Functional Adaptations of Marine Organisms

PHYSIOLOGICAL ECOLOGY

A Series of Monographs, Texts, and Treatises

EDITED BY

T. T. KOZLOWSKI

University of Wisconsin
Madison, Wisconsin

T. T. KOZLOWSKI. Growth and Development of Trees, Volumes I and II – 1971

DANIEL HILLEL. Soil and Water: Physical Principles and Processes, 1971

J. LEVITT. Responses of Plants to Environmental Stresses, 1972

V. B. YOUNGNER AND C. M. MCKELL (Eds.). The Biology and Utilization of Grasses, 1972

T. T. KOZLOWSKI (Ed.). Seed Biology, Volumes I, II, and III – 1972

YOAV WAISEL. Biology of Halophytes, 1972

G. C. MARKS AND T. T. KOZLOWSKI (Eds.). Ectomycorrhizae: Their Ecology and Physiology, 1973

T. T. KOZLOWSKI (Ed.). Shedding of Plant Parts, 1973

ELROY L. RICE. Allelopathy, 1974

T. T. KOZLOWSKI AND C. E. AHLGREN (Eds.). Fire and Ecosystems, 1974

J. BRIAN MUDD AND T. T. KOZLOWSKI (Eds.). Responses of Plants to Air Pollution, 1975

REXFORD DAUBENMIRE. Plant Geography, 1978

JOHN G. SCANDALIOS (Ed.), Physiological Genetics, 1979

BERTRAM G. MURRAY, JR. Population Dynamics: Alternative Models, 1979

J. LEVITT. Responses of Plants to Environmental Stresses, 2nd Edition.
Volume I: Chilling, Freezing, and High Temperature Stresses, 1980
Volume II: Water, Radiation, Salt, and Other Stresses, 1980

JAMES A. LARSEN. The Boreal Ecosystem, 1980

SIDNEY A. GAUTHREAUX, JR. (Ed.), Animal Migration, Orientation, and Navigation, 1981

F. JOHN VERNBERG AND WINONA B. VERNBERG (Eds.), Functional Adaptations of Marine Organisms, 1981

Functional Adaptations of Marine Organisms

Edited by

F. JOHN VERNBERG

Belle W. Baruch Institute
for Marine Biology and Coastal Research
University of South Carolina
Columbia, South Carolina

WINONA B. VERNBERG

College of Health
University of South Carolina
Columbia, South Carolina

1981

ACADEMIC PRESS

A Subsidiary of Harcourt Brace Jovanovich, Publishers
New York London Toronto Sydney San Francisco

ACADEMIC PRESS, INC.
111 Fifth Avenue, New York, New York 10003

United Kingdom Edition published by
ACADEMIC PRESS, INC. (LONDON) LTD.
24/28 Oval Road, London NW1 7DX

Library of Congress Cataloging in Publication Data
Main entry under title:

Functional adaptations of marine organisms.

(Physiological ecology series)
Includes bibliographies and index.
1. Marine fauna--Physiology. 2. Marine flora--Physiology. 3. Adaptation (Physiology) I. Vernberg, F. John, Date. II. Vernberg, Winona B., Date. III. Series.
QL121.F86 574.5'2636 80-1684
ISBN 0-12-718280-2

PRINTED IN THE UNITED STATES OF AMERICA

81 82 83 84 9 8 7 6 5 4 3 2 1

Contents

3 Marine Bacteria

L. Harold Stevenson and Thomas H. Chrzanowski

4 Zooplankton

Donald R. Heinle

5 Meiofauna

Winona B. Vernberg and Bruce C. Coull

6 Benthic Macrofauna

F. John Vernberg

7 Pelagic Macrofauna

Malcolm S. Gordon and Bruce W. Belman

8 Functional Adaptations of Deep-Sea Organisms

Robert Y. George

List of Contributors

Numbers in parentheses indicate the pages on which the authors' contributions begin.

Bruce W. Belman (231), Department of Biology, University of California, Los Angeles, California 90024

Thomas H. Chrzanowski (71), Department of Biology, Belle W. Baruch Institute for Marine Biology and Coastal Research, University of South Carolina, Columbia, South Carolina 29208

Bruce C. Coull (147), Belle W. Baruch Institute for Marine Biology and Coastal Research, University of South Carolina, Columbia, South Carolina 29208

Randolph L. Ferguson (9), National Marine Fisheries Service, NOAA, Southeast Fisheries Center, Beaufort Laboratory, Beaufort, North Carolina 28516

Robert Y. George (279), Institute of Marine Biomedical Research, University of North Carolina at Wilmington, Wilmington, North Carolina 28401

Malcolm S. Gordon (231), Department of Biology, University of California, Los Angeles, California 90024

*Donald R. Heinle** (85), Chesapeake Biological Laboratory, University of Maryland Center for Estuarine Environmental Studies, Solomons, Maryland 20688

Theodore R. Rice (9), National Marine Fisheries Service, NOAA, Southeast Fisheries Center, Beaufort Laboratory, Beaufort, North Carolina 28516

L. Harold Stevenson (71), Department of Biology, Belle W. Baruch Institute for Marine Biology and Coastal Research, University of South Carolina, Columbia, South Carolina 29208

Gordon W. Thayer (9), National Marine Fisheries Service, NOAA, Southeast Fisheries Center, Beaufort Laboratory, Beaufort, North Carolina 28516

*Present address: CH2M Hill, Bellevue, WA 98004

F. John Vernberg (1, 179), Belle W. Baruch Institute for Marine Biology and Coastal Research, University of South Carolina, Columbia, South Carolina 29208

Winona B. Vernberg (1, 147), College of Health, University of South Carolina, Columbia, South Carolina 29208

Preface

To function is to live. To the scientist concerned with understanding how organisms function in their environment, the sea offers diverse marine biota which are subject to very different sets of environmental factors, thereby providing an unequalled source of experimental material. The marine biota offer an opportunity to study organismic responses to "natural" sets of environmental factors as well as man-made sets of perturbations, whether they be the introduction of foreign substances, dredging, off-shore drilling, or overfishing.

This book is written to provide an insight into some of the functional adaptations of marine organisms to both natural and man-made sets of factors. It is organized into chapters representing an ecological orientation. The physiology of plants is presented in terms of both primary producers and decomposers, while functional adaptations of animals are discussed in relation to the major ecological divisions of the sea: zooplankton, meiofauna, benthic macroinvertebrates, and pelagic and deep-sea organisms. We have also written for both marine and nonmarine scientists who have broad interests in acquiring knowledge about the adaptations of organisms in changing environments. We hope that students, our hope for future progress, will be particularly stimulated, learn what has been accomplished, and then, with enthusiasm, seek answers both to persistent and to new problems.

F. John Vernberg
Winona B. Vernberg

Introduction 1

F. J. Vernberg and W. B. Vernberg

I. Introduction

In recent years society has shown an unprecedented interest in the dynamic impact of environment on plants and animals. Not only has this ecological awareness stirred the thoughts and emotions of the layman, but also scientists from diverse scientific disciplines have become professionally involved. As would be expected, this diversity of scientific input inevitably has led to a greater understanding of the interrelationships between organisms and their environment. At the same time, the boundary lines of older scientific disciplines have become blurred, and new ones have emerged. Physiological ecology is an excellent example of a newly forged discipline which has come to the forefront in recent years by drawing on the expertise of scientists who often come from traditional scientific backgrounds. This field is concerned with understanding the basic mechanisms of organismic response to the surrounding environmental complex. Physiological ecologists attempt to interpret physiological responses in terms of their adaptive environmental significance, viewing each organism as a highly integrated system of multiple functional components that may be differentially influenced by environmental factors. The functioning of these component parts must be integrated to ensure that the intact organism can survive and perpetuate the species.

Although the principal unit of study is the individual organism, it is obvious that the organism is both part of a population and of the ecosystem containing this population. Thus, physiological ecologists are concerned with the physiological attributes of populations and communities based on responses of individuals. This continuum of interest over a range of responses from the molecular to the community level of biological organization is necessary to gain insight into the adaptive significance of these responses. The breadth of interest in ecology is restricted only by the vision of the individual investigator and not by artificially

FUNCTIONAL ADAPTATIONS OF MARINE ORGANISMS

ISBN 0-12-718280-2

imposed discipline boundaries. Thus, while physiological ecologists may vary in their background and training, they are unified by their interest in understanding a very fundamental question; how and why do organisms live where they do?

Each organism is exposed to an external environment that consists of a complex combination of interesting factors. Internally the functional machinery must respond to these factors in such a way that the organism can survive. For our convenience in analyzing this environmental–organismic interaction, the external environment can be divided into abiotic and biotic factors. Abiotic factors are numerous and include temperature, salinity, geomagnetic forces, multiple chemical substances, gases, hydrostatic pressure, light, and currents. In the biotic category are such factors as predator–prey interaction, commensalism, and competition. Both biotic and abiotic factors can interact significantly to influence an organism. For example, low oxygen concentrations may adversely affect the ability of a prey species to escape from its predator. Of interest to the theme of this book is the observation that the dynamic equilibrium between the organism and its environment is constantly changing with time and fluctuation in the several components acting individually or in concert. A combination of factors that results in the death of an organism is termed the zone of lethality. When studying a single factor, there is typically a lethal point at a high and low expression of this factor. Between these "high" and "low" lethal points, there is a broad range of sublethal environmental combinations that influence the organism, known as the zone of compatibility. Within the zone of compatibility an organism may survive, but its ability to function efficiently may be markedly reduced. For example, a sublethal temperature may allow the organism to move about and feed, but curtail the reproductive potential to such an extent that the species cannot survive.

The oceanic biota represents a particular challenge to the physiological ecologist—what are the functional ploys that the diverse species from markedly different habitats utilize to survive and reproduce? Briefly, let us now consider some of the multiplicity of habitats within the marine environment.

II. Oceanic Habitats

The sea, with a volume of approximately 315 million cubic miles, covers over 71% of the earth's surface. Although tremendously vast, the unity of the sea environment is demonstrated in that the entire seawater mass is continuous, so that a drop of water could potentially make its way to any part of the total sea. In recent years, the topographical density of the ocean floor has been established by mapping. We now know that there are sea mountains with peaks that break the ocean surface to form small islands, tremendously deep trenches that cleave the ocean floor (such as the 35,800 foot Mindanao Trench), and a submerged mountain range (mid-Atlantic ridge) that extends for 10,000 miles in the North Atlantic Ocean. The ocean floor is now considered to be a dynamic system owing to the concepts of continental drift, seafloor spreading, and plate tectonics. On a geological time scale, a habitat at one specific geographical location changes constantly, and the populations of organisms that live there must adapt to new environmental stresses in an evolutionary sense. An understanding of the physiological ecology of these organisms is a great challenge.

Numerous classifications of marine environments have been proposed, but the one that has been most widely excepted is graphically represented in Fig. 1. This classification was proposed by the Committee on Marine Ecology and Paleoecology. The two major divisions of this classification are benthic and pelagic, and these may be further subdivided. Each assemblage contains a characteristic fauna and flora. Benthic environments extend from high ground to the ocean depths and have been divided into six different zones; (1) supralittoral, (2) littoral or intertidal, (3) sublittoral, (4) bathyal, (5) abyssal, and (6) hadal. The regions of the deep sea are the least well defined, although this area is the largest benthic habitat type, occupying nearly 90% of the ocean floor. The pelagic region can be divided into two main areas; the neretic zone, which comprises the water mass over the Continental Shelf, and the oceanic zone, which includes the main mass of seawater. The oceanic region may further be divided into four sub-regions; (1) the epipelagic, (2) mesopelagic, (3) bathypelagic, and (4) the abyssopelagic.

Marine environments may also be described ecologically. A good illustration are those terms that refer to the amount of light since light steadily decreases in intensity with increasing depth. The terms aphotic and euphotic generally denote regions of darkness or light. Other bases have been proposed to describe marine environments by referring to a general region of the sea, such as divisions of shore and shallow water seas, or to taxonomic groups, such as the level bottom molluscan community or the laminarian intertidal zone. Marine organisms may also be classified on the basis of the habitat type in which they live. For example, benthic organisms are associated with the bottom substrata; those species associated with the surface of the bottom are the epifauna while those that dig or are buried in the substratum are the infauna. Benthic organisms may be further divided into groups based on size; macrobenthos are organisms too large to pass through a 1 mm mesh sieve, meiobenthos are organisms smaller than the macrobenthos, but which are retained by 0.1 mm mesh sieve, and the microbenthos are organisms that are so small that they pass freely through a 0.1 mm mesh sieve. Free-moving organisms that inhabit the water column are called pelagic species, those that can control direction and speed of locomotion are called the nekton, and those primarily dependent on water movement for location or locomotion are called plankton. In turn, plankton are generally recognized as being either phytoplankton (the plant species) or zooplankton (the animal species). Nannoplankton are small plankton ranging in length from 5 to 60 μm,

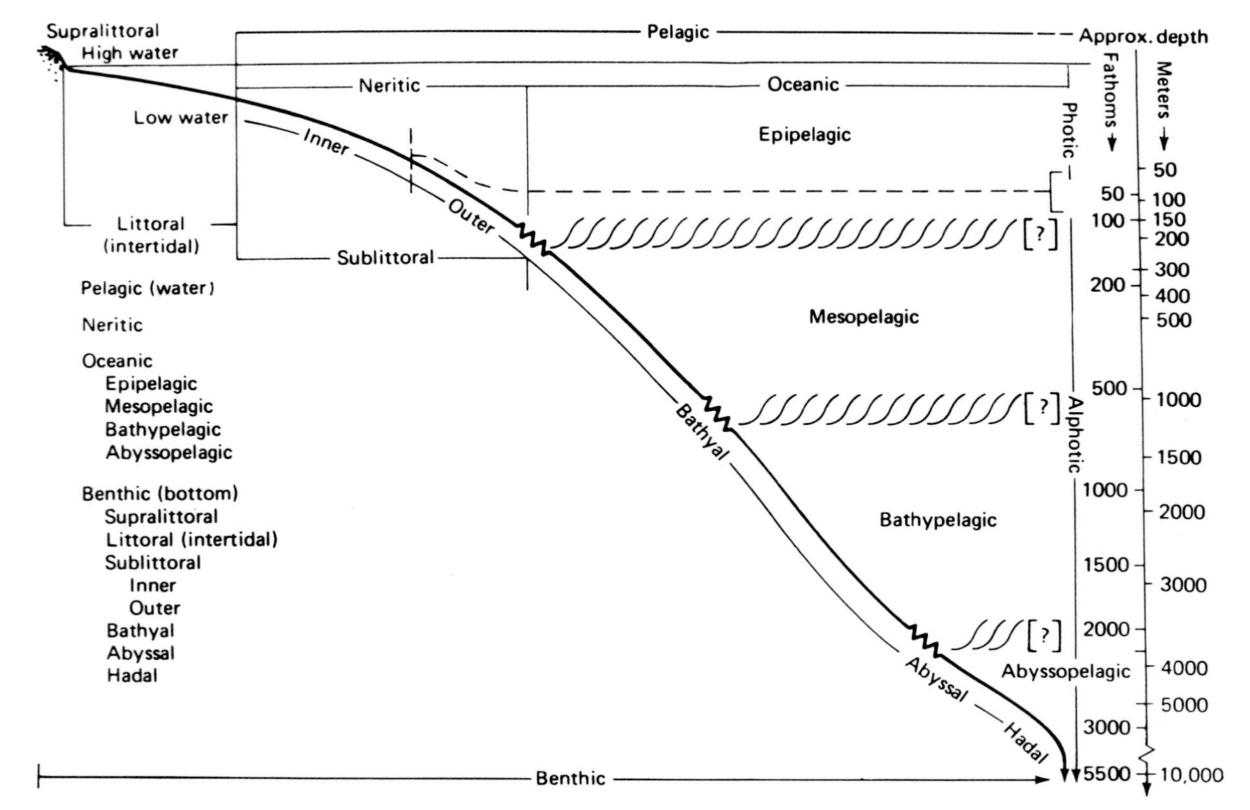

Fig. 1. Classification of marine environments. From J. W. Hedgpeth, 1957. *In* "Treatise on Marine Ecology and Paleoecology" (J. W. Hedgpeth, ed.), Vol. 1, Geol. Soc. Am. Mem. No. 67, New York, pp. 17–28.

and the ultraplankton are smaller than 5 μm. The complex terminology resulting from various classifications of plankton also include some terms that are relevant to physiological ecology. For example, holoplankton species are those in which all of the developmental stages are part of the plankton, while in other species certain life history stages are associated with the benthic environment. If nektonic species regularly occur in the benthic regions they are described as demersal; if plankton are found in the benthic regions they are said to meroplanktonic.

III. The Intertidal Zone

The intertidal zone is a narrow fringe between the ocean and land, and organisms living there are alternately exposed to air and covered by water. This zone is not homogeneous in structure or in physical, chemical, or biotic characteristics. There are, however, three major types of habitats: rocky shores, sandy beaches, and mud-flats. These habitats may be found on the edge of the open ocean and bordering protected regions such as harbors and estuaries.

Characteristically animals living in the intertidal zone are subjected to widely fluctuating environmental conditions, particularly in regions of noticeable tidal change. In such regions, animals may be exposed to wide ranges of seasonal and temporal temperature fluctuations, to salinity changes, and to desiccation. Thus it is not surprising that animals of the intertidal zone tend to be more eurythermal, euryhaline, and resistant to desiccation than other organisms in other marine environments. Many also experience periods of oxygen stress and often must function anaerobically for varying periods of time.

Mechanisms through which intertidal zone animals are able to feed, select their environment, or find a mate may be correlated with their way of life. A burrowing animal, such as a polychaete, will obviously utilize different physiological responses than will a semiterrestrial intertidal-zone crab.

A common characteristic of intertidal zone animals is their tendency to have pelagic larval stages so that their reproductive cycles must be geared to ensure release of gametes into a favorable watery environment.

Thus it can be seen that the intertidal zone offers a nonhomogenous and fluctuating environment with a multiplicity of habitats and microenvironments.

IV. Estuaries

The estuary is a dynamic environment where freshwater mixes with the sea. One of the chief characteristics of this environment is fluctuation in salinity. Tidal changes affect salinity by increasing salinity on a flood tide and decreasing it on an ebb tide, and typically the water is fresher at the surface than at the bottom. Changes in salinity can be sudden and dramatic; e.g., a change in estuarine salinity from 30 to 10‰ within a one-hour period is not atypical. It is also common for the salinity of estuarine waters to vary seasonally, so that the water may be nearly fresh during winter and spring rains, but strongly saline in summer. Although not as great as in the intertidal zone, temperature changes in an estuarine system often are greater than those of either offshore waters or the open ocean, since inflowing river waters tend to be colder than seawater in winter and warmer in summer. Thus, organisms living an an estuarine environment must be able to tolerate a wide range of salinity and, to a lesser degree, change in temperature.

Estuaries are almost always turbid because of the amount of silt in the water. Light cannot penetrate to as great a depth as it does in coastal and open waters, and the animals living in an estuary must depend on sensory cues other than vision to carry out such functions as locating food or avoiding predators. Siltation in the water column is often a major factor affecting primary production, and it also covers the substratum so that mud is the most common type of bottom. These bottom muds are rich in organic detritus (derived primarily from the vegetation along upper tidal levels) and therefore provide a ready source of food for many estuarine organisms. However, this often means that conditions are anoxic a few centimeters beneath the surface of the substrate.

Thus, estuaries offer a murky environment in which salinities and temperature fluctuate markedly, but where food supplies are abundant. While estuarine species are few in number, those that have become adapted to this environment are present in large numbers.

V. Coastal and Open Ocean Waters

Coastal and open ocean waters offer more stable environmental conditions than do either the intertidal zone or an estuarine system. Salinity and temperature are much less variable, although coastal waters will vary in chemical composition more than open ocean water and the range of variation in physical factors also typically is greater in coastal water. These differences are due in large part to some run-off from land masses and the shallowness of coastal areas (less than 200 m). Wave action stirs up bottom sediments and allows mixing and recycling of nutrients and other organic matter.

In contrast to the murk of the estuaries, where light often penetrates only a short distance, the clarity in the upper layers of coastal and open ocean waters permits light penetration to considerable depth. Not surprisingly, light is one of the most important environmental factors in these waters. The depth of the photic zone depends on latitude, season, and amount of particulate matter. The quality of light is variable in the photic zone because of differential penetration of the wavelengths. As a general rule, the yellow-green wavelengths penetrate inshore waters, and the blue-green wavelengths reach deep oceanic water. Vertical migrations of many oceanic species appear to be related to light intensity, and others use the sun as an orienting or navigational device.

Unlike the estuarine environment, the pelagic zone of coastal and open ocean waters have been characterized as a nutritionally dilute environment, for although food supply is essentially limitless, it may be so thinly dispersed that the problem of getting enough to eat is formidable. Consequently, animals living here have per force evolved diverse and efficient methods to feed. Plankton tend to dominate this environment; phytoplankton are the primary producers and zooplankton are the primary herbivores.

Thus, coastal and open ocean waters offer a more stable salinity and thermal environment than do estuarine systems, and light is a major factor in the upper layers of oceanic waters.

VI. The Deep Sea

For the most part, the deep-sea environment (depths below 2000 m) is relatively stable; the salinity is approximately 34.8%, the temperature may vary from 3.6° to 0.6° C, the oxygen concentration is typically high and constant, and no light is present (except for bioluminescence). Food supplies are generally low and organisms exhibit feeding mechanisms characteristic of this type of environment. One unique feature of the deep sea is increased hydrostatic pressure. Since the hydrostatic pressure is increased one atmosphere with every increase of 10 m, it is one variable experienced by deep sea organisms. An organism at 2000 m is subjected to 200 atm while an individual living at 6000 meters experiences a pressure of 600 atm.

With this brief overview as background, we can now proceed to examine some of the physiological adaptations that permit marine organisms to live in such diverse environments.

The following chapters are organized both on an ecological basis of trophic function and habitat. Chapters 2 and 3 deal with primary producers and decomposers, respectively, while the remaining chapters discuss zooplankton, meiofauna, benthic macrofauna, pelagic macrofauna, and deep-sea organisms.

Marine Primary Producers

2

Randolph L. Ferguson, Gordon W. Thayer, and Theodore R. Rice

I. Introduction

Primary producers synthesize organic matter that ultimately becomes the substance and energy source of all forms of life. Not all of the primary production is available to man since a portion is utilized in metabolic processes by the plants and organisms of successively higher trophic levels. Until man can simulate photosynthesis *in vitro,* even he must depend on photosynthetic production of organic matter by plants (Westlake, 1963). The function of plants in a marine ecosystem, however, is not limited to the production of energy rich organic material, since they also cycle nutrients, including trace metals, reduce the effects of winds and tides, stabilize sediments, and provide substrates and habitats for other organisms. Many marine plants also are important to man since they are used as sources of food stabilizers, preservatives, drugs, fertilizer,

FUNCTIONAL ADAPTATIONS OF MARINE ORGANISMS

ISBN 0-12-718280-2

fodder, and for insulation and other building materials. In addition, many marine plant communities, e.g., *Spartina* marshes, because of their assimilative capacity, are highly effective natural "filters" for sewage treatment (Teal and Valiela, 1973; Gosselink *et al.,* 1974).

The marine environment includes open ocean, coastal waters, and estuarine areas, which together represent three-fourths of the earth's surface, approximately 13.5×10^7 km^2 (Odum, 1971). Although the biomass of marine primary producers is small compared to the biomass of terrestrial plants, the relative contribution of these two systems to the earth's fixed carbon budget is similar (Ryther, 1959). Assuming net production by plants is 50% of their gross production, production by marine plants is about 2.2×10^{16} g C/year and by terrestrial plants about 2.9×10^{16} g C/year (Odum, 1971).

The focus of this chapter is on the physiological ecology of phytoplankton, submergent seagrasses, and emergent salt marsh vegetation. Microbenthic algae, macrobenthic algae, and mangroves also are important in estuarine and shallow coastal regions but will be discussed only briefly. For more extensive discussions of microbenthic plants, see Hustedt (1955), Grøntved (1960), Gargas (1970), and Steeman-Nielsen (1975). For a more complete coverage of macrobenthic algae see Chapman (1957, 1966, 1970a), Conover (1958), Biebl (1962), Earle and Humm (1964), den Hartog (1967), and Stewart (1974). For additional material on mangroves see Davis (1940), Scholander *et al.* (1962, 1966), Atkinson *et al.* (1967), Macnae (1967), Chapman (1970b), Kuenzler (1974), and Walsh (1974).

Phytoplankton dominates the open ocean in terms of both biomass and production and consists primarily of diatoms, dinoflagellates, and coccolithophores, although blue-green algae and green flagellates on occasion can be abundant. Regional differences in the occurrence of species and in the production of phytoplankton are a result of many conditions. Three of the more important ones are the supply of nutrients, the quantity of light, and the ambient temperature. Within each region there also are seasonal changes and gradients in these factors with depth. Mean net production values have been found to be 500 mg C/m^2/day for neritic waters, 170 mg C/m^2/day for inshore waters, 100 mg C/m^2/day for subpolar oceanic regions and waters of equatorial divergence, 70 mg C/m^2/day for transition waters between subtropical and subpolar zones, and 35 mg C/m^2/day for oligotrophic waters of central portions of subtropical regions (Koblentz-Mishke *et al.,* 1970).

In shallow coastal waters and in estuaries, macroscopic vegetation dominates the biomass and often the total production as well (Mann, 1972; Thayer *et al.,* 1975b; McRoy and Helfferich, 1977; Phillips and McRoy, 1980). These macroscopic plants include submergent seagrasses and algae and emergent marsh grasses and mangroves. Thus, coastal regions provide a rich variety of organic substrates, both of microphytic and macrophytic origin. The occurrence of macrophytic species varies geographically and rates of production vary in response to regional and seasonal changes in nutrient supply, quantity of light, and temperature. Estimates of the net organic production by marine macrophytes range from 500 to 5000 mg C/m^2/day.

The primary producers of the marine environment possess the capacity, through physiological and morphological adaptability, not only to persist during environmental extremes, but also to thrive over a wide range of environmental conditions. The differences in the form and function of these plant groups are the result of adaptive responses to their specific environments. The ecological success of the various primary producer groups within the marine environment results in high levels of organic production. Marine plants cannot, however, always cope with changes brought about by man's increasing use of the marine environment.

The scientific literature on marine primary producers is extensive, and this chapter is not intended to be an exhaustive review of this group. On the other hand, rates of production and differences in form and function of the various plant groups comprising the marine producers must be responses to identifiable factors in the environment. Man's diversified and still accelerating use of the marine environment makes it imperative that those factors regulating the function (physiological ecology) of major producer groups be understood. This is a prerequisite to the development of successful management practices. For example, while the strategy of fisheries management in the past has been based on the analysis of catch–effort data for the commercially-exploited fish stocks, scientific evidence now confirms that the size of these stocks often is controlled by food availability. In many instances, cause and effect can be traced to the productivity of primary producers.

The phytoplankton and submergent and emergent vascular plants in the marine environment are discussed in this chapter in the three major sections on primary production, adaptations, and human intrusion. This review is based on a search of the literature conducted in late 1976. Some references published subsequently have been incorporated into the text. In the first section, the geographic distribution of representative species, the primary production of dominant plant groups, and factors influencing distribution and production are discussed. In the second section, the physiological and morphological adaptations of phytoplankton, seagrasses, and marsh grasses are reviewed. In the third section, actual and potential impacts of man's activities on the functioning of these three major groups are considered.

II. Primary Production in the Marine Environment

A. PHYTOGEOGRAPHY

1. Overview

Algae and vascular plants, the primary producers of the marine environment, have become adapted to the salinity range of marine waters. Free-floating unicellular algae, the phytoplankton, are ubiquitous. Only a few multicellular algae such as *Sargassum* have adapted to a planktonic existence. Benthic algae and vascular plants are restricted to relatively shallow water or intertidal bottoms. The vascular plants can be subdivided into submergent and emergent species. Phytoplankton are the dominant plant life in the oceans while submergent vascular plants, the seagrasses, are limited to approximately 1% of the sea floor (Drew, 1971). It is only in the narrow band along the coast of islands and continents that sufficient light for the growth of rooted marine plants reaches the bottom. The emergent vascular plants occur in intertidal water or where saline water has saturated the sediments in the coastal zone. Emergent marine vascular plants occupy the transition zone between the marine and freshwater environments and between the aquatic and terrestrial environments.

2. *Phytoplankton*

Phytoplankton are not distributed uniformly throughout the oceans either horizontally or vertically. The species of phytoplankton present in a geographical area at a given time will consist of endemic species and those that have been transported into the area by movement of the water. Species transported into an area may be preadapted to existing environmental conditions or in the process of becoming adapted. Any species that does not become adapted can persist for a considerable period of time and constitute a significant part of the biomass while contributing little to primary production (Watt, 1971). Of a total phytoplankton flora of 150 species for Long Island Sound, less than 10% were considered endemic (Riley, 1967). Of the endemic species, nine were diatoms and four were dinoflagellates.

The species present in a given area and their abundance are in a constant state of flux. Diatoms and dinoflagellates are the most important in relation to the numbers of marine species. Diatoms form an important part of every collection of phytoplankton. The numbers of species of diatoms occurring in samples collected throughout the oceans by many investigators are shown in Fig. 1 (Guillard and Kilham, 1977), and the distribution of several species from the Gulf of Maine to Cape Hatteras is shown in Table I (Watling *et al.,* 1975). The number of diatom species varies from 2 to 400 at the stations sampled. Dinoflagellates are a diverse group of primary producers. Some are animal-like, ingesting particulate food, and some are saprophytic. Species of dinoflagellates at certain times divide rapidly, increasing in numbers to form a "bloom," which can be both spectacular and toxic (Taylor and Seliger, 1979) and for some species is referred to as "red tide" (Joyce and Roberts, 1975; Smith 1975). Coccolithophores, which are small silicoflagellates, make up a significant fraction of phytoplankton in some open ocean areas. These flagellates can occur in large numbers and constitute a significant portion of the biomass of tropical and subtropical oceans. The Cryptophyceae are small red-brown, bluish, or colorless

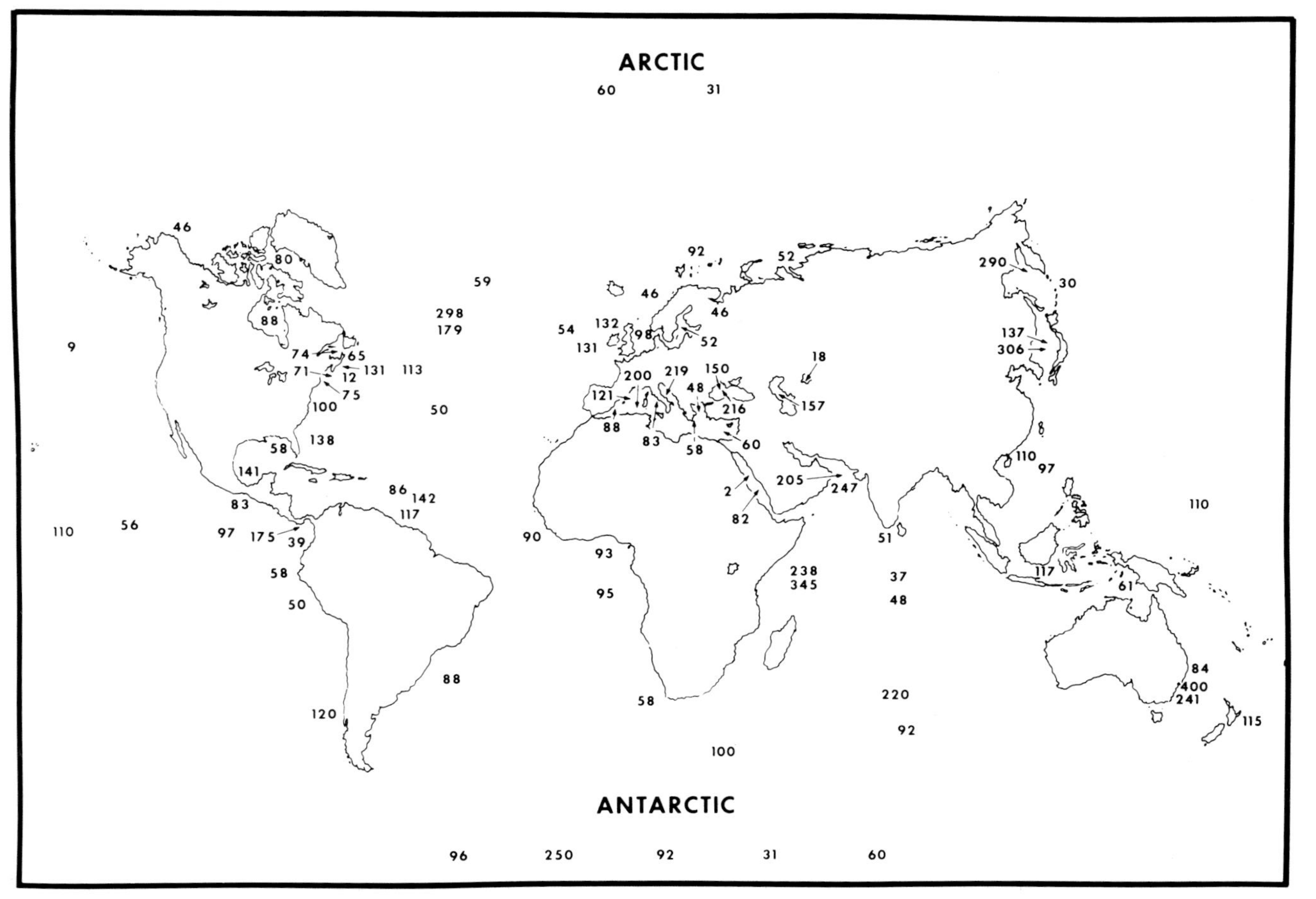

Fig. 1. Numbers of diatom species occurring in samples collected throughout the ocean. (Based on Guillard and Kilham, 1977.)

TABLE I

Representative Species of Diatoms Occurring along the North and Middle Atlantic Coasts of the United States[a]

Species	Range
Asterionella japonica	Massachusetts to Cape Hatteras
Chaetoceros deblis	Gulf of Maine
C. compressus	Gulf of Maine to Long Island Sound
C. decipiens	Gulf of Maine to Cape Hatteras
C. socialis	Gulf of Maine to Cape Hatteras
Coscinodiscus centralis	Gulf of Maine to Rhode Island
Leptocylindricus danicus	Gulf of Maine to Long Island Sound
Melosira sulcata	Gulf of Maine to Cape Hatteras
Nitzschia seriata	Massachusetts to Delaware
N. closterium	Massachusetts to Delaware
Rhizosolenia fragilissima	Gulf of Maine to Cape Hatteras
Skeletonema costatum	Gulf of Maine to Cape Hatteras
Thalassionema nitzschioides	Gulf of Maine to Cape Hatteras
Thalassiosira decipiens	Gulf of Maine to Long Island Sound
T. nordenskioeldii	Gulf of Maine to Long Island Sound
T. gravida	Gulf of Maine to Cape Hatteras

[a] Modified from Watling *et al.* (1975).

flagellates which at times occur in large numbers in inshore waters. The Chlorophyceae are mostly flagellates that occur chiefly in temperate coastal regions, especially in late summer and early winter. Blue-green algae appear to be important primarily in brackish waters or in warm and nutrient poor oceanic waters. Shallow coastal waters around the Gulf of Mexico often contain blue-green algal mats (Sorenson and Conover, 1962) and the frequency of occurrence becomes greatest in the harsh, dry coastal region of southern Texas. Mats occasionally can be found along the east coast of the United States as far north as New Jersey (Pomeroy, 1959) and occasionally are observed in California, Hawaii, and Puerto Rico. Microalgae are found on and within sediments in intertidal and shallow subtidal zones. In areas of turbulence these species are often resuspended and become part of the phytoplankton community. Diatoms dominate the microbenthic algae, although dinoflagellates, green algae, and blue-green algae can be abundant on occasion (Round, 1965; Steeman-Nielsen, 1975). The diatom community of shallow benthic habitats is diverse. For example, on the intertidal mud flats at Beaufort, North Carolina, Hustedt (1955) found 369 species and 19 varieties of diatoms belonging to 63 genera. Seventy-four species of *Navicula,* 48 species of *Nitzschia,* 32 species of *Amphiprora,* 19 species of *Mastogloia,* and 13 species of *Diploneis* were recorded. *Nitzschia* was most characteristic of the area, and 18 of the *Mastogloia* species found in Beaufort appear to be restricted to the Atlantic coasts of subtropical and tropical America.

3. *Macroscopic Plants*

Vascular aquatic plants are those angiosperms and pteridophytes which grow in soil saturated with water or covered with water during a major portion of the year (Weaver and Clements, 1938; Penfound, 1956). The geographic distribution of submergent seagrasses has been reviewed by den Hartog (1970), who reported that 2 families, 12 genera, and 48 species of submerged aquatic angiosperms have successfully colonized the marine environment. The two families, Potamogetonaceae and Hydrocharitaceae are in the class Monocotyledonae of the division Anthophyta. It is remarkable that there are few parts of the world's coastal zone where one or more of these species of seagrasses have not adapted (Fig. 2) (Phillips, 1974).

With few exceptions, seagrasses can be subdivided into tropical–subtropical and temperate and into New and Old World species (Table II, Fig. 2) (Sculthorpe, 1967; den Hartog, 1970). Genera of Potamogetonaceae occur from the tropics through the temperate zones, but no one species extends over this range. On the other hand, the Hydrocharitaceae are tropical. Of the temperate species, only *Zostera marina* has an extensive range, while the distribution of *Posidonia* is discontinuous between the Mediterranean coast of southern Europe and the coast of Australia. *Phyllospadix* has two major centers of occurrence, one on the Pacific coast of North America and the other in eastern Asia. The tropical–subtropical species occur either in the Indo-Malaysian region or in the American tropical region. Within these two regions, the species present have discontinuous distributions, and in many instances species of the tropical Americas are highly localized.

The distribution of species of seagrasses is also stratified vertically in the tidal zone (den Hartog, 1970). For example, the species *Holodule* and *Halophila* generally extend from the upper intertidal to the lower subtidal zone. *Zostera* generally is distributed from the lower intertidal to the lower subtidal zone (Fig. 3). Both *Thalassia* and *Cymodocea* are distributed throughout the lower intertidal and upper subtidal zones, but *Posidonia* and *Syringodium* are restricted to subtidal habitats.

Macroscopic algae are also important submergent plants and are distributed along both horizontal and vertical gradients of the marine environment. Dawson (1966) reported that the green algae, Chlorophyceae, have their maximum evolutionary development in the tropics; the brown algae, Phaeophyceae, are dominant in cold temperate waters, and the red algae, Rhodophyceae, occur at all latitudes. The distribution of some of the species of benthic macroalgae is shown in Figs. 4 and 5. The geographic range of many species of macroalgae depends on suitable water temperatures. The Gulf Stream influences water temperatures on the east coast of North America and favors the occurrence of many tropical species at least as far north as Beaufort, North Carolina (Fig. 4). *Macrocystis,* the giant kelp, occurs along the west coast of North and South America, except

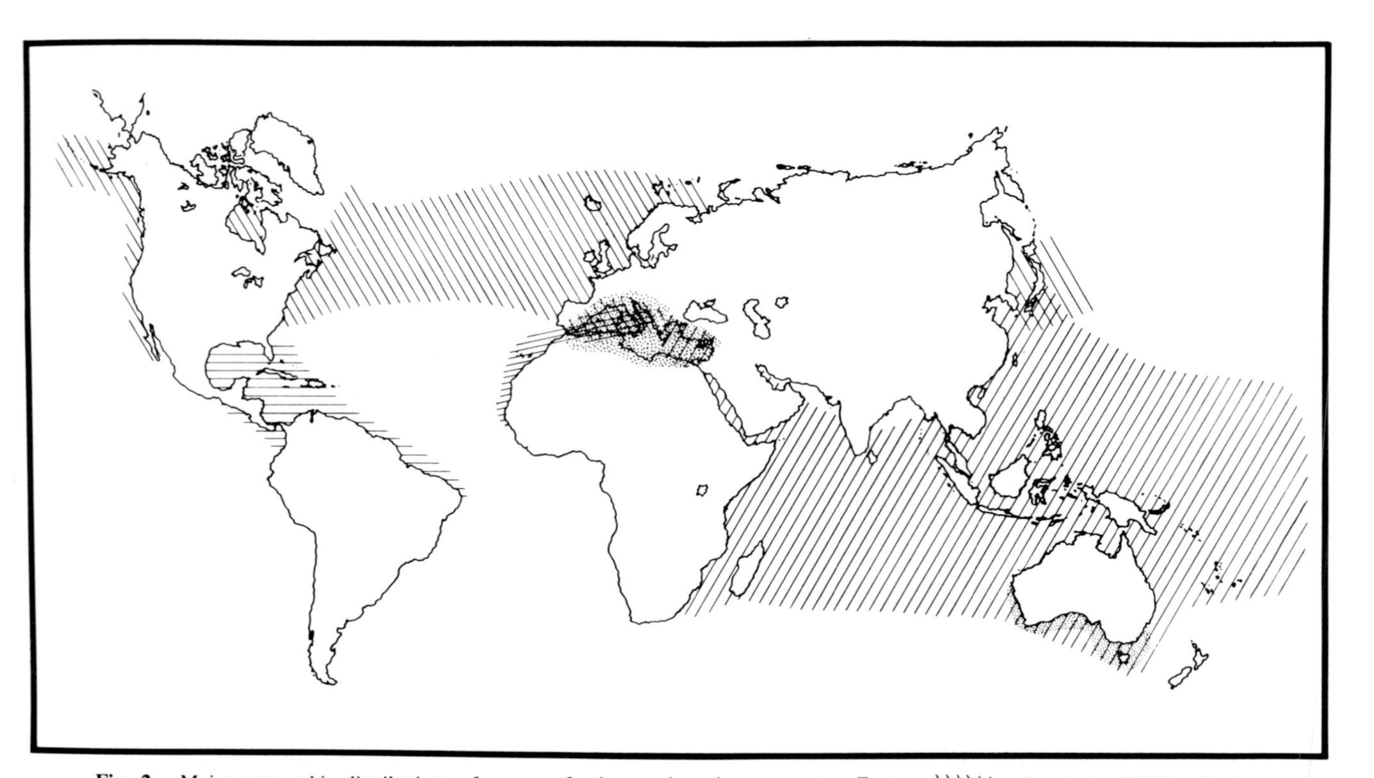

Fig. 2. Major geographic distributions of genera of submerged marine seagrasses: *Zostera*, \\\\\\ ; *Posidonia*. ::::::: ; *Thalassia* and *Halophila*, ≡≡≡ ; *Cymodocea*, ≡≡≡ ; *Syringodium*, *Thalassia*, *Enhalus*, *Halodule*, and *Cymodocea*, //////. (Composite of Fig. 1–10 in den Hartog, 1970.)

TABLE II

Distribution of Seagrasses by Climatic Zone[a]

Climatic zone	Potamogetonaceae	Hydrocharitaceae
Temperate	*Zostera*	
	Phyllospadix	
	Posidonia	
Tropical–warm temperate	*Halodule*	
Tropical–Subtropical	*Cymodocea*	
	Syringodium	
	Amphibolis	
	Thalassodendron	
Tropical		*Enhalus*
		Thalassia
		Halophila

[a] Based on den Hartog (1970).

between Baja, California, and Peru (Fig. 5). The occurrence along the South American coast is a result of the upwelling of cold water. The geographical and seasonal distribution of many other algal species varies with latitude because of the effect of currents on local temperature patterns. For example, the cold water current originating in the Antarctic and traveling up the west coast of Africa is responsible for water temperatures that are suitable for the growth of *Ecklonia,*

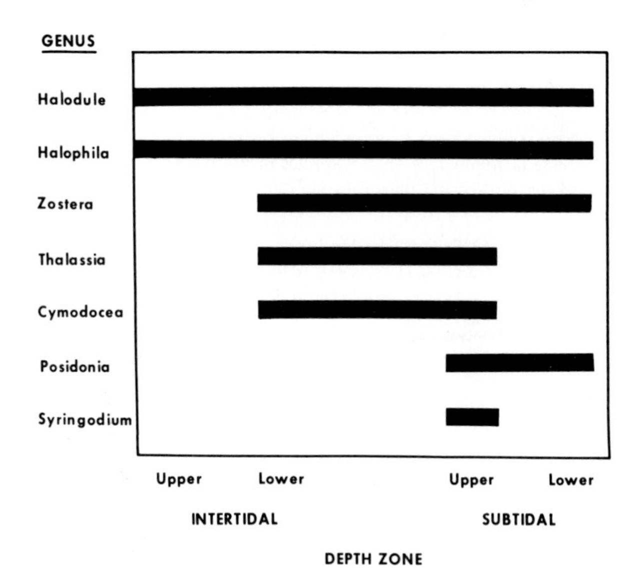

Fig. 3. Vertical distribution of submergent seagrass genera.

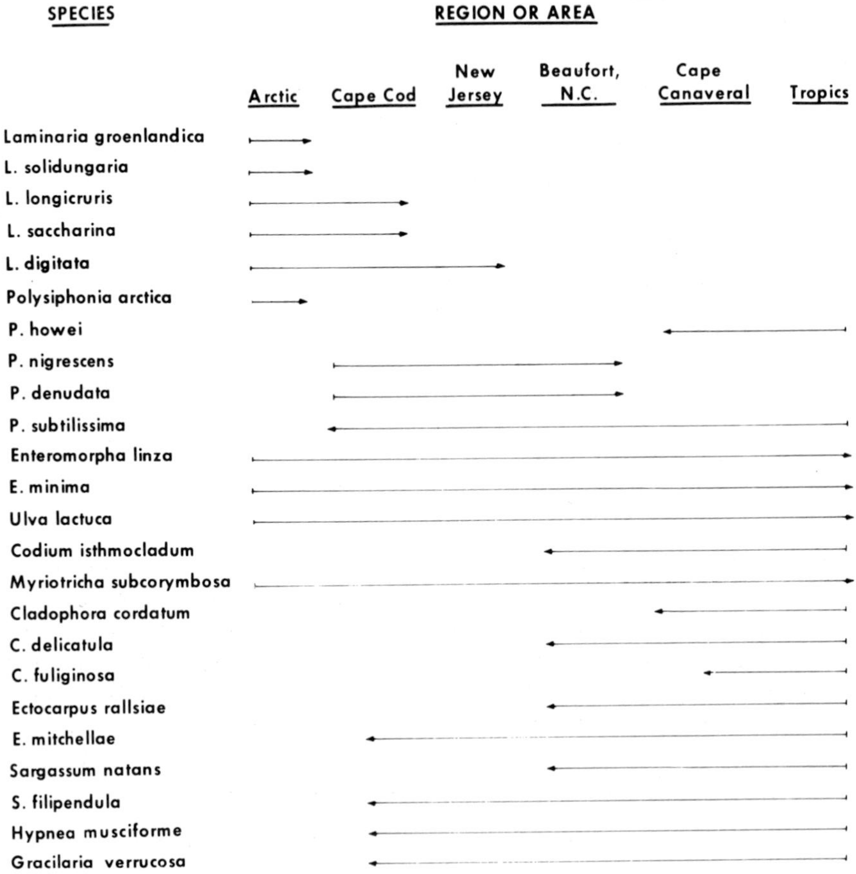

SPECIES

REGION OR AREA

Arctic | Cape Cod | New Jersey | Beaufort, N.C. | Cape Canaveral | Tropics

Laminaria groenlandica
L. solidungaria
L. longicruris
L. saccharina
L. digitata
Polysiphonia arctica
P. howei
P. nigrescens
P. denudata
P. subtilissima
Enteromorpha linza
E. minima
Ulva lactuca
Codium isthmocladum
Myriotricha subcorymbosa
Cladophora cordatum
C. delicatula
C. fuliginosa
Ectocarpus rallsiae
E. mitchellae
Sargassum natans
S. filipendula
Hypnea musciforme
Gracilaria verrucosa

Fig. 4. Geographic distribution of several species of macroscopic benthic algae along the east coast of the United States.

Laminaria, and *Macrocystis* at relatively low latitudes (Fig. 5). The Gulf Stream warms high latitude coastal waters of Great Britian, which are thus favorable for growth of the tropical brown alga, *Cystoseira* (Chapman, 1962). Stratification of marine algae generally indicates that red algae are dominant in deep water, brown algae at intermediate depths, and green algae in shallow water. Taylor (1960), however, reported that in Florida at a depth of 30 m there were 56 species of Chlorophyceae, 15 species of Phaeophyceae, and 60 species of Rhodophyceae, but at 90 m there were 12, 2, and 9 species of these same taxa.

Salt marshes and mangrove swamps are two broad habitat types dominated by species of emergent vascular plants. Salt marshes dominate the intertidal shores of mid- and high-latitude regions, while in the tropics and subtropics they largely

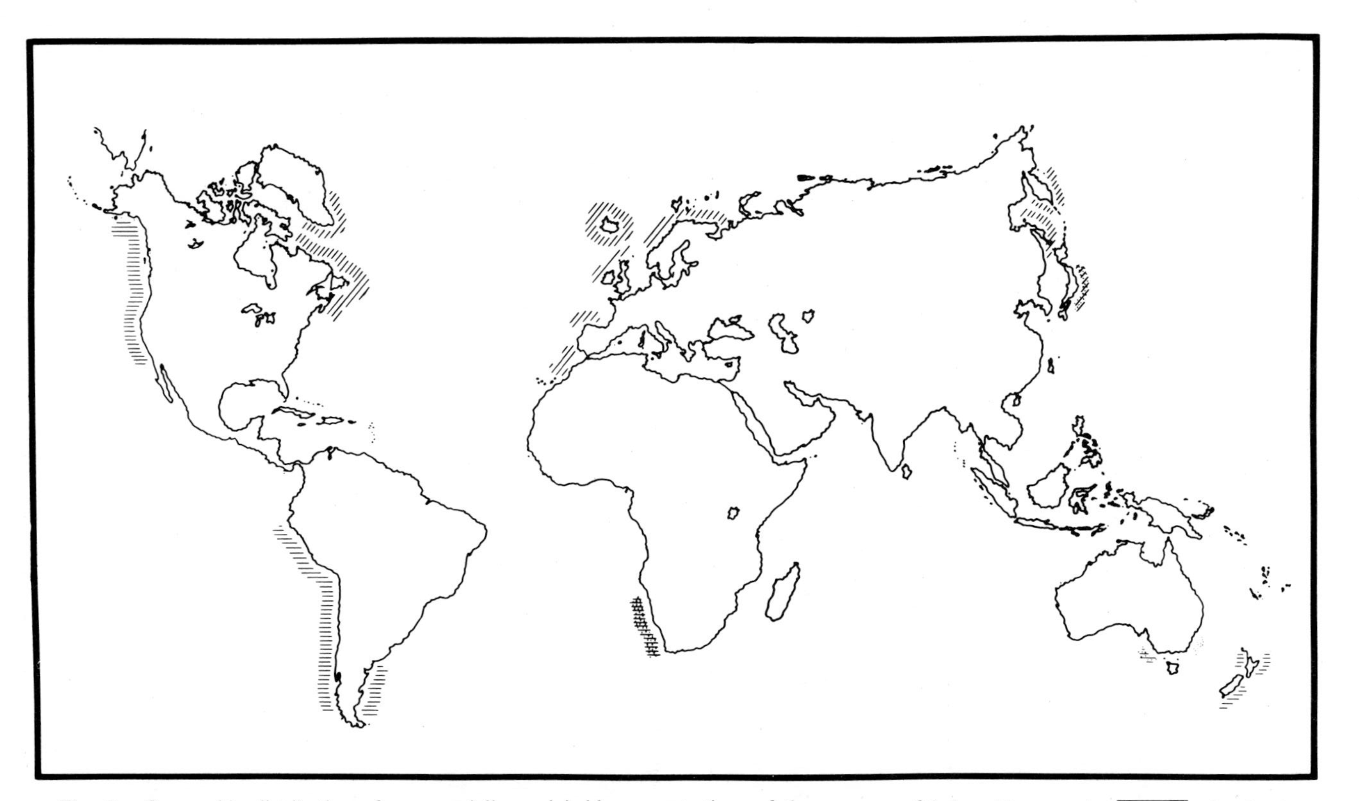

Fig. 5. Geographic distribution of commercially exploitable concentrations of three genera of kelps: *Macrocystis*, [horizontal hatching]; *Laminaria*, [diagonal hatching]; and *Ecklonia*, [dotted pattern]. (Modified from Mann, 1972, after Chapman, 1970a.)

are replaced by mangrove swamps. Vegetation in salt marshes includes herbaceous plants and low shrubs but is generally dominated by grasses or reeds. Benthic green, red, and brown macroscopic algae, benthic microalgae and phanaerograms, such as members of the Chenopodiaceae, e.g., *Salicornia,* also are important and may be locally dominant. Most salt marshes are remarkably uniform in appearance due to the dominance of one or a few species. The list of representative or dominant salt marsh species of the eight geographical regions of the world includes at least 13 genera (Table III). Of the 347 species of halophytes occurring along the Gulf and Atlantic coasts of the United States, 32 species occur in all coastal states (Duncan, 1974). *Spartina alterniflora* is the overwhelming dominant species in the intertidal zone of this area. It extends from mean sea level to mean high tide and is displaced only at very low salinities by brackish or freshwater species. The zone just at or above normal high tide is dominated by *Spartina patens* and *Juncus* spp., generally with *J. gerardi* to the north and *J. roemerianus* to the south of Chesapeake Bay. Other cosmopolitan or

TABLE III

Major Salt Marsh Regions of the World and Representative or Dominant Species in the Lower and Upper Littoral Zones[a]

Region	Lower littoral (intertidal)	Upper littoral (supratidal)
Arctic	*Puccinellietum phryganodis*	*Carex* spp.
North European	*Salicornia* spp. *Puccinella maritima*	*Juncus gerardi*
Mediterranean	*Salicornia* spp. *Limonium* spp.	*Juncus acutus* *Juncus maritimus*
Western Atlantic	*Spartina alterniflora*	*Spartina patens* *Distichlis spicata* *Juncus gerardi* or *Juncus roemerianus* *Salicornia* spp.
Pacific-American-Australasia	*Spartina foliosa*	*Salicornia* spp. *Leptocarpus simplex* *Arthrocnemum* spp. *Distichlis spicata* *Juncus maritimus*
Sino-Japanese	*Triglochin maritima* *Limonium japonicum*	*Zoysia macrostachya*
South American	*Spartina brasiliensis* *Spartina montevidensis* *Heterostachys ritteriana*	*Distichlis spicata* *Juncus acutus* *Juncus maritimus*
Old World Tropics		*Salicornia* spp. *Zoysia pungens*

[a] Based on Chapman (1960, 1974).

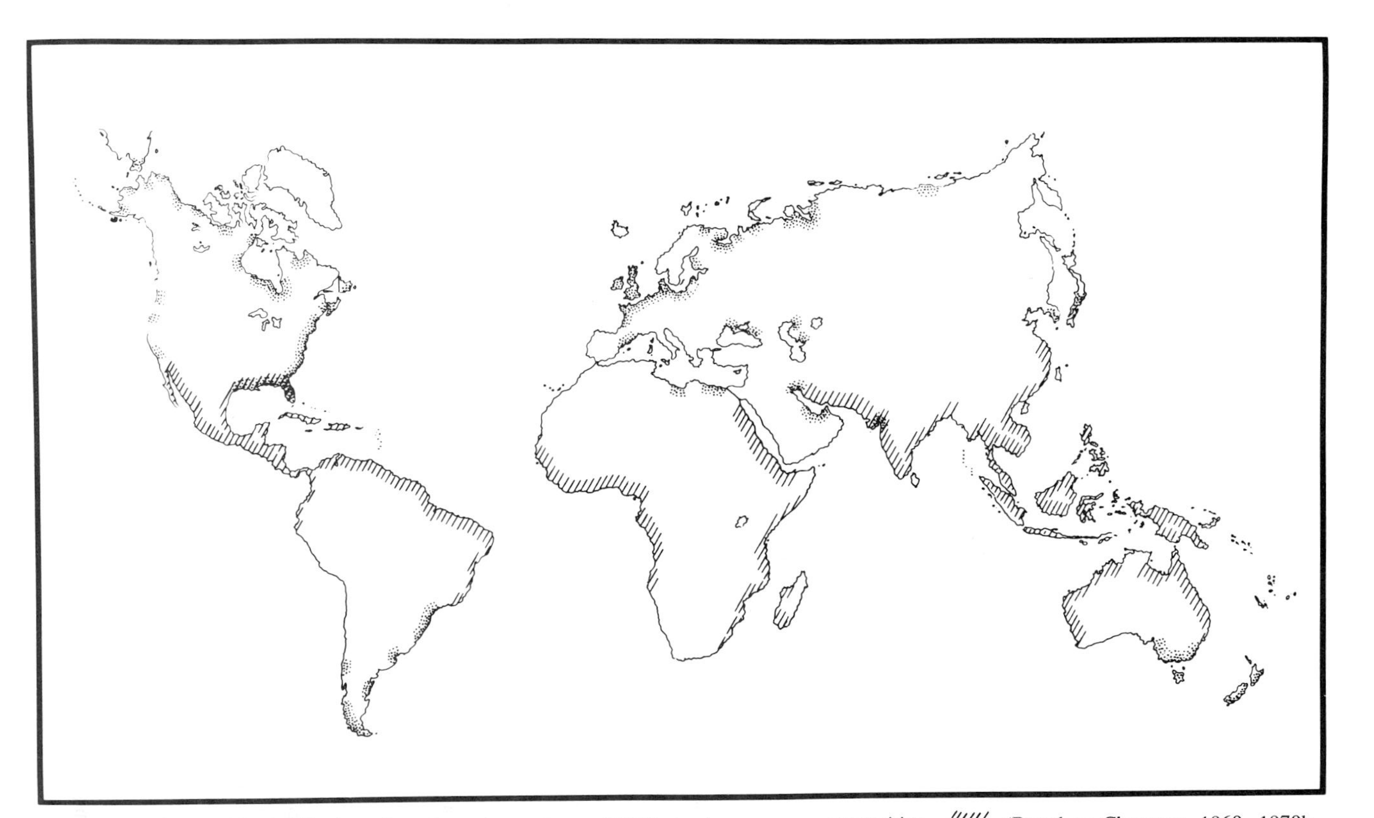

Fig. 6. Geographic distribution of marine salt marshes, ⁘⁘⁘, and mangrove communities, ////. (Based on Chapman, 1960, 1970b, and 1974.)

locally important species in this zone include *Distichlis spicata* and *Salicornia* spp. Pacific coast marshes are limited in distribution and development and are less studied than those on the Atlantic and Gulf coasts (MacDonald and Barbour, 1974). Where present, the lower littoral marsh is dominated by *Spartina foliosa* (*leiantha*) above which are found species of *Salicornia,* accompanied by *Batis marina* in southern California. Species of *Salicornia* are dominant in the upper littoral zone, but are mixed with numerous other species. The *Salicornia* marsh grades into *Dechampsia caespitosa* at higher elevations in Oregon and Washington.

Mangrove swamps are dominated by mangrove trees or shrubs that grow from the high water mark of spring tides to just above mean sea level (Macnae, 1968). The geographical distribution of dominant species of mangrove swamps has been reviewed by Walsh (1974). Mangroves are limited to the area between the Tropics of Cancer and Capricorn (Fig. 6) and occupy an estimated 60 to 75% of the tropical coastlines of the world (McGill, 1958). Mangrove species are classified into 11 families, 4 of which are distributed throughout the tropics and contain 41 of the 55 most common genera (Table IV). In the United States, mangroves occur continuously from Jacksonville on the Atlantic coast to St. Petersburg on the Gulf coast, and are most extensive in southern Florida (Thayer

TABLE IV

Distribution of Mangroves in the Tropics[a]

Family	Genus	Total species	Distribution: Unrestricted	Distribution: Indo-Pacific Region
Rhizophoraceae	*Rhizophora*	7	X	
	Bruguiera	6		X
	Ceriops	2		X
	Kandelia	1		X
Avicenniaceae	*Avicennia*	11	X	
Myrsinaceae	*Aegiceras*	2		X
Meliaceae	*Xylocarpus*	10	X	
Combretaceae	*Laguncularia*	1	X	
	Conocarpus	1	X	
	Lumnitzera	2		X
Bombacaceae	*Camptostemon*	2		X
Plumbaginaceae	*Aegiatilis*	2		X
Palmae	*Nypa*	1		X
Myrtaceae	*Osbornia*	1		X
Sonneratiaceae	*Sonneratia*	5		X
Rubiaceae	*Scyphiphora*	1		X

[a] Based on Chapman (1970b) and Walsh (1974).

and Ustach, in press). On the northern Gulf coast and in the Laguna Madre of Texas only scattered thickets of *Avicennia* bushes are found (Price, 1954). Mangroves reach their best development in the brackish water of estuaries, but they are also well developed in protected regions dominated by seawater or along rivers and streams where saltwater is only occasionally present (Kuenzler, 1974).

B. PRODUCTIVITY

1. Overview

The open ocean area of approximately 326×10^6 km^2 is dominated by phytoplankton production. The area of diverse primary producers, the estuaries, amounts to only 1.8×10^6 km^2 on a worldwide basis. Of this, approximately 1.4×10^6 km^2 is open water and 3.8×10^5 km^2 is marsh and/or mangrove communities (Woodwell *et al.*, 1973). Although higher plants growing on land maintain a biomass more than 1000 times greater than plants in the marine environment, the annual amount of organic production on land and in the sea is about the same (Ryther, 1959). The annual gross production for the world is approximately 10^{17} g C of which 44% occurs in the marine environment (Odum, 1971). If it is assumed that 50% of this production is utilized in catabolic processes of the plants, then net annual primary production in the marine environment is 2.2×10^{16} g C. The gross productivity of marine and terrestrial ecosystems ranges from 20 to 2000 g C/m^2/yr (Table V).

2. Phytoplankton and Benthic Microalgae

Each phytoplankton cell is a complete photosynthesizing unit and no material is expended in the formation of roots, stalks, conductive elements, or other nonphotosynthesizing structures. The growth that takes place between one cell division and the next represents an approximate doubling of organic matter, and all growth goes into the formation of new cells capable of photosynthesis. As a consequence, the annual rate of net organic production by phytoplankton in the oceans, $1.2–1.5 \times 10^{10}$ tons of carbon per year (Smayda, 1966), is approximately four times the standing crop (Ryther, 1960). The production of organic matter on an areal basis by open ocean phytoplankton is about half that of coastal phytoplankton, averaging about 50 g C/m^2/yr for the open ocean and 100 g C/m^2/yr for coastal waters (Ryther, 1969).

Phytoplankton do not divide at a constant rate because of fluctuations in environmental conditions. Thus, the rates of production vary from one geographic area to another and seasonally in the same geographic area. There is a well-defined seasonal periodicity in biomass and productivity in temperate waters, with a maximum generally in spring followed by a decline during summer and a second, but smaller peak in fall. In Arctic and Antarctic waters, there

TABLE V

Major Ecosystems and Their Annual Gross Production[a]

Ecosystem	Area (10^6 km^2)	Productivity (g C/m^2/year)	Total production (10^{16} g C/year)
Marine			
Open ocean	326.0	100	3.3
Coastal zones	34.0	200	0.7
Upwelling zones	0.4	600	0.02
Estuaries and reefs	2.0	2000	0.4
Subtotal	362.4		4.4
Terrestrial			
Deserts and tundras	40.0	20	0.1
Grasslands and pastures	42.0	250	1.1
Dry forests	9.4	250	0.2
Boreal coniferous forests	10.0	300	0.3
Cultivated lands	10.0	300	0.3
Moist temperate forests	4.9	800	0.4
Wet subsidized agriculture	4.0	1200	0.5
Wet tropical and sub-tropical forest	14.7	2000	2.9
Subtotal	135.0		5.8
Total for biosphere	497.4	200	10.2

[a] Modified from Odum (1971).

is a unimodal distribution of biomass and production with a peak during the summer. The unimodal distribution in the tropics, however, peaks during the winter. Biomass and productivity maxima in estuaries often do not follow trends occurring in adjacent oceanic areas. For example, Williams (1966) and Thayer (1971) noted little change in phytoplankton cell numbers between cool and warm periods and a unimodal distribution in production with maximum rates during summer in the warm–temperate estuarine system near Beaufort, North Carolina. This example demonstrates that in estuaries, peak biomass and productivity rates do not necessarily occur simultaneously. Thayer (1971) reported a range, 0.1–19.5 mg/m^3, of individual measurements of chlorophyll concentration. More recent observations in the shallow (average depth at mean low water ~0.8m) Newport River portion of this same estuarine system indicate that brief phytoplankton blooms occur during early spring in response to increasing temperature of the water. On March 2 and 3, 1978, water temperature averged 7°C and chlorophyll *a* concentrations were uniformly low. A bloom with a peak chlorophyll *a* concentration of 104 mg/m^3 at a salinity of 10–15‰ was observed on March 14 and 15, 1978, at a water temperature of 16°C (Palumbo. 1980).

Average daily air temperature had increased gradually from 3° to 15°C over the time period of March 5–16 and then averaged 13°C during March 17–27. Maximum chlorophyll *a* concentrations of about 60 mg/m^3 were observed on March 17 and March 24 but then decreased to less than 15 mg/m^3 by March 27, when the phytoplankton peaked upstream in lower salinity ($<7‰$) water. Peaks in net production rates of phytoplankton are maintained only for comparatively short periods of time except perhaps in areas of continuous upwelling. The single highest daily rate found in the literature, 16.4 g C/m^2, occurred in the Duplin River Estuary, Georgia (Ragotzkie, 1959). Other high daily rates of 11.2 g C/m^2 in the upwelling area off Peru (Ryther, 1969) and 6.4 g C/m^2 in the Arabian Sea (Ryther and Menzel, 1965) have been recorded, but, in general, maximum rates of net production are 1–4 g C/m^2/day (Table VI). The highest sustained production rate found in the literature occurred off the Georgia coast, where Thomas (1966) observed an annual rate of 546 g C/m^2 or a daily average over the year of 1.5 g C/m^2. Average daily rates based on long-term studies usually range from 0.1 to 0.5 g C/m^2 (Table VII).

Microalgae also occur in bottom sediments of the photic zone and their production contributes to the fixed carbon budget of shallow waters. Lowest rates of production occur in sediments underlying deeper water and highest rates occur in protected intertidal and shallow subtidal sediments of estuaries. Measurements of annual gross production range from 100 to 200 g C/m^2 in intertidal estuarine and salt marsh sediments (Table VIII). Similar production rates have been observed in shallow subtidal sediments of the Øresund, Denmark (Gargas, 1970). Measurements of subtidal benthic microalgal productivity using relatively undisturbed cores incubated *in situ* are somewhat lower. In the Newport River estuary, annual gross production in photic zone sediments was 67 g C/m^2 and on total area was 34 g C/m^2 (Bigelow, 1977). Productivity in the lower estuary followed seasonal changes in insolation and at water depths ≤ 1 m, could equal or exceed phytoplankton productivity in the overlying water. In the upper estuary, reduced light penetration to the bottom due to high turbidity contributed to observed low

TABLE VI

Representative Marine Areas and Rates of Production of Phytoplankton

Area	Net production (g C/m^2/day)	Reference
Newport River estuary	1.6	Williams (1966)
Polluted estuary, Long Island	0.9	Barlow *et al.* (1963)
Grand Banks	3.2	Ryther (1959)
Continental Shelf	3.7	Ryther and Yentsch (1958)
Sargasso Sea	0.8	Menzel and Ryther (1960)

TABLE VII

Representative Daily Rates of Net Production Obtained from Yearly Totals

Area	Production ($g\ C/m^2$)	Reference
Sargasso Sea	0.2	Menzel and Ryther (1960)
Continental slope (off Long Island)	0.3	Ryther and Yentsch (1958)
Long Island coast	0.5	Ryther and Yentsch (1958)
Long Island Sound	0.5	Riley (1956)
Estuaries, Beaufort, North Carolina	0.1–0.2	Williams (1966), Thayer (1971)
Wacassassa Estuary, Florida	0.3	Putnam (1966)
Patuxent River estuary, Maryland	0.8	Stross and Stottlemyer (1965)
Altamaha River mouth, Georgia	1.5	Thomas (1966)

productivity. The low productivity observed in sandy beach sediments in Scotland, 4–6 g $C/m^2/yr$, is probably due to the scouring action of waves (Steele and Baird, 1968). Seasonal trends in production rate occurred in the nearshore subtidal sediments in Alaska; 0.5 in winter and 57 mg $C/m^2/hr$ in August (Matheke and Horner, 1974). However, a seasonal trend was not apparent in a Georgia salt marsh (Pomeroy, 1959), where tidal innundation appeared to be an important factor. Average meridian intensity in winter is greater than 9×10^4 lux and greater than 16×10^4 lux in summer in the southeastern United States (Rice and Ferguson, 1975). Production of benthic microflora increases about 10% per degree temperature increase between 4° and 20°C, but may be inhibited at tem-

TABLE VIII

Gross Production of Benthic Microflora

Area	Sediment	Production	Reference
		($g\ C/m^2/year$)	
Georgia	Salt marsh	200	Pomeroy (1959)
Washington	Tidal flat	144	Pamatmat (1968)
Dutch Wadden Sea	Tidal flat	101	Cadee and Hegeman (1974)
Scotland	Sandy beach	4–9	Steele and Baird (1968)
		($mg\ C/m^2/hr$)	
Eastern Long Island Sound	Shallow subtidal estuarine	0–165	Marshall *et al.* (1971)
North Carolina	Shallow subtidal estuarine	67	Bigelow (1977)
Alaska	Nearshore subtidal	0.5–57	Matheke and Horner (1974)
Caribbean Sea	To depth of 60 m	0–31.6	Bunt *et al.* (1972)

peratures above 22°C. Also, production is light saturated at 1×10^4 lux and photoinhibited at higher intensities (Colijn and van Buurt, 1975). Therefore, the depressed microfloral production in Georgia (Pomeroy, 1959) probably resulted from light and thermal inhibition. The depression of photosynthesis was most evident in summer and resulted in similar rates of production throughout the year.

3. Macroscopic Plants

Submergent seagrasses have a high rate of organic production; i.e., 300–600 g C/m^2/year (Table IX). Values of 500–1500 g C/m^2/year have been estimated for *Thalassia testudinum,* the most important tropical–subtropical species, while the dominant temperate species, *Zostera marina,* produces 150–300 g C/m^2/year. These production ranges indicate a ratio of production of leaves to their average biomass, or turnover, of about 6 times per year for *Thalassia* and between two and three times per year for *Zostera*. Although the areal production of *Zostera marina* is less than that of *Thalassia testudinum,* it represents a significant portion of total production in some temperate regions. For example, in the estuaries near Beaufort, North Carolina, *Zostera* accounts for 62% of the total production, phytoplankton 27%, *Spartina alterniflora* 7%, and benthic microalgae 3% (Williams, 1973; Ferguson and Murdoch, 1975; Thayer, *et al.,* 1975a). The production data do not include root production, and hence, are conservative estimates. The biomass of the root–rhizome system ranges from 50–90% of the total biomass of the plant (Burkholder *et al.,* 1959; Penhale, 1976). Production of roots and rhizomes is significant. For example, *Zostera marina,* starting from a single seed, can cover an area of 30 cm^2 by vegetative growth during the first year, 1 m^2 during the second, and 2 m^2 during the third (Arasaki, 1950).

The leaves of submergent angiosperms provide a substrate for the attachment of epiphytic organisms, which support a grazing food chain. In the tropics, the biomass of this epiphytic community may equal that of the macrophyte leaf (Wood *et al.,* 1969). In temperate waters, the epiphytic community on *Zostera* can range from 1% (Wood *et al.*, 1969) to 24% (Penhale, 1976) of the total biomass of the leaf plus epiphytes. The epiphytes on eelgrass blades near Beaufort had a mean annual production of about 70 g C/m^2/yr, approximately 18% of the mean annual *Zostera* plus epiphyte production (Penhale, 1976).

Benthic macroalgae also contribute to the productivity of the coastal zone, and on an areal basis the productivity of some species can equal or exceed that of submerged seagrasses (Mann, 1972, 1973). Net production rates for *Laminaria* beds in north temperate areas can exceed 1500 g C/m^2/yr. Other brown algae, the fucoids and kelps, contribute about 1000 g C/m^2 annually to some areas of the north temperate zone. New tissue produced each year by some kelps in Canada, e.g., *Laminaria longicruris, L. digitata,* and *Agarum criorosum,* averaged nine times the initial biomass of the blade (Mann, 1973).

TABLE IX

Organic Production by Marine Macrophytes[a]

Macrophyte	Production (g/m²/year)	Location	Reference
Seagrasses			
Zostera	5–26	Massachusetts	Conover (1958)
	140	Denmark	Grøntved (1958)
	180–320	Alaska	McRoy (1966)
	90–540	Washington	Phillips (1974)
	680[b]	North Carolina	Dillon (1971)
	330	North Carolina	Penhale (1976)
Cymodocea	150	Mediterranean	Gessner and Hammer (1960)
Posidonia	750	Malta	Drew (1971)
Thalassia	900	Florida	Jones (1968)
	580	Texas	Brylinsky (1967)
	500–1500	Indian Ocean	Mann (1972)
Seaweeds			
Laminaria	1200–1900	Atlantic Ocean	Mann (1972)
Macrocystis	400–820	California	Mann (1972)
	1000–2000	Indian Ocean	Mann (1973)
Fucus	640–800	Canada	Mann (1972)
High salt marsh			
Juncus roemerianus	280–612	North Carolina	Williams and Murdoch (1972), Waits (1967), Foster (1968), Stroud and Cooper (1968)
	425	Everglades, Florida	Heald (1969)
Spartina patens	500	Long Island	Harper (1918)
	650	North Carolina	Waits (1967)
S. patens together with *Distichlis spicata* and *Scirpus robustus*	670	North Carolina	Waits (1967)
Fresh and brackish marsh			
Spartina alterniflora	1950	Louisiana	Kirby (1971)
S. cynosuroides	500	Georgia	Odum and Fanning (1973)
Phragmites communis	675	Long Island	Harper (1918)
Low salt marsh			
Spartina alterniflora	220–2000	East and Gulf Coasts, United States	Morgan (1961), Odum (1959), Williams and Murdoch, (1969), Smalley (1959), Kirby (1971), Teal (1962), Odum and Fanning (1973)
Mangroves (leaf litter)			
Rhizophora mangle	400	Florida and Puerto Rico	Heald (1969) Golley *et al.* (1962)
R. brevistylis	375	Panama	Golley *et al.* (1969)

[a] When necessary, dry weight values have been converted to carbon, assuming a carbon/dry weight ratio of 50%.

[b] Includes the production of *Ectocarpus* and *Halodule*.

Emergent coastal marshes and swamps also are among the most productive natural habitats. The vascular plant flora dominate total primary production in these communities. In the salt marshes of Sapelo Island, Georgia, for example, *Spartina alterniflora* accounts for about 80% of the total vascular plant plus algal net production (Teal, 1962). Productivity values ranged from 200 to 2000 g C/m^2/yr for marshes and averaged about 375 g C/m^2/yr for mangroves (Table IX). Measurements of the production of vascular plants in natural salt marshes generally are limited to estimates of increases in above ground biomass through the year. Those estimates for above-ground production for which losses of leaf material have been taken into account are summarized in Table IX. The exception to this are estimates of annual leaf litter production by mangrove trees, which are the most readily measurable rates of production. These estimates of production are conservative because growth of below-ground portions of the plants are not included. The root and rhizome biomass is at least half of the total biomass of emergent marsh species, and below-ground production of *Spartina alterniflora* and *S. patens* may be six times that of the above-ground production (Valiela *et al.*, 1976).

Production of submergent and emergent vascular plants varies seasonally because of many factors, including climate and characteristics of the species. Vascular plants have a growing season which, in general, is from spring through fall. Initiation of growth in the spring and its termination in the fall depends on the response of the species to changing temperatures and/or day lengths. Growth may be reduced during the warmest summer periods and may continue through winter in milder climates.

Important aspects of submergent and emergent vascular plant dominated habitats are the time of death of leaves and other above-ground portions of the plant, the addition of this material to the detrital component of the habitat, and the exchange of matter between these habitats and adjacent open waters. In the submergent plant, *Zostera,* leaves mature and age on the plant before dying and dropping. This is a continuous process, with maximum defoliation at Beaufort, North Carolina during the fall (G. W. Thayer, unpublished data). Dead leaves of the submergent tropical species *Thalassia* also enter the detrital pool during this period (Zieman, 1975b). Zieman *et al.* (1979) have demonstrated that significant quantities of seagrasses are exported from Tague Bay, Virgin Islands. Surface export of *Syringodium filiforme* represented 60–100% of the leaf production of this plant, whereas surface export of *Thalassia testudinum* represented only 1% of its production. In Florida, red mangroves, *Rhizophora mangle,* drop leaves year-round, but the peak period in June represents 60% of the year's total (Heald and Odum, 1970). For low salt marsh species along the eastern seaboard of the United States, the peak standing crop occurs in late summer to early fall with a gradient from north to south. For example, peak standing crop occurs from mid-August to early September near the Virginia–Maryland border (Keefe and Boynton, 1973). In the Beaufort, North Carolina area, *Spartina alterniflora* enters the estuarine system as partially decomposed material primarily from late fall through winter (Thayer, 1974). For high salt marsh species such as *Juncus roemerianus,* the occurrence of peak standing crops is difficult to establish because of the presence of large amounts of standing dead material (Williams and Murdoch, 1972). In low salt marshes and even more so in high salt marshes, the export of plant matter to the adjacent estuarine and coastal ecosystem is both seasonal and episodic, a function of spring and storm tides which dislodge and transport dead and dying plant matter. In a review of marsh–estuarine interactions, Nixon (1980) indicates that few studies have been adequate to make an accurate assessment of the exchange of organic matter and other nutrients between salt marshes and adjacent open waters of estuaries. He concludes, however, that where sufficient data do exist, tidal marshes appear to export dissolved and particulate organic carbon, dissolved organic nitrogen, and inorganic phosphorus.

C. ENVIRONMENTAL FACTORS INFLUENCING DISTRIBUTION AND ABUNDANCE

Physical, chemical, and biological factors and their interactions affect the distribution and abundance of marine primary producers. Proximity to dry land or to the bottom has a profound impact on the characteristics of the water and thus, on primary productivity. Depth of the water moderates the influence of both edaphic and climatic factors. Water movements, currents, upwelling, downwelling, tides, and wave action also have important influences on productivity.

Light intensity and temperature are important physical factors that vary geographically and with depth of the water. Illumination at the surface varies geographically and seasonally in terms of intensity throughout the day and also in length of the daily light period. Both quantity and quality of light vary with depth due to selective absorption of different wavelengths by the water and its dissolved constituents and to absorption and scattering of light by suspended particles (Jerlov, 1951). Estuarine and coastal water are often turbid and yellow-brown in color and are most transparent to light at a wavelength of 580 nm. Clearest oceanic water is most transparent to light at a wavelength of 480 nm and is deep blue in color. The depth of penetration of 1% of surface illumination ranges from a maximum of about 100–150 m in oceanic water to a minimum of between 1 and 4 ms in turbid estuaries. In estuaries of high and variable turbidity, illumination at different depths is a function of turbidity. Thus, seasonal changes in day-length are directly significant to phytoplankton, whereas seasonal changes in surface illumination affect phytoplankton indirectly by increasing water temperature (Ferguson, 1972; Ferguson *et al.,* 1976). Temperature also varies geographically, seasonally, and diurnally. The high specific heat of water moderates seasonal and diurnal changes, particularly in the ocean. The shallowness of coastal water and the tidal exposure of the littoral zone allows larger and more frequent temperature oscillations. In the Newport River estuary, North Carolina, for example, hourly gradients of up to 6.2°C have been observed (Rice and Ferguson, 1975). The north–south gradients in daylight and temperature affect morphological and physiological traits of emergent vascular plants along the east coast of the United States (Table X).

Chemical characteristics of freshwater and seawater are dependent on the dissolved constituents (Table XI). While the major solutes of flowing freshwater and seawater are similar, their concentrations and ratios are different. The concentrations of specific ions in freshwater vary from place to place. For example, while the bicarbonate content of natural fresh waters ranges from 0.0001 to 0.4 molal, the elemental composition of oceanic water, including the bicarbonate content, is remarkably constant and may have been so for at least the past 100 million years (Stumm and Morgan, 1970). The pH of freshwater generally ranges from 6–9 to an occasional low of 4–5 in rivers draining freshwater swamps and marshes, e.g., in eastern North Carolina (R. L. Ferguson, W. Sunda, and A. V. Palumbo, unpublished data; R. T. Barber and W. W. Kirby-Smith, unpublished data). A pH as low as 7 has occurred at a salinity of one-third strength seawater in the Pamlico River estuary of North Carolina during periods of freshwater flushing of upstream marshes (Hobbie, 1974). A pH as high as 9 has been measured at a similar salinity in the Newport River estuary of North Carolina during a phytoplankton bloom (Palumbo, 1980). On the other hand, the pH of seawater is less

TABLE X

Studies on Morphological and Physiological Characteristics of *Spartina alterniflora* along the Latitudinal Gradient of the Atlantic and Gulf Coasts of the United States[a,b]

Dependent variable	Phytotron and regional seeding studies	North Carolina seeding studies (second year data)
	response from North to South	
Flowering	initiated later	initiated later
Vegetative growth period	adapted to longer growing season	initiated earlier and terminated later
Dependence of growth rate on photoperiod	less influence	not applicable
Basal culm diameter and leaf widths	increasing	increasing[c]
Production	increasing	increasing[c]

[a] Based on Seneca (1974).

[b] Seeds were collected throughout the coast and planted in a phytotron under simulated regional conditions or under actual environmental conditions in the region or in North Carolina.

[c] Exclusive of Gulf coast group.

TABLE XI

Major Solutes, Bicarbonate Concentrations, and pH of Freshwater and Seawater[a]

Ion	Freshwater (molar)	Seawater (molar)
Total ionic strength	0.00015–0.02	0.65
Cations		
Sodium	0.00027	0.47
Potassium	0.000059	0.01
Calcium	0.00038	0.01
Magnesium	0.00034	0.054
Anions		
Chloride	0.00022	0.55
Sulfate	0.00012	0.038
Bicarbonate	0.0027[b]	0.0024
pH	7.5[c]	8.2

[a] Based on Stumm and Morgan (1970).

[b] Range of various freshwater streams, groundwater, and lakes; 0.0001–0.4.

[c] Range of pH of freshwater is generally 6–9 or occasionally as low as 4 or 5 in rivers draining freshwater swamps and marshes; e.g., Newport and South Rivers, North Carolina (R. L. Ferguson, W. Sunda, and A. V. Palumbo, unpublished data, and R. T. Barber and W. W. Kirby Smith, unpublished data).

variable. Maximum changes in pH in ocean water are associated with upwelling. For example, off the coast of Oregon, a range of pH from 7.7 to 8.3 has been observed (Park, 1968).

Freshwater and seawater mix in the littoral and shallow neritic zones. Salinity in coastal areas fluctuates with freshwater runoff but can be greater than that of seawater during periods of high evaporation. In parts of the Laguna Madre of the Texas Gulf coast, salinity may exceed by 2.2–2.8 times the salinity of oceanic seawater (Conover, 1964). During periods of low tide and high evaporation, water trapped in depressions on the shore may reach saturation concentrations resulting in the formation of salt crystals. Also, the rate of change of salinity may be rapid in some estuaries. In the Newport River estuary, North Carolina, for example, the maximum salinity change observed over a period of 1 hr was 20‰ (Rice and Ferguson, 1975).

The concentrations of nutrients other than inorganic carbon are not the same in freshwater, intermediate salinity water, and oceanic water. Two of the most important are nitrogen and phosphorus. Nitrogen is present in the water as dissolved gas and organic nitrogen and as fixed inorganic nitrogen, nitrate, nitrite, and ammonia. Phosphorus is present in a variety of ortho- and condensed organic and inorganic phosphates. In oceans the N:P molar ratio (inorganic fixed nitrogen:orthophosphate) is generally 15:1 (Redfield *et al.*, 1963). The concentration of nutrients in surface ocean water generally is not as high as it is in bottom sediments or in terrestrial run off. Thus, nitrogen and phosphorus levels tend to be higher in less saline and in shallower water. In estuaries the N:P ratio often is far below 15 (Thayer, 1971), but the total fixed inorganic nitrogen is seldom less than 0.5 m*M* and is normally at least several times that concentration (Williams, 1972). The range of phosphate content for most United States rivers is 0.3–3 m*M* (Stumm and Morgan, 1970) and less than 2.8 m*M* in unpolluted estuaries (Hobbie, 1974). Nitrogen most often is the nutrient that limits phytoplankton growth in estuaries (Thayer, 1974; Rice and Ferguson, 1975), and added nitrogen often stimulates growth of *Spartina alterniflora* (Table XII). The rate of production and the generally high levels of trace metals and nutrients in submerged sediments suggest that seagrasses may not be nutrient limited. Orth (1977) demonstrated, however, a significant increase in the length, biomass, and total number of shoots of *Zostera marina* in experimental field plots to which commercial fertilizers had been introduced to the substrate. Submerged aquatic plants are capable of utilizing nutrients from both the water column and the sediments (McRoy and McMillan, 1977). It also appears that seagrasses utilize different forms of nitrogen either directly or indirectly subsequent to microbial transformation (McRoy and McMillan, 1977; Harlin, 1980; Klug, 1980; Smith, Hayasaka and Thayer, 1980).

Three micronutrients—silicon, iron, and copper—illustrate both the di-

TABLE XII

Relationships of Above-Ground *Spartina alterniflora* Productivity to Observed Factors of Marshes and Responses of Natural Marshes and Marshes Planted on Dredge Soil to Added Nutrients[a,b]

	Soil Factor			Nutrient Added	
Marsh type	Salinity	Sulfur	Phosphorus	Phosphorus	Nitrogen
Natural muddy	–	–	+		
Tall form	–	–	+	0	0
Short form	–	–	+	+	+
Natural sandy	–	–	+	+	+
Planted soil	–	–	+	+	+

[a] Based on Woodhouse *et al.* (1974).

[b] (–), negative correlation between factor and observed productivity; (+) positive correlation between factor and observed productivity; (0) no correlation between factor and observed productivity.

vergence of concentrations of dissolved nutrients in fresh and saltwater and the physicochemical problems associated with dissolved and particulate fractions of these elements (Stumm and Morgan, 1970). The dissolved concentrations in freshwater are about 1–2 orders of magnitude greater than in seawater (Table XIII). Suspended clays contain relatively large amounts of these elements, being 5.1% iron and 0.03% copper by weight (Cronan, 1969), in contrast to their minute dissolved concentrations. While the proportion of dissolved and particulate fractions is highly pH dependent, it also is affected by biotic factors.

The size of the standing crop of primary producers at any given time is a direct function of primary production and mortality rates. Physical factors also may concentrate and disperse plants and biotic factors will influence primary production indirectly by their interactions with physical and chemical factors. Primary producers absorb light and reduce its penetration into deeper water. Some microorganisms have the capacity to fix atmospheric nitrogen while others are denitrifyers which decrease the levels of fixed nitrogen. Macroscopic attached forms reduce the speed of water movement and increase the deposition of suspended material (Thayer *et al.*, 1975b). Benthic forms of both macroscopic and microscopic primary producers tend to stabilize the sediments (Wood *et al.*, 1969; Thayer *et al.*, 1975b; Zieman, 1975b). Relative rates of photosynthesis and respiration by the biota affect pH and the dissolved oxygen content of the water. Production of dissolved and particulate organic matter increases rates of heterotrophic activity. This activity affects nutrient regeneration and the chelation of dissolved metals which will influence their availability for uptake or potential for

TABLE XIII

Concentration of Dissolved Silicon, Iron, and Copper in Water[a]

Element	Freshwater (μM)	Seawater (μM)
Si	65–386[b]	0.05–46[c]
Fe	0.2–25[b]	0.01–0.03[d]
Cu	0.1–0.4[e]	0.005–0.03[d]

[a] Particulate fraction of iron and silicon is highly variable and may exceed dissolved fraction by up to 2 or 3 orders of magnitude.

[b] Livingstone (1963).

[c] Range of averages for surface waters of the major oceans. Concentration increases with depth to a range of averages of 43–150 μM (after Armstrong, 1965).

[d] Chester and Stoner (1974). The range of copper in near surface open ocean waters is now estimated to be 1–3 nM (Boyle *et al.*, 1977; Bruland *et al.*, 1979).

[e] Range for dissolved metal in 16 drainage basins in the United States (after Kopp and Kroner, 1969).

toxicity. Abundance of organic matter, decreased penetration of light, and reduced circulation of water or permeability of bottom sediments all tend to produce low oxygen levels or anoxic conditions.

III. Adaptation to the Marine Environment

A. OVERVIEW

All organisms adapt to a range of each environmental factor. Above or below these ranges, physiological stress or death can occur. Adaptations vary from subtle to highly detectable and can take place rapidly or slowly. Individual organisms are limited by their genetic potential under existing environmental conditions. Populations can adapt genetically in terms of the relative frequency of resistive, physiological, and morphological capacities. Many phytoplankton and marine macrophyte species are capable of rapidly adjusting to hyper- and hypoosmotic conditions and to changes in temperature. Marine macrophytes also adapt, often in a short period of time, to atmospheric exposure. Such environmental factors as temperature, salinity, and light quality and quantity, are important to all plants. In addition, macrophytes must adapt to tidal and wave action and conditions of the sediment. These factors affect not only the geographic distribution and vertical zonation of plants within each habitat, but also the morphological and physiological condition of individuals of a species. Variations in form and function within these plant groups are the result of their adaptive responses.

B. PHYTOPLANKTON

One of the major morphological adaptations to a planktonic existence is a reduction of size, which increases the surface area to volume ratio. This adaption facilitates flotation, uptake of nutrients, and absorption of light energy. Other morphological adaptations influence the relative sinking rates of different species, of individuals of the same species in different growth stages, and of the same species found in waters of different densities. There are four major structural adaptations exhibited by diatoms: (1) the bladder type cell, e.g., *Coscinodiscus,* in which the protoplasm and cell wall form a thin layer inside a relatively large test with the majority of the cell volume being enclosed in an inner vacuole; (2) the needle type cell, e.g., *Rhizosolenia,* in which the cell is elongated along one axis; (3) the ribbon type cell, e.g., *Fragilaria,* in which the cell is flattened along one plane; and (4) the branched type cell, e.g., *Chaetoceros,* in which the cell exhibits projections. The presence of oil also facilitates flotation. Dinoflagellates, along with some other groups of phytoplankton, possess flagella that provide for limited locomotion as well as maintenance of position within the water column. Projections are found in dinoflagellates and also provide an additional flotation mechanism.

A number of phytoplankton species show regional and seasonal variation in cell structure, presumably in response to water density and viscosity. These are adaptations that decrease sinking rates. For example, *Chaetoceros decipiens,* a branched, chain-forming diatom, has a thin cell wall, long, thin projections, and large interstices between cells during the summer. In the winter, however, when density and viscosity of water are greater, the cell walls are thicker, the projections are shorter and more substantial, and the interstices are smaller (Raymont, 1963). Pelagic diatoms usually are thinner shelled than littoral forms. Dinoflagellates also possess larger projections or flaring structures in warmer waters.

Sinking rates are dependent not only on morphology, but also on the physiological condition of the cells (Eppley *et al.,* 1967). Control of vertical position within the photic zone is a primary requirement for survival of phytoplankton in nature. Physiologically less active cells sink faster in culture media, but the ecological significance of this phenomenon is not completely understood at this time. Those cells most active physiologically, however, are often the closest to being neutrally buoyant and the least likely to change their vertical position. Although sinking out of surface waters might be considered of potentially greater significance to oceanic than to estuarine forms, this is not necessarily correct. Hulburt (1970) suggests that the dominant species in shallow estuaries tend to sink more slowly than coastal and oceanic species, and thus avoid extinction through entrapment in the low turbulence water overlying the sediment. The size and structural modifications of the cell also may play significant roles in the vulnerability of phytoplankton to filter feeding herbivores (Raymont, 1963).

Marine phytoplankton exhibit resistive physiological and morphological adaptations to environmental factors. Adaptations often are best observed by following the response of an isolated phytoplankton species in controlled laboratory experiments. For instance, it has been shown that the response of isolates to many environmental factors are correlated with the ranges of those factors which characterize their natural habitat (Braarud, 1961; Fogg, 1965; Rice and Ferguson, 1975). Relative adaptability of estuarine phytoplankton is greatest to low salinity but least to low nutrient concentrations, whereas oceanic species are adapted to high salinity but low nutrients (Table XIV). The response of neritic species generally is intermediate to those from estuaries and open oceans.

Phytoplankton cells adapt physiologically to salinity by adjusting their internal salt concentration to a level slightly higher than that of the growth medium (Guillard, 1962). This establishes an osmotic pressure which favors the flow of water into the cell and maintains cell turgor. Thus, the salinity range at which cell

TABLE XIV

Physiological Adaptation of Phytoplankton from Different Marine Environments to Salinity and Nitrate[a]

	Salinity		Nitrate	
Marine environment	Number of species tested	Minimum allowing half-maximum division rate (‰)	Number of species tested	Mean half-saturation uptake concentration (μMoles N/liter)
Oceanic	10	18–24	6	0.3
Neritic	6	0.5–20	15	2.3
Estuarine	6	0.5–8	4	5.1

[a] After Rice and Ferguson (1975).

division occurs and the rate of cell division depends on the metabolic rates as affected by altered internal salt concentrations. Adaptive adjustment of the internal salt concentration must proceed rapidly, since cell division rates of euryhaline diatoms at different salinities do not change as a result of conditioning to these salinities (Williams, 1964). Isolates of oceanic species divide at over half maximum rates at salinities above 18–24‰. Estuarine forms, on the other hand, divide at over half maximum rates at salinities above 0.5–8‰. The response of neritic species is intermediate to that of oceanic and estuarine forms. Some estuarine species can divide in freshwater but most freshwater species will divide only at salinities of less than 2‰ (Chu, 1942). *Chlamydomonas moewusii,* a freshwater species, however, will divide at salinities of up to 29‰ (Guillard, 1960).

While cell division can be limited by the uptake and assimilation of some nutrients at low concentrations, uptake of other nutrients may exceed requirements even at very low concentrations. For example, the rate of uptake of phosphorus by *Phaeodactylum* at low phosphate concentrations (0.01–0.17 μM P) is greater than that needed for maximal division rate (Kuenzler and Ketchum, 1962). Most frequently nitrogen is the limiting nutrient in estuarine waters (Williams, 1973; Thayer, 1974), and its concentration can affect the rate of cell division. Under conditions of nitrogen limitation, *Isochrysis galbana* divides more rapidly in continuous culture with an increased flow rate of the medium (Caperon, 1968). Thus, the rate of supply of a nutrient also determines the rate of cell division.

Phytoplankton adapt to low levels of nitrogen by changing the assimilation number, i.e., the productivity per unit chlorophyll. Nitrogen-deficient cells have lower assimilation numbers than non-deficient cells since with progressive nitrogen deficiency, the chlorophyll content of cells will decrease more rapidly than the nitrogen content. Also, this chlorophyll is less active than that in non-deficient cells (Bongers, 1956; Fogg, 1959). With the addition of nitrogen to the culture medium, cells will recover from their deficiency; i.e., their chlorophyll content will increase and higher assimilation numbers will occur (Syrett, 1962). Phytoplankton adapt to the nutrient availability of their environment not only by changing their assimilation number, but also by changing the organic content in the cells. For example, in *Chlorella* the proportions of carbohydrate, protein, and lipid vary in relation to nutrient availability and the length of time the cells have grown in nutrient-depleted medium (Milner, 1948). Carbohydrate content varied from 15 to 37.5%, protein from 7.9 to 46.4%, and lipid from 20.2 to 77.1%, as a result of shifts in metabolic end products in response to nutrient deficiency. This metabolic shift has an adaptive advantage since nitrogen-deficient cells take up ammonia-nitrogen at rates of four to five times those of normal cells (Bongers, 1956; Harvey, 1953). This ammonia-nitrogen uptake results in an increased respiration rate and a reduced carbohydrate reserve in the cells (Syrett, 1953).

Phytoplankton adapt to temperature by changing their rate of division (Eppley, 1972) and productivity per unit chlorophyll (Talling, 1955; Williams and Murdoch, 1966; Ichimura, 1968). If temperatures are within the range favorable for growth, division rates of cells in culture generally increase by a factor of two to four times with a 10°C increase in temperature (Fogg, 1965). For example, cells of *Thalassiosira pseudonana* estuarine clone 3H divide twice as fast at 20°C as at 10°C (Ferguson, 1971). Both division rate and assimilation number increase with temperature until unfavorably high temperatures occur, since metabolic rates, including the dark reaction rates of photosynthesis, are temperature dependent. Phytoplankton also adapt to temperature with a change in the organic composition of their cells. *Skeletonema costatum* cells show increases in photosynthetic enzymes and organic matter at low temperatures and double their carbon content per cell as temperature decreases from 20° to 7°C (Jørgensen, 1968). This increase in cell carbon and carbon per unit chlorophyll *a* with decreasing temperature is characteristic of marine phytoplankton (Eppley and Sloan, 1966; Jørgensen, 1968). An increase in amount of cellular enzymes per cell to some degree offsets the decrease of enzyme activity with decrease in temperature, since cell division and dark reaction rates of photosynthesis depend on rates of enzymatic processes. The evidence for adaptation of species to the thermal regime of the water from which they were isolated is not entirely consistent (Braarud, 1961). Many species have been shown to have temperature optima which approximate that temperature observed during periods when they are abundant. However, laboratory determinations of temperature optima in some species, e.g., *Asterionella japonica* and *Thalassiosira nordenskioeldi,* appear to be at variance with field observations, which indicate lower temperature optima than those observed in culture (Kain and Fogg, 1958; Braarud, 1961). Although Braarud

(1961) correctly criticized the lack of suitable controls in work limited to non-bacterial free cultures, later studies have established the additional problem of the interaction of temperature with other environmental factors (McCombie, 1960; Ukeles, 1961; Jitts *et al.*, 1964; Maddux and Jones, 1964; Ferguson *et al.*, 1976). In particular, the very high nutrient concentrations that characterize media used in cultures of limited volume often interact with temperature to result in higher apparent temperature optima.

The rate of cell division is light dependent, since it is related to the supply of photosynthetically produced carbon. The rate of production is equal to the rate of respiration at the compensation intensity. At even lower light intensities, cellular carbon is used faster than it is produced. Increases in division rate occur with increases above the compensation intensity until unfavorably high light intensities are reached or until some other factor becomes limiting. Phytoplankton also adapt to changes in light intensity by changing the amounts of pigments or the amounts of photosynthetic enzymes in the cells. *Chlorella pyrenoidosa* responds by decreasing the amount of pigment per cell as the light intensity increases (Steeman-Nielsen and Jørgensen, 1968). *Cyclotella meneghiniana,* on the other hand, adapts to increasing light intensity by increasing the maximum rates of enzymatic reactions (Jørgensen, 1964). Decreasing the chlorophyll *a* content of the cell increases the assimilation number and decreases the compensation intensity. These adaptive changes allow cells to utilize light of lower intensity than cells that have not been adapted.

The evidence of genetic adaptation within a species is sparse and conflicting. Clones of the same species isolated from different waters often show significantly different half-saturation concentrations (Ks) for nitrate uptake. Clones isolated from the Sargasso Sea where nitrogen content was $\leq 0.5\ \mu M$, however, retained their relatively low Ks values even after having been cultured for 12 years in a high nitrate medium (Carpenter and Guillard, 1970). Different clones of the same species of *Peridinium micans* and *P. trichoideum* have demonstrated differing temperature optima according to the temperature regime of the water from which they were isolated (Braarud, 1961). Similar observations, which are also conflicting, have been noted for the genetic adaptation of metabolic rates to internal salt concentrations. For example, two different clones of the species *P. trichoideum,* one isolated from water of 20‰ and the other from 37‰, responded in a similar manner to a variation in salinity (Braarud, 1961). Oceanic, neritic, and estuarine clones of *Thalassiosira pseudonana,* however, showed half maximum division rates at salinities corresponding to their environmental ranges (Guillard and Ryther, 1962; Guillard, 1963).

Studies on the excretion of soluble, extracellular substances by phytoplankton have been concerned with the chemical nature of the excreted products, the conditions of the environment and of the cells which control the production of these substances, and their ecological significance. A metabolite may act as an accessory growth substance, as an inhibitory substance, as a chelator of heavy metals, or it may have no effect on the growth of a given species. In culture, many species of phytoplankton produce metabolites which serve as direct or accessory growth substances for other species. The growth rates of *Scenedesmus obliquus* and *Dictyosphaerum ehrenbergianum* have approximately doubled when grown in medium containing water extracts of disrupted phytoplankton cells (Hartman, 1960). Jørgensen (1956) has found a stimulation of the growth of *Scenedesmus quadricauda* when grown in filtrates from cultures of *Chlorella pyrenoidosa* and *Nitzschia palea*.

Phytoplankton also can produce metabolites which inhibit their own division rate as well as that of other species. Pratt and Fogg (1940) grew cultures of *Chlorella vulgaris* in inorganic media for different periods of time and until different population sizes had been obtained. The cells then were removed by filtering the culture medium. After pH adjustment this filtered medium was used in different proportions with fresh medium to prepare new cultures. The growth of *Chlorella* in these media was slower than when no conditioned medium had been added. Worthington (1943) found that when an *Asterionella* bloom was disappearing, the water of Lake Windermere, England, was unsuitable for the preparation of culture medium. This fact suggests that *Asterionella* produces an inhibitory substance. Phytoplankton also produce metabolites which, when released into the water, often inhibit the growth of other species. The growth rates of *Chlorella vulgaris* or *Nitzschia frustulum* are less when grown together in mixed cultures than when grown alone, depending on the size of the populations used (Rice, 1954). An increase in the initial concentration of *Nitzschia* reduced the growth rate of *Chlorella* and similarly, a decrease in the growth rate of small populations of *Nitzschia* occurred when the initial concentration of *Chlorella* was increased.

The presence in the water of organic substances that chelate toxic heavy metal ions may reduce the availability of the metal to the cells and therefore reduce the apparent toxicity of the metal (Steeman-Nielsen and Wium-Andersen, 1970; Sunda, 1975; Sunda and Guillard, 1976). These organic chelators may be excreted by the phytoplankton growing in the media (Steeman-Nielsen and Wium-Andersen, 1971) or be included in the synthetic media (Provasoli *et al.*, 1957). In the oceans, deep water and newly upwelled water is relatively rich in inorganic nutrients, due to heterotrophic activity in the aphotic zone. This activity has its impact on the levels of both dissolved and particulate organic matter. Although some deep and newly upwelled water still can have sufficient quantities of dissolved organic chelators to limit the toxicity of the trace metals present, some do not (Barber *et al.*, 1971). Suitability of upwelled water in the equatorial Pacific for the growth of phytoplankton can increase with time at the surface despite the reduction of such nutrients as nitrate, phosphate, and silicate. This fact suggests that the water is being conditioned through the scavaging of trace metals or the excretion of chelators by the phytoplankton (Barber and Ryther, 1969; Barber, 1973). Direct resistance to trace metal toxicity or release of chelators by phytoplankton species can be of adaptive significance in the marine environment. Copper concentrations greater than 10 μg/liter decreased productivity and biomass of the phytoplankton enclosed in plastic containers moored in Saanich Inlet, British Columbia; however, a return to initial levels occurred after 2–3 weeks, with a shift of dominant species from the diatom *Chaetoceros* spp. to small microflagellates (Thomas and Siebert, 1977; Thomas *et al.*, 1977). Relative sensitivity of phytoplankton is to a certain extent related to taxonomic groups of which dinoflagellates and cyanophytes are generally the most sensitive and green flagellates the least sensitive species (Erickson *et al.,* 1970). Huntsman and Sunda (in press) have recently reviewed the role of trace metals in regulating phytoplankton growth in natural waters, and they further detail biological adaptations of phytoplankton to trace metal stress.

C. SUBMERGENT VASCULAR PLANTS

Marine submergent vascular plants inhabit intertidal and comparatively shallow subtidal regions of estuaries and the nearshore coastal zone that are characterized by large predictable and unpredictable fluctuations in environmental factors. These plants generally are tolerant of environmental stresses and, in fact most of the dominant species can withstand fluctuations in environmental extremes greater than those normally experienced. The marine vascular plants, and vascular plants in general, are morphologically adaptive in terms of somatic organization, structural modifications, and growth characteristics. The ecological success of seagrasses in this fluctuating environment in part is a manifestation of their inherent morphological plasticity.

The seagrasses possess two morphological adaptations to an existence in an aquatic environment subject to wave and tidal action and shifting sediments. These characteristics are linear, grasslike leaves and an extensive root and rhizome system. Linear leaves also are characteristic of many brackish water species, e.g., *Ruppia* and *Althenia,* and several freshwater species, but the combination of this type leaf plus an extensive root system is unique to submerged marine and estuarine plants. The leaves and stems of submergent angiosperms possess structural modifications, which are adaptations to life in flowing water, restricted and differential penetration of light of different wavelengths into the water column, and reduced rates of gas diffusion in water relative to air. The leaves and stems generally are thin, possess reduced structural tissue, and have an extensive system of lacunal air spaces. Extensive mechanical support in leaves with high surface to volume ratios would be detrimental in flowing water. Buoyancy is provided by the lacunae. These adaptations facilitate both light penetration and diffusion of dissolved gases.

Adsorption and diffusion of gases and nutrients also are facilitated by thin cellulose walls of epidermal, mesophyll, and cortical cells and by large areas of these cells abutting on the extensive lacunar system. The seagrasses also lack stomata and possess a thin, perforated cuticle. There apparently is no physiological barrier to reduce water loss in seagrasses (Gessner, 1971). In contrast, there is a loss of only 1.5% of recently fixed ^{14}C as dissolved organic carbon from dessicated *Zostera marina* (Penhale, 1976).

Adaptations in morphology are apparent even within a particular habitat. *Zostera marina* and *Thalassia testudinum* have narrow, short leaves in shallow waters and wider, longer leaves in deeper waters (Marmelstein *et al.,* 1968; Biebl and McRoy, 1971; Phillips, 1972). Biebl and McRoy (1971) have demonstrated a physiological adaptation associated with the short-leaved forms of *Zostera.* The intertidal form is less sensitive to increased temperatures in terms of respiration and photosynthetic rate than is the subtidal, long-leaved form.

Most submergent leaves lack differentiation of the mesophyll layer and possess chloroplasts throughout the mesophyll of the leaves and the outer cortex of the stems. In common with many shade-adapted dicots, seagrass leaves and stems possess chloroplasts in the epidermis in higher concentrations than in any other tissue. Thus, the epidermis is the principle site of photosynthesis (Sculthorpe, 1967). The amount of chlorophyll in this tissue, however, is not uniform throughout the blade and the absolute amount of chlorophyll *a* and the ratio of this pigment to accessory pigments decreases from tip to base in blades of *Zostera* (Stirban, 1968). An increase in *in vivo* chlorophyll *a* fluorescence from tip to base also was observed. Stirban (1968) suggested that this increase was an adaptation to the decrease of total light and light of longer wavelength with increasing water depth. Accessory pigments serve to absorb light and transfer energy to chlorophyll *a,* particularly near the base of the blades.

The primary functions of the roots are anchorage and absorption of nutrients. In general, the rhizomes possess bundles of longitudinal sclerenchyma fibers throughout the inner and outer cortex. Mechanical strength also may be provided by collenchyma tissue as in *Cymodocea* and *Zostera* or by sclerenchyma as in *Posidonia* (Sculthorpe, 1967). The roots and rhizomes of the submergent angiosperm also have extensive lacunae, which may be continuous with those of the stems and leaves. The large air space to volume ratio of the roots minimizes the respiratory oxygen demand of the roots (Williams and Barber, 1961). Numerous investigators have shown opposing linear gradients in oxygen and carbon dioxide concentration in submergent plants, oxygen decreasing from the leaves to the roots. This suggests that the roots derive their oxygen supply from photosynthetic organs of the leaves and stems, with the gas diffusing to the roots through the lacunar system. The extent of this lucunar system provides submergent plants the capability to anchor oxygen-requiring root tissues in anaerobic sediments. *Halophila decipiens* is able to recover its photosynthetic capacity completely after 5–12 hr of anaerobic exposure, but is unable to recover after 24 hr of the same. On the other hand, *Thalassia testudinum* showed complete photosynthetic recovery after 32 hr of oxygen deprivation (Hammer, 1969). These responses are explained in part by the relatively larger lucunar system in the leaves and roots of *Thalassia* compared to *Halophila*.

The importance of the roots in mineral uptake and water transport had been a subject of controversy (Sculthorpe, 1967). The roots of submergent plants have reduced vascular systems relative to other vascular plants. However, the high rate of organic production by seagrass leaves (Table IX) that occurs in nutrient-poor water is primarily due to nutrient and trace metal uptake from the sediments by the roots. Uptake of nutrients by the roots of marine seagrasses is significant, and indeed, the majority of nutrient uptake is accomplished by the roots (McRoy and Barsdate, 1970; McRoy *et al.*, 1972; McRoy and Goering, 1974). For example, uptake of phosphorus by the leaves of *Zostera marina* was insignificant compared to that of the roots (Penhale, 1976), It is only through root uptake and subsequent transfer that the phosphorus demands of the plant can be satisfied.

There is an evolving literature demonstrating the interaction of seagrasses and the physical and chemical characteristics of the substrate (see reviews and literature cited in McRoy and Helfferich, 1977; Thayer *et al.*, 1979; Phillips and McRoy, 1980). In North Carolina, the allocation of productivity to roots and leaves and total production of *Zostera marina* is in part a function of substrate (Kenworthy, 1980). In sediments which are coarse and low in both organic matter and in total and exchangeable combined inorganic nitrogen, a larger proportion of the productivity is allocated to root production than to leaf production. In sediments which are finer and have higher levels of nitrogen, however, a greater proportion is allocated to leaf production. On a per shoot basis, total plant production was higher in the fine texture, nutrient-enriched sediments.

D. SALT MARSH PLANTS

Salt marsh plants possess some of the morphological features characteristic of submergent seagrasses. In general, many species exhibit extensive lacunae which are continuous between above and below-ground portions of some species (Anderson, 1974). Teal and Kanwisher (1966) have demonstrated opposing linear gradients in oxygen and carbon dioxide concentrations between the stems and roots of *Spartina alterniflora*. Most emergent marsh plants also exhibit an extensive root–rhizome system. As opposed to seagrasses, the leaves of emergent plants have stomata, the above-ground portions generally are structurally rigid, and some species have woody fibrous tissues.

Two fundamental characteristics of the salt marsh are saline and anaerobic soils. All plants adjust their internal osmotic condition to maintain osmotic flow of water into the plant (Caldwell, 1974). The necessity for plants to remain hyperosmotic to their environment restricts their ability to limit total internal ion concentration. Salt tolerant plants differ from salt intolerant plants in the degree to which they tolerate high concentrations of internal ions and avoid potentially toxic ion imbalances. Selective ion absorption by the roots (Rains, 1972) and the need for only exceedingly small quantities of sodium and chloride ions (Rains and Epstein, 1967) does not prevent the accumulation of considerable quantities of salt by coastal marsh plants. Emergent marine plants avoid toxic ion imbalances through selective ion absorption, synthesis of organic ions, and selective excretion of salts from salt glands.

Salt tolerant plants excrete salt from salt glands, many of the cells of which have large numbers of mitochondria. The structure and function of plant glands has been reviewed by Lüttge (1971). The role of mitrochondria in some salt glands may be to supply ATP for active ion transport. Salt glands located in the leaves or in the roots are physiologically and morphologically analogous. Excretion by salt glands in the leaves, however, may be coupled to photosynthetic electron flow or reducing power as in *Atriplex spongiosa* (Lüttge *et al.*, 1970). In either case, the stoichiometry of salt excretion by plants has not been established. Nevertheless, the movement of ions out of the plant requires energy directly or indirectly translatable to terminal phosphate bonds of ATP. The energy for this process ultimately is derived through photosynthesis.

Three types of carbon dioxide fixation in vascular plants are C_3 or reductive pentose phosphate, C_4 or C_4-dicarboxylic acid, and CAM or crassulacean acid metabolism (Black, 1973). Characteristics associated with these pathways are summarized in Table XV. The data indicate that C_4 plants may be well adapted to warm, arid environments. Caldwell (1974) indicated that both C_3 and C_4 plants thrive in salt desert environments and show similar rates of growth, but that much

TABLE XV

Characteristics of Salt Marsh Plants Associated with Different Primary Types of Carbon Dioxide Fixation[a,b]

Primary type of CO_2 fixation	CO_2 compensation (ppm)	Transpiration ratio (g H_2O/g dry wt)	Optimum day temp. for growth (°C)	Maximum growth rate (g dry wt/m^2/day)
C_3	30–70	450–950	20–25	19.5 ± 3.9
C_4	0–10	250–350	30–35	30.3 ± 13.8
CAM	0–5 dark 0–200 daily rhythm	50–55	~ 35	*

[a] Modified from Black (1973).

[b] (*) Data not available but ratio of growth relative to C_4 and CAM plants on a per leaf area basis is less than 2% and 0.4%, respectively.

work is needed to unravel the ecological ramifications of the C_4 photosynthetic pathway. One point is clear, however; C_4 plants exhibit a longer growth season and produce more dry weight per unit of transpired water loss (Caldwell, 1974).

The primary adaptations of C_4 plants may be in response to the selective pressures prevalent in xeric and saline environments such as tropical savannas, river plains, and seasonally arid valleys where adaptations for water retention are important. The C_4 pathway is polyphyletic and of relatively recent origin. All C_4 plants also exhibit C_3 photosynthesis. Hough (1974) pointed out that to date there are no known C_4 submergent species. The C_4 system may not be of adaptive value for submergents, since photorespiration is less important to them, in contrast to emergents, because of the generally lower maximum oxygen, light, and temperature levels to which submergents are subjected.

CAM plants are adapted to environments of more or less constant aridity, whereas C_4 plants are found primarily in regions of intermittent aridity, where they apparently evolved. Aerial portions of CAM plants have low surface area/volume ratios that provide optimum geometry for water retention but not for efficient gas exchange. C_4 plants, however, are adapted to survive during periods of drought, yet they compete with rapidly growing mesophytes during periods when water is not limiting. This latter adaptation is a structural compromise which restricts water loss yet permits efficient uptake and fixation of carbon dioxide. The presence of a C_4-dicarboxylic acid system in emergent marsh plants may be of energetic significance since C_4 plants generally are more productive under conditions of high rates of photorespiration than are C_3 plants. An example of the possible energetic significance of refixation of respired carbon dioxide is given below.

Spartina alterniflora is a C_4 plant whose productivity in natural salt marshes approaches that of the most productive energy subsidized cultivated crops (Table IX). This occurs in a saline environment with an anaerobic substrate. Oxygen necessary for root respiration is supplied by diffusion through hollow air tubules, whick serve as ducts connecting lacunae in the leaves to those in the roots (Teal and Kanwisher, 1966). Measurements and calculated estimates of oxygen flow in these plants exceed respiratory requirements of the roots by 30 to 200%. This has led to speculation about possible benefits to the plant of an oxidized microhabitat surrounding the root hairs (Teal and Kanwisher, 1970). As oxygen diffuses to the roots, carbon dioxide diffuses from the roots to the leaves. It is possible that refixation in the leaves of carbon dioxide generated by respiration within the entire plant may be of considerable energetic significance to *Spartina alterniflora*. Based on experimental studies, Teal and Kanwisher (1970) indicated that for *S. alterniflora,* transpiration is of little significance as a means of maintaining the leaves below lethal temperatures, even in direct sunlight; transpiration occurs during periods of carbon dioxide diffusion from the air into the leaves at a rate of

TABLE XVI

Relationship of Energy Loss to Ion Excretion as a Function of Carbon Dioxide Source

	Atmospheric CO_2			CO_2 from plant respiration[d]
Atmospheric condition	Molar ratio of transpired water loss to CO_2 intake[a]	Molar ratio of cation intake to CO_2 intake[b]	Energy loss to cation excretion (%)[c]	
Cool humid	70	0.6	3	1
Hot dry	460	4.0	22	9
Average for growing season	200	1.7	9	4

[a] Data from Teal and Kanwisher (1970).
[b] Na^+ content of seawater of 0.47 molal (Stumm and Morgan, 1970).
[c] Assuming 6 moles ATP gained per mole CO_2 fixed and ⅓ mole ATP lost per mole Na^+ excreted.
[d] Ratio based on Ferguson and Williams (1974).

from 70 to 460 moles H_2O lost per mole of CO_2 gained. Water to be transpired is obtained by the roots and enters the plant with its dissolved ion complement. Thus, at a given salinity, the molar ration of ion intake by the roots to carbon dioxide intake by the leaves can be estimated. The sodium content of seawater is 0.47 molal (Table XI), and thus, the ratio of moles of sodium to moles of water is about 0.0085.

As reported earlier, the stoichiometry of the various suspected mechanisms of salt excretion is unknown. However, for sodium transport in mammalian erythrocytes it is approximately three sodium atoms per ATP consumed, and this ratio is consistent over a wide range of concentration differentials (Lehninger, 1970). Using this ratio as a first approximation, the advantage of refixing carbon dioxide respired within the plant is substantial (Table XVI). Assuming that *Spartina alterniflora* obtains 60% of the carbon dioxide it fixes from respiration within the plant (Ferguson and Williams, 1974), the energy savings over a growing season is approximately 5% of gross photosynthesis or 20% of net photosynthesis (Teal, 1962; Ferguson and Williams, 1974). Selective advantage may rest, therefore, with an internal gas transport system that is optimal for carbon dioxide retention and flow, although it is somewhat excessive for oxygenation of the roots.

IV. Human Intrusion

A. OVERVIEW

The major populations of the world have lived on marine coastlines or on major rivers not far from the sea. Historically, man has developed technologies for fishing, boat building, defense and warfare, excavated navigable channels and constructed dykes, all in an effort to live near and to utilize the sea (Rice, 1971). From 1930 to 1960, the population of the coastal counties of the United States grew 78%, while the national population grew 45%. These coastal counties represent only 15% of the land area of the contiguous United States. Presently 40% of all industrial facilities are located along the coast (Thayer, 1975), and by the year 2000, it is estimated that 70% of the nation's population will reside in the coastal region, including the coastal area of the Great Lakes (Ketchum and Tripp, 1972).

This coastal environment is a complex, important, and unique system, which must be maintained for the benefit of all. Yet, man has used this habitat for generations without any planning (Rice, 1971). Man's use of the oceans, estuaries, and marshes is multifaceted and includes demands on these areas for food, energy development, transportation, waste disposal, living space, recreation, and as a source of aesthetic pleasure. Not all of these uses are compatible with each other, and in many instances they are mutually exclusive (Fig. 7). The

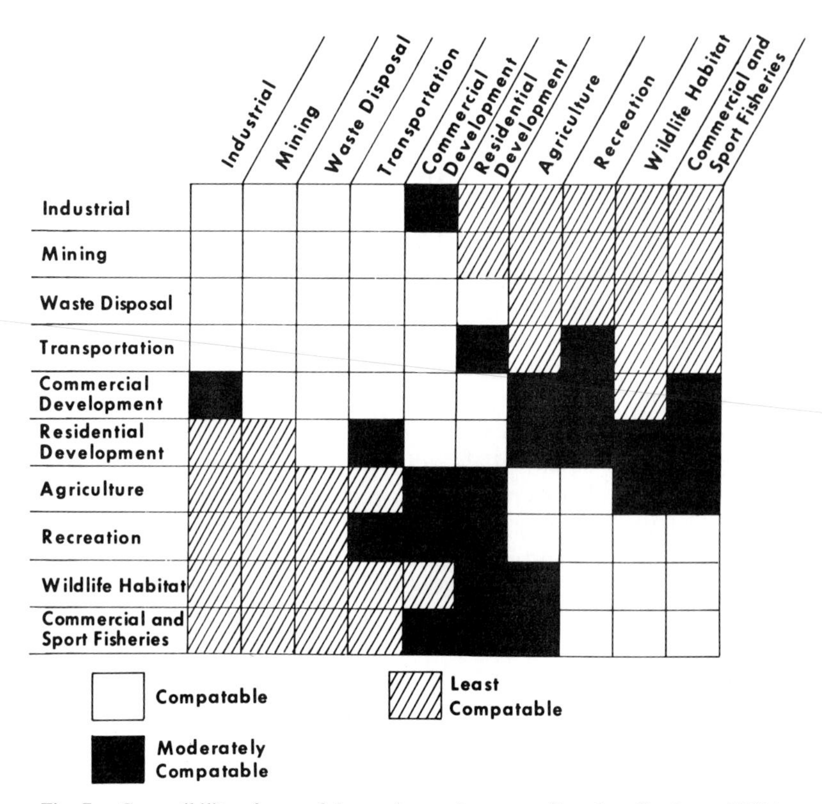

Fig. 7. Compatibility of uses of the marine environment. (Based on Ketchum, 1969.)

uses may be beneficial, detrimental, or they may have no direct impact on the quality of the environment (Fig. 8), and in turn, on the productivity of primary producers (Table XVII). Present demands now may exceed the regenerative capacity of living marine resources and the needs of future generations must also be considered.

The continued availability to man of the biological resources of the marine environment depends on either a maintenance of natural genetic and species diversity or a controlled reduction of this diversity in parts of the marine environment in a manner analogus to terrestrial agriculture. As opposed to energy subsidized agricultural crops, however, the productivity of natural marine systems is free to us at no expense except for that associated with the harvest and prevention of contamination or destruction of the environment. The biological productivity of any natural system is a function of the availability of nutrients and light energy and of the temperature. The stability of natural systems is a function of both the stability of these environmental factors and the genetic and species diversity of the organisms present. Levels and fluctuations in levels of nutrients,

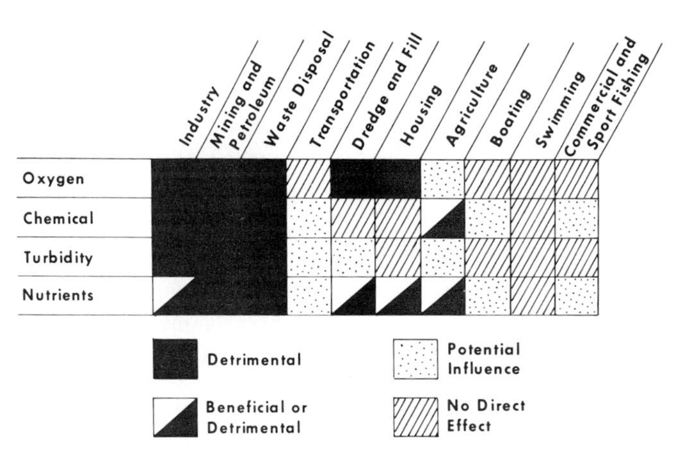

Fig. 8. Actual and potential impacts of man's use of the marine environment on water quality. (Modified from Ketchum, 1969.)

light, ambient temperature, and toxic materials are influenced by man's activities. Although many man-induced factors can increase short term productivity, these often lead to the elimination of genotypes and species (Teal *et al.*, 1972) and thus, a simplification of the natural system. Simplification reduces the stability of natural systems and can result in a reduction in productivity or in a conversion of this productivity to species less desirable to man.

TABLE XVII

Man's Activities and Their Potential Impact on Productivity of Marine Plants[a]

Man's activities	Factor	Phytoplankton	Vascular Plants: Submergents	Vascular Plants: Emergents
Dredging and filling (Erosion)	Silt	+ or −	0 or −	+ or −
Water impoundment (Channelization)	Salinity	+ or −	+ or −	+ or −
Power generating	Temperature and	+ or −	+ or −	+ or −
	Radioactivity	0 or −	0 or −	0 or −
Offshore drilling (Transportation)	Petroleum	0 or −	0 or −	0 or −
Sewage disposal and agricultural drainage	Nutrients	+ or −	+ or −	+ or −
	Toxicants	0 or −	0 or −	0 or −

[a] Potential influence dependent on concentration. (+) Increase in productivity; (−) Decrease in productivity; (0) No change in productivity.

B. PHYTOPLANKTON

The abundance, species composition, diversity, and species succession of phytoplankton can be influenced by man's activities (Patrick, 1973). Man alters the marine environment, particularly estuaries, by dredging, filling, and channelization, which change the direction and rate of water flow. These activities modify the shape and reduce the total area of many estuaries, increase turbidity, and change salinity patterns. In addition, man adds many materials to the marine environment. Among these are toxic substances, nutrients, and radioactive materials. Also, primarily through power generating plants, man adds large quantities of heated water to the marine environment.

Silt suspended in estuaries or in sea water may exert no effect on primary productivity or it can reduce or increase production (Table XVII). When the bottom is within the euphotic zone, production is high, even with high turbidities (Odum and Wilson, 1962). When the depth of the mixing layer exceeds the euphotic zone or when cell respiration exceeds photosynthesis due to turbidity, the production of the phytoplankton population is reduced. Additions of nutrient-rich suspended material can stimulate photosynthesis (Odum and Wilson, 1962). In instances when respiration exceeds photosynthesis due to light limitation, inorganic nutrients accumulate in the water. This nutrient accumulation stimulates photosynthesis when light conditions again become suitable.

A correlation exists between temperature and the distribution of phytoplankton. Temperature influences the rates of chemical processes of all phytoplankton, but each species grows best within a certain temperature range. The effect of an increase in temperature on survival and growth is dependent on the temperature to which the phytoplankton are acclimated and how near this temperature is to the upper lethal limit (Gurtz and Weiss, 1972). Potentially, thermal pollution can lead to eutrophication. Less drastic effects, which can occur and be of long-term significance, are shifts in species dominance or moderate alterations in the biomass or productivity of the phytoplankton community (Cairns, 1971; Coutant and Goodyear, 1972).

Toxic substances released into the marine environment can have an adverse effect on phytoplankton or their predators. This occurs because phytoplankton cells have the capacity to take up toxic substances at rates that exceed dilution within the cytoplasm of the cell due to growth and division. High initial uptake of a toxic substance will increase its concentration within the cells to high levels. Through uptake, the substance can be reduced to levels that are not measurable either in the water or eventually, if growth continues, in the cells. It is probable that the initially high uptake rate and high internal concentration of a toxic substance will be most likely to affect cell division. If a cell does not accumulate a toxicant, deleterious effects may not occur. If the toxicant is accumulated by the phytoplankton but the phytoplankton is not affected by or adapts to the presence of the toxicant inside the cell, there will be no response and, therefore, no deleterious effect. Given the ability of phytoplankton to tolerate or adapt to increased levels of toxic materials within their cells and the known capacity of aquatic animals to accumulate such materials from their plant foods, the tolerance of the feeders may be, in some cases, the most suitable criterion for limiting toxic waste releases to the environment (Stockner and Antia, 1976). Some toxic substances, however, inhibit the growth of phytoplankton at very low concentrations. For instance, copper (Davey *et al.*, 1973) and mercury (Harris *et al.*, 1970) have been found to inhibit the growth of phytoplankton at levels of 1 part per billion. Pesticides also are toxic at very low levels. Wurster (1968) reported that DDT at 10 ppb can inhibit the photosynthesis of coastal and oceanic phytoplankton.

The addition of sewage to an estuary was found to greatly increase the phosphorus and nitrogen content of the water and was responsible, in part, for the rich growth of diatoms (Braarud and Hope, 1952). Nutrients contained in sewage or agricultural drainage into estuarine and coastal waters may increase or decrease the growth rate of phytoplankton and change species dominance, depending on the concentration of fertilizer and the length of time it is available (Ryther and Dunstan, 1971). Run off from duck farms along the tributary streams of Moriches Bay in Long Island Sound, New York, resulted in the development of dense algal blooms of small green flagellates where previously diatoms dominated (Ryther, 1954).

Fuel oils of different origin and components of oil vary in their toxicity to phytoplankton (Parker, 1974). For example, water soluble fractions from Baytown and Montana fuel oils are lethal to the blue-green algae, *Agmenellum quadruplicatum* and *Nostoc* sp., and water soluble fractions from New Jersey fuel oil are lethal to the green algae, *Dunaliella tertiolecta* and *Chlorella autotrophica,* but not to the two blue-green species. *Thalassiosira pseudonana* is not greatly influenced by water soluble fractions of these three fuel oils, but 15 mg/liter of hydrocarbon derived from American Petroleum Institute #2 fuel oil inhibits photosynthesis. Low molecular weight hydrocarbons also differ in their influence on phytoplankton. Aliphatic hydrocarbons, e.g., pentane and hexane, depress the photosynthetic rate of mixed ocean phytoplankton cultures and cultures of *Skeletonema costatum* considerably more than do aromatic hydrocarbons, e.g., benzene, at a given concentration. Concentrations of petroleum hydrocarbons toxic to phytoplankton species recognized as food of the oyster, however, were far in excess of levels previously demonstrated to be lethal to adult and larval oysters (Mahoney and Haskin 1980). In addition, no strong evidence for major damage to phytoplankton populations as a result of an oil spill have been documented (Michael, 1977). Off the coast of Brittany, France, in waters affected and unaffected by the *Amoco Cadiz* oil spill for example, the oil spill did not appear to be a factor in observed distributions of phytoplankton (Spooner, 1978).

Changes in salinity resulting from water impoundment, channelization, dredge and fill, and mosquito ditching can be important in controlling the species present. It was found that the quantity of diatoms present in Osaka Bay, Japan, could be expressed as a function of salinity and temperature (Uyeno, 1957). The lowest diversity of species was reported by Patten *et al.* (1963) to occur in Chesapeake Bay on the eastern side, where the salinity was highest.

C. SUBMERGENT VASCULAR PLANTS

As a result of their shallow sublittoral and, for some species, intertidal existence (Fig. 3), seagrasses are subject to many of the stresses imposed by man's use of the coastal environment (Fig. 8, Table XVII). The growth form of seagrasses is unique for marine plants and thus, as a group, they are subject to perturbations both of the aquatic medium and of the sediment. Thayer *et al.* (1975b) and Zieman (1975b) have discussed much of the literature available at that time concerning the impact of stresses on temperate and tropical seagrasses. Much of the discussion here will be drawn from their papers.

Dredge and fill operations potentially have the most damaging impacts to seagrass beds and have resulted in the destruction of more grass bed habitats than any other form of stress imposed by man. Several factors are involved. Dredging not only removes sediment but also increases turbidity, accelerates sediment deposition, and alters the oxidation–reduction (redox) potential of the sediments. These modifications result in decreased light transmission, smothering of both the grass and its associated fauna, and release of nutrients from the suspended sediments. Seagrasses appear dependent on an anaerobic subsurface sediment for root and rhizome growth, and hence, oxidation of these sediments may retard growth. Propeller cuts made in grass beds, trawling over grass beds, or the commercial harvest of bay scallops also result in disturbance of the grass and of the sediments.

The input of sewage or other nutrient-enriched water into estuarine and coastal waters in the vicinity of seagrass beds may have both positive and negative impacts on these systems, depending on the degree and duration of the stress. These inputs not only alter nutrient availability but also influence the turbidity of the overlying water. Since seagrasses can absorb nutrients through the blades as well as through the root system, increases in nutrients may stimulate productivity. In this regard, Odum (1963) indicated that in a *Thalassia* area adjacent to but not smothered by silt from a dredge operation, there was increased production and chlorophyll *a* content of the grass, possibly resulting from increased nutrient availability. Increases in nutrients in the water column, however, probably influence phytoplankton and epiphytic production more directly. If there is a significant increase in the growth of these producers, there also will be diminished light transmission through the water and a concommitant decrease in the productivity of seagrasses. For example, in Hillsborough Bay, Florida, light transmission to the grass blades was so reduced by epiphytic and planktonic algae that there was a reduction in the biomass of *Thalassia* (Taylor *et al.*, 1973). Sewage input has been implicated in species shifts from *Thalassia* to other seagrasses or filamentous green algae in some tropical areas (McNulty, 1970; Nichols *et al.*, 1972).

Release of oil can influence seagrasses by reduction in light transmission to the plants and by coating of blades, thereby reducing gas and light absorption. Spilled oil can consolidate the sediment and increase its buoyancy, which makes the sediment more susceptible to dislodgement by waves (Zieman, 1975b). Diaz-Piferrer (1962) reported a loss of both *Thalassia* and about 3000 m^3 of sediment from the coast of Puerto Rico following a crude oil spill. Following the Santa Barbara oil spill in 1969, intertidal forms of *Phyllospadix* were damaged (Foster *et al.*, 1971), and crude oil was implicated in the demise of *Zostera* populations in the English Channel in the 1930s (Duncan, 1933). Sublethal effects of oil on photosynthesis have been reported (McRoy and Williams, 1977).

Seagrasses generally are considered eurybionts and most undergo seasonal fluctuations in production and abundance which are associated with seasonal temperature regimes. Alterations in temperature or salinity may significantly influence both production and distribution of these plants. Phillips (1960) reported that the optimum temperature range for *Thalassia testudinum,* perhaps the single most important tropical seagrass, is between 20°–30°C, and that above or below this range there is leaf mortality. Maximum net organic production values occur between 28°–30°C and decrease rapidly on either side of this range (Zieman, 1975a). The rhizome system, however, imparts regenerative capacity to the plants, and even if the leaves are lost as a result of thermal stress, there appears to be rapid regeneration of new leaves from rhizomes. Wood and Zieman (1969), however, warn that prolonged heating, especially during the summer, may sufficiently heat the substrate to destroy the rhizome system and that recovery of the bed would take several years, even if the stress were discontinued. Although the impact of thermal discharges on temperate seagrasses has not been documented, these species also exhibit specific thermal ranges and may be seriously affected by man-induced changes in temperature. *Posidonia,* a generally warm-temperature seagrass, has a more restricted thermal range than many of the tropical and temperate genera, being generally restricted to average temperatures of 10°–20°C (den Hartog, 1970). Optimal temperatures for *Posidonia* are 17°–20°C, and temperatures above about 22°C result in extensive leaf mortality. *Zostera marina,* a temperate species, grows over a wide range of temperatures with extremes of 0°–40°C (Phillips, 1972), but with an effective range of 5°–27°C (Thayer *et al.*, 1975b). Setchell (1929) noted that vegetative growth occurred between 10° and 15°C, and sexual reproduction between 15° and 20°C in *Zostera marina* at Mt. Desert Isle, Maine. Changes in thermal regimes from heated water discharges, therefore, could disrupt the normal sexual and vegetative reproductive cycle of this seagrass.

Seagrasses exhibit wide variations and ranges of salinity tolerance from about 1 to 70‰, but generally the lower limit for successful growth is near 15‰ (McMillan and Mosely, 1967; McMahan, 1968; McMillan, 1974). The inability of marine seagrasses to tolerate freshwater and extremely low salinities for protracted periods may result in part from a reduced inorganic carbon supply relative to seawater (Ogata and Matsui, 1965; Hammer, 1969). For example, *Posidonia* exhibits drastically reduced photosynthetic capacity in freshwater or low salinity water, but when high concentrations of bicarbonates are supplied, photosynthesis is greater than in full strength seawater (Hammer, 1969). Although we have found no evidence for the impact on seagrasses from man-induced salinity changes, Zieman (1970) reported stresses on *Thalassia* communities resulting from temperature–salinity interactions. A decrease in the plant populations occurred under conditions of high temperatures and low salinities. Other temperature–salinity combinations did not result in such decreases. This suggests that stream channelization may be detrimental to seagrasses during periods of normally high ambient water temperatures because of increased input of freshwater into estuaries and coastal waters.

D. EMERGENT VASCULAR PLANTS

The extent and productivity of marine emergents and the composition and succession of species can be influenced by man's activities. In general, man's pertinent activities can be classified into physical alteration and chemical and thermal pollution Dredging and land filling alter the shape, circulation patterns, tidal regime, and the rate of silt deposition in intertidal and shallow water. Emergent plant marshes and swamps can be eradicated by direct dredging and land filling associated with development of the coastal zone. The deposition of silt is less likely to smother emergent plants than submergent species, and in some instances the accretion of sediment stimulates both areal production and the extent of marshes or swamps (Ranwell, 1964). Alterations that decrease tidal amplitude also can result in reduced areal extent, reduced height at maturity, and reduced productivity of *Spartina* marshes (Table XVIII). Alterations that increase the salinity of water over the marshes also can reduce the productivity of *Spartina alterniflora* (Adams, 1963) and *S. foliosa* (Phleger, 1971). The growth of *S. foliosa* was reduced 4% for each 8‰ increase. Increases in both salinity and the degree of tidal inundation, however, favor the growth of *S. alterniflora* over competing species (Woodhouse *et al.*, 1974).

In a 2-year study in eastern North Carolina, the effects of mosquito-control ditching on high marshes dominated by *Juncus roemerianus* were not apparent beyond the immediate removal of *Juncus* from ditches and its smothering under

TABLE XVIII

Effect of Tidal Irrigation on Relative Height at Maturity and Annual Net Areal Production Rates of *Spartina* at Sapelo Island, Georgia[a]

Position in marsh	Relative grass height at maturity	Tidal irrigation	Net production[b] (g C/m²/year)
Low	Tall	Frequent/vigorous	2000
Intermediate	Medium	Frequent/gentle	1650
High	Short	Infrequent/gentle	375

[a] Based on Odum (1974).
[b] Assuming C/dry wt ratio of 50%.

spoil piles along the ditches (Kuenzler and Marshall, 1973). A continuing invasion by bushy vegetation, *Baccharis* and *Iva,* on some of the spoil piles and in previously pure stands of *Juncus* may indicate accelerated succession to a habitat dominated by more terrestrial species. Whether ditching increases the frequency and extent of tidal inundation or reduces the moisture level of high marsh soils may vary according to local conditions, thereby causing seemingly inconsistent effects. In North Carolina *Juncus* marshes, the productivity and abundance of *Juncus* between ditches did not seem to be affected (Kuenzler and Marshall, 1973), but then neither was the mosquito population. Another invasion of the salt marsh encouraged by man has been by domesticated herbivores, particularly in European marshes. In general, grazing of marshes reduces total production by up to 70% and deleteriously affects dominant species, favoring those species less susceptible to trampling (Reimold, 1976; Reimold *et al.,* 1975).

Chemical pollutants include oil, pesticides, heavy metals, and domestic sewage. ZoBell's (1964) statement that only at its worst does oil pollution appear to be injurious to animal and plant life in the sea is a generalization that can be misleading. The effect of oil on salt marsh plants varied with its source, its degree of weathering, and with the species affected. Fresh oil, in general, has serious effects on marine plants at 0.01–0.1% concentrations and kills them at 0.5–5% (Nelson-Smith, 1967). *Distichlis spicata, Spartina alterniflora,* and young *Avicennia* appear to be more readily harmed by oil pollution than are, for example, oysters. As little as 25 ml of oil per square foot retained by wooden bulkheads was detrimental to marsh grasses (Mackin, 1950). On the other hand, most salt marsh plants at Cotes du Nord, Brittany, survived all but the heaviest contamination of oil greater than 10 cm thick during the Torrey Canyon incident (Stebbings,1970). This oil had been weathered at sea 14–18 days before coming ashore. The most susceptible species were *Salicornia perennis* and *Beta maritima*. *Juncus gerardi* and *Puccinella maritima* were particularly resistant and the marshes appeared to have recovered 16 months later. *Salicornia* and *Sueda* seedlings reinvaded the previous *Salicornia perennis* zone. There also was exceptionally vigorous growth by some species, e.g., *Juncus maritimus,* growing up through cracks in the residual dried oil (Stebbings, 1970). Although species of salt marsh plants are susceptible to high levels of oil, many species, e.g., *Spartina townsendii* and *Puccinella maritima,* recover from acute exposure by the following year (Baker, 1970). However, in this case *Juncus maritimus* only occasionally recovered (Baker, 1973). Production of *Spartina alterniflora* was not affected by additions of oil up to 32 liters /m^2 in field studies in the Barataria Basin, Louisiana, although injury to the plant may have occurred if oil contacted the leaf blade (DeLaune *et al.,* 1979). Mangrove seedlings on the other hand, were found to be especially damaged in a mangrove community in Panama which had been heavily impacted by Bunker C and diesel fuel oil (Rutzler and Steerer, 1970).

There is little data on the concentration of pesticides in tissues of salt marsh plants. Although DDT was concentrated in the leaves of salt marsh plants by a factor of up to 1500 times that appearing in the water 1 month after application, this pesticide was not retained by the plants beyond the first month and the level subsequently decreased (Croker and Wilson, 1965). The concentration factor of toxaphene in leaves of *Spartina alterniflora* relative to marsh sediments was approximately one (Reimold, 1974). Neither author reported any observable detrimental effects on the plants by these pesticides.

Many heavy metals tend to accumulate in salt marsh soils due to chemical precipitation or to sedimentation of particulate matter, although this amount may be less than 25% of the total of some metals entering the estuary (Dunstan *et al.,* 1975). Williams and Murdoch (1969) reported the significance of *Spartina alterniflora* as a mechanism for the return of iron, zinc, and manganese from marsh sediments to the aquatic environment. More recently, Banus *et al.* (1974) reported a similar mechanism of flow of lead contamination in marshes. When lead was added to salt marshes dominated by *Spartina alterniflora,* increased levels appeared in the emergent parts of the plant. Some heavy metals are toxic to *Spartina alterniflora* seedlings exposed for 8 weeks to 100 ppm concentrations (Dunstan *et al.,* 1975). The responses to copper, lead, and cadmium were death in 2 weeks, 50% mortality in 8 weeks with surviving plants showing only minimal growth, or no effect in 8 weeks. However, these exposures were higher than environmental levels, and as for their potential as toxicants, *Spartina* appears to be resistant to moderate increases in the levels of copper, lead, and cadmium in the environment. Although *Spartina* also may be an important mechanism for movement of methyl-mercury from marsh sediments to the overlying water, this toxicant does not appear to accumulate in *Spartina* or inhibit its growth (Gardener *et al.,* 1975).

We have already reviewed the apparent increase in salt marsh production resulting from nitrogen fertilization. Thus, it is not surprising that domestic sewage pollution in fact can stimulate production of salt marshes. Teal and Valiela (1973) termed *Spartina* marshes as effective "living filter" tertiary treatment systems that respond to sludge application with increased growth by the plants, reduced rates of nitrogen fixation, and increased rates of denitrification. *Spartina* marshes were able to process up to twice the amount of sludge per unit area per year as were terrestrial living filter systems. Although emergent species in general appear to be resistant or are able to recover from chemical pollutants, resultant changes in species composition and in the level of contaminants presented to higher trophic levels in salt marsh plant tissues and detritus could significantly affect higher trophic levels in the marine ecosystem.

The high productivity of coastal and some oceanic systems is rivaled only by highly mechanized agricultural crops and tropical and subtropical rainforests (Table V). Argricultural crops are efficient and productive but they are neither diverse nor stable. Energy and material expenditures are required to maintain their structure and productivity. The stability of the production of marine and estuarine areas results from their natural complexity. This stability is a self-sustaining process, but is very susceptible to intervention by man. Man's extensive and often incompatible resource needs make it imperative that we understand those factors which regulate the structure and function of coastal and oceanic biological communities. Eventually, man must manage these systems wisely to minimize conflicts and maximize benefits. The first step in the management of our biological marine resources is to maintain the natural habitats, the productivity, and the diversity of marine primary producers.

Acknowledgment

The authors express their appreciation to Mrs. Marianne B. Murdoch, Mrs. Ann B. Hall, and Mr. Curtis A. Oden for their assistance in the literature search and compiling of the references; to Mrs. Margaret L. Rose for typing; and to Mr. Herbert R. Gordy for his assistance in preparing the figures. Contribution number 80-41B, Southeast Fisheries Center, National Marine Fisheries Service, NOAA, Beaufort, N.C. 28516.

References

Adams, D. A. (1963). Factors influencing vascular plant zonation in North Carolina salt marshes. *Ecology* **44**, 445–456.

Anderson, C. E. (1974). A review of structure in several North Carolina salt marsh plants. *In* "Ecology of Halophytes" (R. J. Reimold and W. H. Queen, eds.), pp. 307–344. Academic Press, New York.

Arasaki, M. (1950). Studies on the ecology of *Zostera marina* and *Zostera nana* (in Japanese). *Bull. Jpn. Soc. Sci. Fish.* **16**(2), 70–76.

Armstrong, F. A. J. (1965). Silicon. *In* "Chemical Oceanography" (J. P. Riley and G. Skirrow, eds.), Vol. 7, pp. 409–432. Academic Press, New York.

Atkinson, M. R., Findlay, G. P., Hope, A. B., Pitnan, M. G., Saddler, H. W., and West, K. R. (1967). Salt regulation in the mangroves, *Rhizophora* Lam. and *Aegealitis annulata* R. Br. *Aust. J. Biol. Sci.* **20,** 589–599.

Baker, J. M. (1970). Oil pollution in salt marsh communities. *Mar. Pollut. Bull.* **1,** 27–28.

Baker, J. M. (1973). Recovery of salt marsh vegetation from successive oil spillages. *Environ. Pollut.* **4,** 223–230.

Banus, M., Valiela, I., and Teal, J. M. (1974). Export of lead from salt marshes. *Mar. Pollut. Bull.* **5,** 6–9.

Barber, R. T. (1973). Organic ligands and phytoplankton growth in nutrient-rich seawater. *In* "Trace Metals and Metal-organic Interactions in Natural Waters" (P. C. Singer, ed.), pp 321–338. Ann Arbor Sci. Publ., Ann Arbor, Michigan.

Barber, R. T., and Ryther, J. H. (1969). Organic chelators: Factors affecting primary production in the Cromwell Current upwelling. *J. Exp. Mar. Biol. Ecol.* **3,** 191–199.

Barber, R. T., Dugdale, R. C., McIsaac, J. J., and Smith, R. L. (1971). Variations in phytoplankton growth associated with the source and conditioning of upwelling water. *Invest. Pesq.* **35,** 171–193.

Barlow, J. P., Lorenzen, C. J., and Myren, R. T. (1963). Eutrophication of a tidal estuary. *Limnol. Oceanogr.* **8,** 251–262.

Biebl, R. (1962). Protoplasmatisch-okologische Untersuchungen an Mangrovealgen von Puerto Rico. *Protoplasma* **55,** 572–606.

Biebl, R., and McRoy, C. P. (1971). Plasmatic resistance and role of respiration and photosynthesis of *Zostera marina* at different salinities and temperatures. *Mar. Biol.* **8,** 48–56.

Bigelow, G. W. (1977). Primary productivity of benthic microalgae in the Newport River estuary. M.S. Thesis, North Carolina State University, Raleigh.

Black, C. C., Jr. (1973). Photosynthetic carbon fixation in relation to net CO_2 uptake. *Annu. Rev. Plant Physiol.* **24,** 253–286.

Bongers, C. H. J. (1956). Aspects of nitrogen assimilation by cultures of green algae (*Chlorella vulgaris,* strain A and *Scenedesmus*). *Meded. Landbouwhogesch. Wageningen* **56,** 1–52.

Braarud, T. (1961). Cultivation of marine organisms as a means of understanding environmental influences on population. *In* "Oceanography" (M. Sears, ed.), Publ. No. 67, pp. 271–298. Am. Assoc. Adv. Sci., Washington D.C.

Braarud, T., and Hope, B. (1952). The annual phytoplankton cycle of a landlocked fjord near Bergen. *Rep. Norw. Fish. Mar. Invest., Rep. Technol. Res.* **9,** 1–26.

Brylinsky, M. (1967). "Energy Flow in a Texas Turtle Grass Community." Class rep. Inst. Ecol., University of Georgia, Athens.

Bunt, J. S., Lee, C. C., and Lee, E. (1972). Primary productivity and related data from tropical and subtropical marine sediments. *Mar. Biol.* **16,** 28–36.

Burkholder, P. R., Burkholder, L. M., and Rivero, J. A. (1959). Some chemical constituents of turtle grass, *Thalassia testudinum*. *Bull. Torrey Bot. Club* **86**(2), 88–93.

Cadee, G. C., and Hegeman, J. (1974). microflora living on tidal flats in the Dutch Wadden Sea. *Neth. J. Sea Res.* **8,** 260–291.

Cairns, J. J. (1971). Thermal pollution—a cause for concern. *J. Water Pollut. Control Fed.* **43,** 55–66.

Caldwell, M. (1974). Physiology of desert halophytes. *In* "Ecology of Halophytes" (R. J. Reimold and W. H. Queen, eds.), pp. 355–378. Academic Press, New York.

Caperon, J. (1968). Population growth response of *Isochrysis galbana* to nitrate variation at limiting concentrations. *Ecology* **49,** 866–872.

Carpenter, E. J., and Guillard, R. R. L. (1970). Intrspecific differences in nitrate half-saturation constants for three species of marine phytoplankton. *Ecology* **52,** 183–185.

Chapman, V. J. (1957). Marina algal ecology. *Bot. Rev.* **23,** 320–350.
Chapman, V. J. (1960). "Salt Marshes and Salt Deserts of the World." Wiley (Interscience), New York.
Chapman, V. J. (1962). "The Algae." St. Martin's Press, New York.
Chapman, V. J. (1966). The physiological ecology of some New Zealand seaweeds. *Proc. Int. Seaweed Symp., 5th, 1965* pp. 29–54.
Chapman, V. J. (1970a). "Seaweeds and Their Uses." Methuen, London.
Chapman, V. J. (1970b). Mangrove phytosociology. *Trop. Ecol.* **2,** 1–19.
Chapman, V. J. (1974). "Salt Marshes and Salt Deserts of the World," 2nd suppl., reprinted ed. Cramer, Lehre.
Chester, R., and Stoner, J. H. (1974). The distribution of zinc, nickel, manganese, cadmium, copper and iron in some surface waters from the world ocean. *Mar. Chem.* **2,** 17–32.
Chu, S. P. (1942). The influence of the mineral composition of the medium on the growth of planktonic algae. Part 1. Methods and culture media. *J. Ecol.* **30,** 284–325.
Colijn, F., and van Buurt, G. (1975). Influence of light and temperature on the photosynthetic rate of marine benthic diatoms. *Mar. Biol.* **31,** 209–214.
Conover, J. T. (1958). Seasonal growth of benthic marine plants as related to environmental factors in an estuary. *Publ. Inst. Mar. Sci., Univ. Tex.* **5,** 97–147.
Conover, J. T. (1964). The ecology, seasonal periodicity, and distribution of benthic plants in some Texas lagoons. *Bot. Mar.* **7,** 4–41.
Coutant, C. C., and Goodyear, C. P. (1972). Thermal effects. *J. Water Pollut. Control Fed.* **44,** 1250–1294.
Croker, R. A., and Wilson, A. J. (1965). Kinetics and effects of DDT in a tidal marsh ditch. *Trans. Am. Fish. Soc.* **94,** 152–159.
Cronan, D. S. (1969). Average abundances of Mn, Fe, Ni, Co, Cu, Pb, Mo, V, Cr, Ti, and P in Pacific pelagic clays. *Geochim. Cosmochim. Acta* **33,** 1562–1565.
Davey, E. W., Morgan, M. J., and Erickson, S. J. (1973). A biological measurement of copper complexation capacity of seawater. *Limnol. Oceanogr.* **18,** 993–997.
Davis, J. H., Jr. (1940). The ecology and geologic role of mangroves in Florida. *Carnegie Inst. Washington Publ.* **517,** Pt. 16, 303–424.
Dawson, E. Y. (1966). "Marine Botany." Holt, New York.
DeLaune, R. D., Patrick, W. H., Jr., and Buresh, R. J. (1979). Effect of crude oil on a Louisiana *Spartina alterniflora* salt marsh. *Environ. Pollut.* **20,** 21–31.
den Hartog, C. (1967). The structural aspect in the ecology of seagrass communities. *Helgol. Wiss. Meeresunters.* **15,** 648–659.
den Hartog, C. (1970). "The Seagrasses of the World." North-Holland Publ., Amsterdam.
Diaz-Piferrer, M. (1962). The effects of an oil spill on the shore of Guanica, Puerto Rico. *Assoc. Isl. Mar. Lab., 4th Meet.* pp. 12–13.
Dillon, R. C. (1971). A comparative study of the primary productivity of estuarine phytoplankton and macrobenthic plants. Ph.D. Thesis, University of North Carolina, Chapel Hill.
Drew, E. A. (1971). Botany. *In* "Underwater Science" (J. D. Woods and J. N. Lythgoe, eds.), pp. 175–233. Oxford Univ. Press, London and New York.
Duncan, F. M. (1933). Disappearance of *Zostera marina*. *Nature (London)* **132,** 483.
Duncan, W. H. (1974). Vascular halophytes of the Atlantic and Gulf coasts of North America north of Mexico. *In* "Ecology of Halophytes" (R. J. Reimold and W. H. Queen, eds.), pp. 23–50. Academic Press, New York.
Dunstan, W. M., Windom, H. L., and McIntire, G. L. (1975). The role of *Spartina alterniflora* in the flow of lead, cadmium, and copper through the salt marsh ecosystem. *In* "Mineral Cycling in Southeastern Ecosystems" (F. G. Howell, J. B. Gentry, and M. H. Smith eds.), CONF-740513, pp. 250–256. U.S. E.R.D.A., Washington, D.C.
Earle, L. C., and Humm, H. J. (1964). Intertidal zonation of algae in Beaufort Harbor. *J. Elisha Mitchell Sci. Soc.* **80,** 78–82.

Eppley, R. W. (1972). Temperature and phytoplankton growth in the sea. *Fish. Bull.* **70,** 1063-1085.

Eppley, R. W., and Sloan, P. K. (1966). Growth rates of marine phytoplankton: Correlation with light adsorption by cell chlorophyll *a*. *Physiol. Plant.* **19,** 47-59.

Eppley, R. W., Holmes, R. W., and Strickland, J. D. H. (1967). Sinking rates of marine phytoplankton measured with a flowmeter. *Mar. Biol. Ecol.* **1,** 191-208.

Erickson, S. J., Lackie, N., and Maloney, T. E. (1970). A screening technique for estimating copper toxicity to estuarine phytoplankton. *J. Water Pollut. Control Fed.* **42,** R 270-278.

Ferguson, R. L. (1971). Growth kinetics of an estuarine diatom: Factorial study of physical factors and nitrogen sources. Ph.D. Thesis, Florida State University, Tallahassee.

Ferguson, R. L. (1972). Population density limitation of an estuarine diatom by illumination and nitrogen source. *J. Elisha Mitchell Sci. Soc.* **88,** 188.

Ferguson, R. L., and Murdoch, M. B. (1975). Microbial ATP and organic carbon in sediments of the Newport River estuary, North Carolina. *In* "Estuarine research" (L. E. Cronin ed.), Vol. 1, pp. 229-250. Academic Press, New York.

Ferguson, R. L., and Williams, R. B. (1974). A growth chamber for the production of ^{14}C-labelled salt marsh plants and its application to smooth cordgrass, *Spartina alterniflora* Loisel. *J. Exp. Mar. Biol. Ecol.* **14,** 251-259.

Ferguson, R. L., Collier, A., and Meeter, D. A. (1976). Kinetic response of an estuarine diatom to illumination, temperature and nitrogen source. *Chesapeake Sci.* **17,** 148-158.

Fogg, G. E. (1959). Nitrogen nutrition and metabolic patterns in algae. *Symp. Soc. Exp. Biol.* **13,** 106-125.

Fogg, G. E. (1965). "Algal Cultures and Phytoplankton Ecology." Univ. of Wisconsin Press, Madison.

Foster, M., Charters, A. C., and Neushul, M. (1971). The Santa Barbara oil spill. Part 1. Initial quantities and distribution of pollutant crude oil. *Environ. Pollut.* **2,** 97-113.

Foster, W. A. (1968). Studies on the distribution and growth of *Juncus roemerianus* in southeastern Brunswick County, North Carolina. M.S. Thesis, North Carolina State University, Raleigh.

Gardener, W. S., Windom, H. L., Stephens, J. A., Taylor, F. E., and Stickney, R. R. (1975). Concentrations of total mercury and methyl mercury in fish and other coastal organisms: Implications to mercury cycling, *In* "Mineral Cycling in Southeastern Ecosystems" (F. G. Howell, J. B. Gentry, and M. H. Smith, eds.), CONF-740513, pp. 268-278. U.S. E.R.D.A., Washington, D.C.

Gargas, E. (1970). Measurements of primary production, dark fixation and vertical distribution of the microbenthic algae in the Oresund. *Ophelia* **8,** 231-253.

Gessner, F. (1971). The water economy of the seagrass *Thalassia testudinum*. *Mar. Biol.* **10,** 258-260.

Gessner, F., and Hammer, L. (1960). Die Primärproduktion in mediterranen *Caulerpa-Cymodocea*-Wiesen. *Bot. Mar.* **2,** 157-163.

Golley, F. B., Odum, H. T., and Wilson, R. F. (1962). The structure and metabolism of a Puerto Rican red mangrove forest in May. *Ecology* **43,** 9-19.

Golley, F. B., McGinnis, J. T., Clements, R. G., Child, G. I., and Deuver, M. J. (1969). The structure of tropical forests in Panama and Colombia. *BioScience* **19,** 693-696.

Gosselink, J. G., Odum, E. P., and Pope, R. M. (1974). "The Value of the Tidal Marsh," Publ. No. LSU-S6-74-03. Center for Wetland Research, Louisiana State University, Baton Rouge.

Grøntved, J. (1958). Underwater macrovegetation in shallow coastal waters. *J. Conserv.* **24**(1), 32-42.

Grøntved, J. (1960). On the productivity of microbenthos and phytoplankton in some Danish fjords. *Medd. Dan. Fisk.-Havunders.* [N.S.] **3,** 55-92.

Guillard, R. R. L. (1960). A mutant of *Chlamydomonas moewusii* lacking contractile vacuoles. *J. Protozool.* **7,** 262-268.

Guillard, R. R. L. (1962). Salt and osmotic balance. *In* "Physiology and Biochemistry of Algae" (R. A. Lewin, ed.), pp. 529–540. Academic Press, New York.
Guillard, R. R. L. (1963). Organic sources of nitrogen for marine centric diatoms. *In* "Symposium on Marine Microbiology," (C. H. Oppenheimer, ed.), pp. 93–104. Thomas, Springfield, Illinois.
Guillard, R. R. L., and Kilham, P. (1977). The ecology of marine plankton diatoms. *In* "The Biology of Diatoms" (D. Werner, ed.), pp. 372–469. Blackwell, Oxford.
Guillard, R. R. L., and Ryther, J. H. (1962). Studies of marine planktonic diatoms. I. *Cyclotella nana* Hustedt and *Detonula confervacea* (Cleve) Gran. *Can. J. Microbiol.* **8,** 229–239.
Gurtz, M. E., and Weiss, C. M. (1972). Field investigations of the response of phytoplankton to thermal stress. Univ. North Carolina, Dept. Environ. Sci. Eng., *ESE Publ.* **321,** 1–152.
Hammer, L. (1969). Anaerobiosis in marine algae and marine phanerogams. *Proc. Int. Seaweed Symp., 6th, 1968* Vol. 7, pp. 414–419.
Harlin, M. M. (1980). Seagrass epiphytes. *In* "Handbook of Seagrass Biology: An Ecosystem Perspective" (R. C. Phillips and McRoy, C. P., eds.), pp. 117–151. Garland STPM Press, New York.
Harper, R. M. (1918). Some dynamic studies of Long Island vegetation. *Plant World* **21,** 38–46.
Harris, R. C., White, D. B., and Macfarlane, R. B. (1970). Mercury compounds reduce photosynthesis by plankton. *Science* **170,** 736.
Hartman, R. T. (1960). Algae and metabolites of natural waters. *In* "The Ecology of Algae" (A. C. Tryon, Jr. and R. T. Hartman, eds.), Spec. Publ. No. 2, pp. 38–55. Pymatuning Lab. Field Biol., University of Pittsburgh, Pittsburgh, Pennsylvania.
Harvey, H. W. (1953). Synthesis of organic nitrogen and chlorophyll by *Nitzschia closterium. J. Mar. Biol. Assoc. U.K.* **31,** 477–487.
Heald, E. J. (1969). "The Production of Organic Detritus in a South Florida Estuary," Sea Grant Program, Sea Grant Tech. Bull. No. 6, University of Miami, Miami, Florida.
Heald, E. J., and Odum, W. E. (1970). The contribution of mangrove swamps to Florida fisheries. *Gulf Caribb. Fish. Inst., Univ. Miami, Proc.* **22,** 130–135.
Hobbie, J. E. (1974). Nutrients and eutrophication in the Pamlico River estuary, N.C., 1971–1973. *Univ. N. C., Water Resour. Res. Inst., Rep.* No. 100, pp. 1–239.
Hough, R. A. (1974). Photorespiration and productivity in submersed aquatic vascular plants. *Limnol. Oceanogr.* **19,** 912–927.
Hulburt, E. M. (1970). Competition for nutrients by marine phytoplankton in oceanic, coastal and estuarine regions. *Ecology* **51,** 475–484.
Huntsman, S. A., and Sunda, W. G. (1980). The role of trace metals in regulating phytoplankton growth in natural waters. *In* "Primary Productivity of Natural Waters: Physiological Approaches to the Ecology of Phytoplankton" (E. M. Morris, ed.). Blackwell Scientific Publications, Oxford, (in press).
Hustedt, F. (1955). Marine littoral diatoms of Beaufort, North Carolina. *Duke Univ. Mar. Stn., Bull.* No. 6, pp. 1–67.
Ichimura, S. (1968). Phytoplankton photosynthesis. *In* "Algae, Man and the Environment" (D. F. Jackson, ed.), pp. 103–120. Syracuse Univ. Press, Syracuse, New York.
Jerlov, N. G. (1951). Report of the Swedish Deep-Sea Expedition. 3. *Phys. Chem.* **1,** 1–59.
Jitts, H. R., McAllister, C. D., Stephens, K., and Strickland, J. D. H. (1964). The cell division rates of some marine phytoplankters as a function of light and temperature. *J. Fish. Res. Board Can.* **21,** 139–157.
Jones, J. A. (1968). Primary productivity by the tropical turtle grass, *Thalassia testudinum* Konig and its epiphytes. Ph.D. Thesis, University of Miami, Coral Gables, Florida.
Jørgensen, E. G. (1956). Growth inhibiting substances formed by algae. *Physiol. Plant.* **9,** 712–726.
Jørgensen, E. G. (1964). Adaptation to different light intensities in the diatom *Cyclotella meneghiniana* Kütz. *Physiol. Plant.* **17,** 136–145.

Jørgensen, E. G. (1968). The adaptation of plankton algae. II. Aspects of the temperature adaptation of *Skeletonema costatum*. *Physiol. Plant.* **21,** 423–427.

Joyce, E. A., Jr., and Roberts, B. S. (1975). Florida Department of Natural Resources red tide program. *In* "Toxic Dinoflagellate Blooms" (V. R. LoCicero, ed.), pp. 95–104. Massachusetts Science and Technology Foundation, Wakefield.

Kain, J. M., and Fogg, G. W. (1958). Studies on the growth of marine phytoplankton. I. *Asterionella japonica* Gran. *J. Mar. Biol. Assoc. U.K.* **37,** 397–413.

Keefe, C. W., and Boynton, W. R. (1973). Standing crop of salt marshes surrounding Chincoteague Bay, Maryland - Virginia. *Chesapeake Sci.* **14,** 117–123.

Kenworthy, W. J. (1981). The interrelationship between seagrasses, *Zostera marina* and *Halodule wrightii,* and the physical and chemical properties of sediments in a coastal plain estuary near Beaufort, N.C. M.S. Thesis, University of Virginia, Charlottesville.

Ketchum, B. H. (1969). An ecological view of environmental management. *Oceanus* **15,** 14–23.

Ketchum, B. H., and Tripp, B. W. (1972). "The Water's Edge: Critical Problems of the Coastal Zone." MIT Press, Cambridge, Massachusetts.

Kirby, C. P. (1971). The annual net primary production and decomposition of the salt marsh grass *Spartina alterniflora* Loisel in the Barataria Bay estuary of Louisiana. Ph.D. Thesis, Louisiana State University, Baton Rouge.

Klug, M. J. (1980). Detritus-decomposition relationships. *In* "An Ecosystem Perspective" (R. C. Phillips and C. P. McRoy; eds.), pp. 225–245. Garland STPM Press, New York.

Koblentz-Mishke, O. J., Valkovinsky, V. V., and Kabanova, J. G. (1970). Plankton primary production of the world ocean. *In* "Proceedings of a Symposium on the Scientific Exploration of the South Pacific" (W. S. Wooster, ed.), pp. 183–193. National Academy of Sciences, Washington, D.C.

Kopp, J. F., and Kroner, R. C. (1969). "Trace Metals in Waters of the United States, a Five-Year Summary of Trace Metals in Rivers and Lakes of the United States (Oct. 1, 1972–Sept. 30, 1967)." U.S. Department of the Interior, Fed. Water Pollut. Control Admin., Div. Pollut. Surveillance, Cincinnati, Ohio.

Kuenzler, E. J. (1974). Mangrove swamp systems. *In* "Coastal Ecological Systems of the United States" (H. T. Odom, B. J. Copeland, and E. A. McMahan, eds.), Vol. 1, pp. 346–371. Conservation Foundation, Washington, D.C.

Kuenzler, E. J., and Ketchum, B. W. (1962). Rate of phosphorus uptake by *Phaeodactylum tricornutum*. *Biol. Bull.* (*Woods Hole, Mass.*) **5,** 134–145.

Kuenzler, E. J., and Marshall, H. L. (1973). Effects of mosquito control ditching on estuarine ecosystems. *Univ. N.C., Water Resour. Res. Inst., Rep.* No. 81, pp. 1–83.

Lehninger, A. L. (1970). "Biochemistry." Worth, New York.

Livingstone, D. A. (1963). Chemical composition of rivers and lakes. *U.S., Geol. Surv., Prof. Pap.* 440–G.

Lüttge, U. (1971). Structure and function of plant glands. *Annu. Rev. Plant Physiol.* **22,** 23–44.

Lüttge, U., Pallaghy, C. K., and Osmond, C. B. (1970). Coupling of ion transport in green cells of *Atriplex spongiosa* leaves to energy sources in the light and in the dark. *J. Membr. Biol.* **2,** 17–30.

McCombie, A. M. (1960). Actions and interactions of temperature, light intensity and nutrient concentration on the growth of the green algae, *Chlamydomonas reinhardi* Dangeard. *J. Fish. Res. Board Can.* **17,** 871–894.

MacDonald, K. B., and Barbour, M. G. (1974). Beach and salt marsh vegetation of the North American Pacific Coast. *In* "Ecology of Halophytes" (R. J. Reimold and W. H. Queen, eds.), Academic Press, New York.

McGill, J. T. (1958). Map of the coastal landforms of the world. *Geogr. Rev.* **48,** 402–405.

Mackin, J. G. (1950). "Effects of Crude Oil and Bleedwater on Oysters and Aquatic Plants," Prog. Rep. Texas A & M University, College Station.

McMahan, C. A. (1968). Biomass and salinity tolerance of shoal grass and manatee grass in lower Laguna Madre, Texas. *J. Wildl. Manage.* **32,** 501–506.

McMillan, C. (1974). Salt tolerance of mangroves and submerged aquatic plants. *In* "Ecology of Halophytes" (R. J. Reimold and W. H. Queen, eds.), pp. 379–390. Academic Press, New York.

McMillan, C., and Mosely, F. N. (1967). Salinity tolerances of five marine spermatophytes of Redfish Bay, Texas. *Ecology* **48,** 403–506.

Macnae, W. (1967). Zonation within mangroves associated with estuaries in North Queensland. *In* "Estuaries" (G. H. Lauff, ed.), Publ. No. 83, pp. 432–444. Am. Assoc. Adv. Sci., Washington D.C.

Macnae, W. (1968). A general account of the fauna and flora of mangrove swamps and forests in the Indo-West Pacific region. *Adv. Mar. Biol.* **6,** 73–270.

McNulty, J. K. (1970). Effects of abatement of domestic sewage pollution on the benthos, volumes of zooplankton and the fouling organisms of Biscayne Bay, Florida. *Stud. Trop. Oceanogr.* **9,** 1–107.

McRoy, C. P. (1966). The standing stock and ecology of eelgrass, *Zostera marina,* Izembek Lagoon, Alaska. M.S. Thesis, University of Washington, Seattle.

McRoy, C. P., and Barsdate, R. J. (1970). Phosphate absorption in eelgrass. *Limnol. Oceanogr.* **15,** 6–13.

McRoy, C. P., and Goering, J. J. (1974). Nutrient transfer between seagrass *Zostera marina* and its epiphytes. *Nature (London)* **248,** 173–174.

McRoy, C. P., Barsdate, R. J., and Nebert, M. (1972). Phosphorus cycling in an eelgrass (*Zostera marina* L.) ecosystem. *Limnol. Oceanogr.* **17,** 58–67.

McRoy, C. P., and Helferich, C., eds. (1977). "Seagrass Ecosystems." Marcel Dekker, New York.

McRoy, C. P., and McMillan, C. (1977). Production ecology and physiology of seagrasses. *In* *"Seagrass Ecosystems"* (C. P. McRoy and C. Helfferich, eds.), pp. 53–87. Marcel Dekker, Inc. New York.

McRoy, C. P., and Williams, S. L. (1977). Sublethal effects on seagrass photosynthesis. pp. 636–673. *In* U.S. NOAA. Outer Continental Shelf Environmental Assessment Program. Environmental assessment of the Alaskan continental shelf. Vol. XII. Effects. Boulder Colorado, U.S. NOAA, Outer Continental Shelf Environmental Assessment Program. 1977.

Maddux, W. S., and Jones, R. F. (1964). Some interactions of temperature, light intensity and nutrient concentration during the continuous culture of *Nitzschia closterium* and *Tetraselmis* sp. *Limnol. Oceanogr.* **9,** 79–86.

Mahoney, B. M., and Haskin, H. H. (1980). The effects of petroleum hydrocarbons on the growth of phytoplankton recognized as food forms for the eastern oyster, *Crassostrea virginica* Gmelin. *Environ. Pollut.* **22,** 123–132.

Mann, K. H. (1972). Macrophyte production and detritus food chains in coastal waters. *Mem. Ist. Ital. Idrobiol.* **29,** Suppl., 353–383.

Mann, K. H. (1973). Seaweeds: Their productivity and strategy for growth. *Science* **182,** 975–981.

Marmelstein, A. D., Morgan, P. W., and Pequegnat, W. E. (1968). Photoperiodism and related ecology in *Thalassia testudinum. Bot. Gaz. (Chicago)* **129,** 63–67.

Marshall, N., Skauen, D. A., Lampe, H. C., and Oviatt, C. A. (1971). Productivity of the benthic microflora of shoal estuarine environments in southern New England. *Int. Rev. Gesamten. Hydrobiol.* **56,** 947–956.

Matheke, G. E., and Horner, R. (1974). Primary productivity of the benthic microalgae in the Chukchi Sea near Barrow, Alaska. *J. Fish. Res. Board Can.* **31,** 1779–1786.

Menzel, D. W., and Ryther, J. H. (1960). The annual cycle of primary production in the Sargasso Sea off Bermuda. *Deep-Sea Res.* **6,** 351–467.

Milner, H. W. (1948). The fatty acids of *Chlorella. J. Biol. Chem.* **177,** 813–817.

Morgan, M. H. (1961). Annual angiosperm production on a salt marsh. M.S. Thesis, University of Delaware, Newark.
Nelson-Smith, A. (1967). Oil, emulsifiers and marine life. *In* Conservation and the Torrey Canyon. *J. Devon Trust Natl. Conserv., July Suppl.* pp. 29–33.
Michael, A. D. (1977). The effects of petroleum hydrocarbons on marine populations and communities. *In* "Fate and Effects of Petroleum Hydrocarbons in Marine Organisms and Ecosystems" (D. A. Wolfe, ed.), pp. 129–137. Pergamon Press, New York.
Nichols, M. M., Sallenger, A., Van Eepoel, R., Grigg, D., Brady, R., Olman, J., and Crena, R. (1972). "Environment, Water and Sediments of Christiansted Harbor, St. Croix," Rep. No. 16 Div. Environ. Health, Caribean Research Institute, College of the Virgin Islands.
Nixon, S. W. (1980). Between coastal marshes and coastal waters - a review of twenty years of speculation and research on the role of salt marshes in estuarine productivity and water chemistry. *In* "Estuarine and Wetland Processes" (P. Hamilton and K. B. MacDonald, eds.), pp. 437–525. Plenum Press, New York.
Odum, E. P. (1959). "Fundamentals of Ecology," 2nd ed., Saunders, Philadelphia, Pennsylvania.
Odum, E. P. (1971). "Fundamentals of Ecology," 3rd ed., Saunders, Philadelphia, Pennsylvania.
Odum, E. P. (1974). Halophytes, Energetics and ecosystems. *In* "Ecology of Halophytes" (R. J. Reimold and W. E. Queen, eds.), pp. 599–602. Academic Press, New York.
Odum, E. P., and Fanning, M. E. (1973). Comparison of the productivity of *Spartina alterniflora* and *Spartina cynosuroides* in Georgia coastal marshes. *Bull. Ga. Acad. Sci.* **31,** 1–12.
Odum, H. T. (1963). Productivity measurements in Texas turtle grass and the effects of dredging on an intercoastal channel. *Publ. Inst. Mar. Sci., Univ. Tex.* **9,** 48–58.
Odum, H. T., and Wilson, R. F. (1962). Further studies on reaeration and metabolism of Texas bays, 1958–1960. *Publ. Inst. Mar. Sci., Univ. Tex.* **8,** 23–55.
Ogata, E., and Matsui, T. (1965). Photosynthesis in several marine plants of Japan as affected by salinity, drying and pH, with attention to their growth habits. *Bot. Mar.* **8,** 199–217.
Orth, R. J. (1977). Effect of nutrient and enrichment on growth of the eelgrass, *Zostera marina,* in the Chesapeake Bay, Virginia, U.S.A. *Mar. Biol.* **44,** 187–194.
Pamatmat, M. M. (1968). Ecology and metabolism of a benthic community on an intertidal sandflat. *Int. Rev. Gesamten Hydrobiol.* **53,** 211–298.
Park, K. (1968). Alkalinity and pH off the coast of Oregon. *Deep-Sea Res.* **15,** 171–183.
Parker, P. L., ed. (1974). "Effects of Pollutants on Marine Organisms." National Science Foundation, Washington, D.C.
Patten, B. C., Mulford, R. A., and Warriner, J. E. (1963). An annual phytoplankton cycle in the lower Chesapeake Bay. *Chesapeake Sci.* **4,** 1–20.
Penfound, W. T. (1956). Primary production of vascular aquatic plants. *Limnol. Oceanogr.* **1,** 92–101.
Penhale, P. A. (1976). Primary production, dissolved organic carbon excretion, and nutrient transport in an epiphyte-eelgrass (*Zostera marina*) system. Ph.D. Thesis, North Carolina State University, Raleigh.
Phillips, R. C. (1960). "Observations on the Ecology and Distribution of Florida Sea Grasses," Prof. Pap. Ser. No. 2. Fla. Board Conserv., Mar. Lab.
Phillips, R. C. (1972). Ecological life history of *Zostera marina* L. (eelgrass) in Puget Sound, Washington, Ph.D. Thesis, University of Washington, Seattle.
Palumbo, A. V. (1980). Dynamics of bacterioplankton in the Newport River estuary. Ph.D. Thesis, North Carolina State University, Raleigh, North Carolina.
Patrick, R. (1973). Use of algae, especially diatoms, in the assessment of water quality. *In* "Biological Methods for the Assessment of Water Quality. (J. Cairns Jr. and K. L. Dickson, eds.). Am. Soc. Test. Mater. Spec. Tech. Publ. 528, 76–95.
Phillips, R. C., and McRoy, C. P. (1980). "Handbook of Seagrass Biology: An Ecosystem Perspective." Garland STPM Press, New York.

Phillips, R. C. (1974). Temperate grass flats. *In* "Coastal Ecological Systems of the United States" (H. T. Odum, B. J. Copeland, and E. A. McMahan, eds.), Vol. 2, pp 244–299. Conservation Foundation, Washington, D.C.

Phleger, C. F. (1971). Effects of salinity on growth of salt marsh grass. *Ecology* **52,** 908–911.

Pomeroy, L. R. (1959). Algal productivity in salt marshes of Georgia. *Limnol. Oceanogr.* **4,** 386–397.

Pratt, R. J., and Fogg, J. (1940). Studies on *Chlorella vulgaris*. II. Further evidence that *Chlorella* cells form a growth-inhibiting substance. *Am. J. Bot.* **27,** 431–436.

Price, W. A. (1954). Shorelines and coats of the Gulf of Mexico. *Fish. Bull.* **55,** 39–65.

Provasoli, L., McLaughlin, J. J. A., and Droop, M. R. (1957). The development of artificial media for marine algae. *Arch. Mikrobiol.* **25,** 392–428.

Putnam, H. D. (1966). Limiting factors for primary productivity in a west coast Florida estuary. *Adv. Water Pollut. Res.* **3,** 121–142.

Ragotzkie, R. A. (1959). Plankton productivity in estuarine waters of Georgia. *Publ. Inst. Mar. Sci., Univ. Tex.* **6,** 146–158.

Rains, D. W. (1972). Salt transport by plants in relation to salinity. *Annu. Rev. Plant Physiol.* **23,** 51–72.

Rains, D. W., and Epstein, E. (1967). Preferential absorption of potasium by leaf tissue of the mangrove, *Avicennia marina;* An aspect of halophytic competence in the coping with salt *Aust. J. Biol. Sci.* **20,** 847–857.

Ranwell, D. W. (1964). *Spartina* salt marshes in southern England. III. Rates of establishment, succession and nutrient supply at Bridgewater Bay, Somerset. *J. Ecol.* **52,** 95–105.

Raymont, F. E. G. (1963). "Plankton and Productivity in the Oceans." Pergamon, Oxford.

Redfield, A. C., Ketchum, B. H., and Richards, F. A. (1963). The influence of organisms on the composition of sea-water. *In* "The Sea" (M. N. Hill, ed.), Vol. 2, pp. 26–77. Wiley (Interscience), New York.

Reimold, R. J. (1974). "Toxaphene interactions in estuarine ecosystems," Tech. Rep. Ser. No. 74-6. Ga. Mar. Sci. Cent., Skidaway Island.

Reimold, R. J. (1976). Grazing on wetland meadows. *In* "Estuarine Processes" (M. Wiley, ed.), Vol. 1, pp. 219–225. Academic Press, New York.

Reimold, R. J., Linthurst, R. A., and Wolf, P. L. (1975). Effects of grazing on salt marsh. *Biol. Conserv.* **8,** 105–125.

Rice, T. R. (1954). Biotic influences affecting population growth of planktonic algae. *Fish. Bull.* **54,** 227–245.

Rice T. R. (1971) Ocean use planning. *Trans. Am. Nucl. Soc.* **14,** 74.

Rice, T. R., and Ferguson, R. L. (1975). Response of estuarine phytoplankton to environmental conditions. *In* "Physiological Ecology of Estuarine Organisms" (F. J. Vernberg, ed.), pp. 1–43. Univ. of South Carolina Press, Columbia.

Riley, G. A. (1956). Oceanography of Long Island Sound, 1952–1954. IX. Production and utilization of organic matter. *Bull. Bingham Oceanogr. Collect.* **15,** 324–344.

Riley, G. A. (1967). Mathematical model of nutrient conditions in coastal waters. *Bull. Bingham Oceanogr. Collect.* **19,** 72–80.

Round, R. E. (1965). "The Biology of the Algae" Arnold, London.

Rutzler, K. and Sterrer, W. (1970). Oil pollution damage observed in tropical communities along the Atlantic seaboard of Panama. *Bioscience* **20,** 222–234.

Ryther, J. H. (1954). The ecology of phytoplankton blooms in Moriches Bay and Great South Bay, Long Island, New York. *Biol. Bull.* (*Woods Hole, Mass.*) **106,** 198–209.

Ryther, J. H. (1959). Potential productivity of the sea. *Science* **130,** 602–608.

Ryther, J. H. (1960). Organic production by planktonic algae, and its environmental control. *In* "The Ecology of Algae" (C. A. Tryon, Jr. and R. T. Hartman, eds.), Spec. Publ. No. 2, pp. 72–83. Pymatuning Lab. Field Biol., Univeristy of Pittsburgh, Pittsburgh, Pennsylvania.

Ryther, J. H. (1969). Photosynthesis and fish production in the sea. Science **166,** 72–76.

Ryther, J. H., and Dunstan, W. M. (1971). Nitrogen, phosphorus and eutrophication in the coastal marine environment. *Science* **171,** 1008–1013.

Ryther, J. H., and Menzel, D. W. (1965). On the production, composition and distribution of organic matter in the western Arabic Sea. *Deep-Sea Res.* **12,** 199–209.

Ryther, J. H., and Yentsch, C. S. (1958). Primary production of continental shelf waters off New York. *Limnol. Oceanogr.* **3,** 327–335.

Scholander, P. F., Himmel, H. T., Hemmingsen, E., and Gary, W. (1962). Salt balance in mangroves. *Plant Physiol.* **37,** 722–729.

Scholander, P. F., Bradstreet, E. D., Himmel, H. T., and Hemmingsen, E. A. (1966). Sap concentrations in halophytes and some other plants. *Plant Physiol.* **41,** 529–532.

Sculthorpe, C. D. (1967). "The Biology of Aquatic Vascular Plants" Arnold, London.

Seneca, E. D. (1974). Germination and seedling response of Atlantic and Gulf coast populations of *Spartina alterniflora*. *Am. J. Bot.* **61,** 947–956.

Setchell, W. A. (1929). Morphological and phenological notes on *Zostera marina* L. *Univ. Calif., Berkeley, Publ. Bot.* **14,** 389–452.

Smalley, A. E. (1959). The role of two invertebrate populations, *Littorina irrorata* and *Orchelium fidicinium* in the energy flow of a salt marsh ecosystem. Ph.D. Thesis, University of Georgia, Athens.

Smayda, T. J. (1966). Phytoplankton. *In* "The Encyclopedia of Oceanography" (R. W. Fairbridge, ed.), pp. 712–716. Van Nostrand-Reinhold, Princeton, New Jersey.

Smith, G. B. (1975). The 1971 red tide and its impact on certain reef communities in the mid-eastern Gulf of Mexico. *Environ. Lett.* **9,** 141–152.

Smith, G. W., Hayasaka, S. S., and Thayer, G. W. (1980). Microbiology of the rhizosphere of seagrass systems, *In* "Annual Report of the Beaufort Laboratory to the U.S. Department of Energy." pp. 120–138. U.S. Department of Commerce, National Oceanic and Atmospheric Administration, National Marine Fisheries Service, Southeast Fisheries Center, Beaufort Laboratory, Beaufort, North Carolina.

Sorenson, L. O., and Conover, Y. T. (1962). Algal mat communities of *Lyngbya confervoides* (C. Agardh) Gomont. *Publ. Inst. Mar. Sci., Univ. Tex.* **8,** 61–74.

Spooner, M. F. (1978). Editorial introduction. *Mar. Pollut. Bull.* **9,** 281–284.

Stebbings, R. E. (1970). Recovery of salt marsh in Brittany sixteen months after heavy pollution by oil. *Environ. Pollut.* **1,** 163–167.

Steele, J. H., and Baird, I. E. (1968). Production ecology of a sandy beach. *Limnol. Oceanogr.* **13,** 14–25.

Steeman-Nielsen, E. (1975). "Marine Photosynthesis," Am. Elsevier, New York.

Steeman-Nielsen, E., and Jørgensen, E. G. (1968). The adaptation of plankton algae. I. General part. *Physiol. Plant.* **21,** 401–413.

Steeman-Nielsen, E., and Wium-Andersen, S. (1970). Copper ions as poison in the sea and in freshwater. *Mar. Biol.* **6,** 93–97.

Steeman-Nielsen, E., and Wium-Andersen, S. (1971). The influence of copper on photosynthesis and growth in diatoms. *Physiol. Plant.* **24,** 480–484.

Stewart, W. D. P. (1974). "Algal Physiology and Biochemistry." Univ. of California Press, Berkeley.

Stockner, J. G., and Antia, N. H. (1976). Phytoplankton adaptation to environmental stresses from toxicants, nutrients, and pollutants–a warning. *J. Fish. Res. Bd. Can.* **33,** 2089–2096.

Stirban, M. (1968). Relationship between the assimilatory pigments, the intensity of chlorophyll fluorescence and the level of the photosynthesis zone in *Zostera marina* L. *Rev. Roum. Biol., Ser. Bot.* **13,** 291–295.

Stross, R. G., and Stottlemyer, J. R. (1965). Primary production in the Patuxent River. *Chesapeake Sci.* **6,** 125–140.

Stroud, L. M., and Cooper, A. W. (1968). Color-infrared aerial photographic interpretation and net primary productivity of a regularly flooded North Carolina salt marsh. *Univ. N.C., Water Resour. Res. Inst., Rep.* No. 14, pp. 1–81.

Stumm, W., and Morgan, J. J. (1970). "Aquatic Chemistry." Wiley (Interscience), New York.

Sunda, W. (1975). The relationship between cupric ion activity and the toxicity of copper to phytoplankton. Ph.D. Thesis, Massachusetts Institute of Technology, Cambridge, and Woods hole Oceanographic Institution, Woods Hole, Massachusetts.

Sunda, W., and Guillard, R. R. L. (1976). The relationship between cupric ion activity and toxicity of copper to phytoplankton. *J. Mar. Res.* **34,** 511–529.

Syrett, P. J. (1953). The assimilation of ammonia by nitrogen-starved cells of *Chlorella vulgaris.* I. The correlation of assimilation with respiration. *Ann. Bot.* (*London*) [N.S.] **17,** 1–19.

Syrett, P. J. (1962). Nitrogen assimilation. *In* "Physiology and Biochemistry of Algae" (R. A. Lewin, ed.), pp. 171–188. Academic Press, New York.

Talling, S. F. (1955). The relative growth rate of three plankton diatoms in relation to underwater radiation and temperature. *Ann. Bot.* (*London*) [N.S.] **19,** 329–341.

Taylor, D. L., and Seliger, H. H. (1979). Toxic Dinoflagellate Blooms. *In* "Developments in Marine Biology," Volume 1. Elsevier/North-Holland, New York.

Taylor, J. L., Saloman, C. H., and Prest, K. W. (1973). Harvest and regrowth of turtle grass (*Thalassia testudinum*) in Tampa Bay, Florida. *Fish. Bull.* **71,** 145–148.

Taylor, W. (1960). "Marine Algae of the Eastern Tropical and Subtropical Coasts of the Americas." Univ. of Michigan Press, Ann Arbor.

Teal, J. M. (1962). Energy flow in the salt marsh ecosystem of Georgia. *Ecology* **43,** 614–624.

Teal, J. M., and Kanwisher, J. W. (1966). Gas transport in the marsh grass, *Spartina alterniflora. J. Exp. Bot.* **17,** 355–361.

Teal, J. M., and Kanwisher, J. W. (1970). Total energy balance in salt marsh grasses. *Ecology* **51,** 690–696.

Teal, J. M., and Valiela, I. (1973). The salt marsh as a living filter. *Mar. Technol. Soc. J.* **7,** 19–21.

Teal, J. M., Jameson, D. L., and Bader, R. G. (1972). Living resources. *In* "The Water's Edge" (B. H. Ketchum ed.), pp. 37–62. MIT Press, Cambridge, Massachusetts.

Thayer, G. W. (1971). Phytoplankton production and the distribution of nutrients in a shallow unstratified estuarine system near Beaufort, N.C. *Chesapeake Sci.* **12,** 240–253.

Thayer, G. W. (1974). Identity and regulation of nutrients limiting phytoplankton production in the shallow estuaries near Beaufort, N.C. *Oecologia* **14,** 75–92.

Thayer, G. W. (1975). The estuary—an area of environmental concern. *In* "Coastal Development and Areas of Environmental Concern" (S. Baker, ed.), Sea Grant Publ., UNC-SG-75-18, pp. 50–72. North Carolina State University, Raleigh.

Thayer, G. W., Adams, S. M., and LaCroix, M. W. (1975a). Structural and functional aspects of a recently established *Zostera marina* community. *In* "Estuarine Research." (L. E. Cronin, ed.), Vol. I, pp. 518–540. Academic Press, New York.

Thayer, G. W., Stuart, H. H., Kenworthy, W. J., Ustach, J. F., and Hall, A. B. (1979). Habitat values of salt marshes, mangroves and seagrasses for aquatic organisms. *In* "Wetland Functions and Values" (P. Greeson, J. R. Clark, and J. E. Clark, eds.), pp. 235–247. American Water Resources Association, Minneapolis, Minnesota.

Thayer, G. W., and Ustach, J. F., (1980). Gulf of Mexico Wetlands: Value, State of Knowledge and Research Needs, Proc. GOMEX Workshop, NOAA/Office of Marine Pollution Assessment, Miami, Florida, Oct. 1979, (in press).

Thayer, G. W., Wolfe, D. A., and Williams R. B. (1975b). The impact of man on seagrass systems. *Am. Sci* **63,** 288–296.

Thomas, W. H. (1966). Effects of temperature and illuminance on cell division rate of tropical oceanic phytoplankton. *J. Physiol.* (*London*) **2,** 17–22.

Thomas, W. H., Holm-Hansen, O., Siebert, D. L. R., Azam, F., Hodson, R., and Takahashi, M.

(1977). Effects of copper on phytoplankton standing crop and productivity: Controlled ecosystem pollution experiment. *Bull. Mar. Sci.* 27, 34-43.

Thomas, W. H. and Siebert, D. L. R. (1977). Effects of copper on the dominance and the diversity of algae: controlled ecosystem pollution experiment. *Bull. Mar. Sci.* **27,** 23-33.

Ukeles, R. (1961). The effect of temperature on the growth and survival of several marine algal species. *Biol. Bull.* (*Woods Hole, Mass.*) **120,** 255-264.

Uyeno, F. (1957). The variation of diatom communities and the schematic explanation of their increase in Osaka Bay in summer. III. Schematic explanation of increase of diatoms in relation to the distribution of chlorinity. *J. Oceanogr. Soc. Jpn.* **13,** 107-110.

Valiela, I., Teal, J. M., and Persson, N. Y. (1976). Production and dynamics of experimentally enriched salt marsh vegetation: Belowground biomass. *Limnol. Oceanogr.* **21,** 245-252.

Waits, E. D. (1967). Net primary productivity of an irregularly flooded North Carolina salt marsh. Ph.D. Thesis, North Carolina State University, Raleigh.

Walsh, G. E. (1974). Mangroves: A Review. *In* "Ecology of Halophytes" (R. J. Reimold and W. H. Queen, eds.), pp. 51-174. Academic Press, New York.

Watling, L., Pembroke, A., and Lind, H. (1975). An evaluation of multipurpose offshore industrial port islands for the Atlantic and Gulf coasts. Draft report prepared for National Science Foundation's Research Applied to National Needs Program, Contract No. GI-4311. University of Delaware, Newark.

Watt, W. D. (1971). Measuring the primary production rates of individual phytoplankton species in natural mixed populations. *Deep-Sea Res.* **18,** 329-339.

Weaver, J. E., and Clements, F. E. (1938). "Plant Ecology," 2nd ed. McGraw-Hill, New York.

Westlake, D. F. (1963). Comparisons of plant productivity. *Biol. Rev. Cambridge Philos. Soc.* **38,** 385-425.

Williams, R. B. (1964). Division rates of salt marsh diatoms in relation to salinity and cell size. *Ecology* **45,** 877-880.

Williams, R. B. (1966). Annual phytoplanktonic production in a system of shallow temperate estuaries. *In* "Some Contemporary Studies in Marine Science" (H. Barnes, ed.), pp. 699-716. Allen & Unwin, London.

Williams, R. B. (1972). Steady-state equilibriums in simple non-linear food webs. *In* "Systems Analysis and Simulation in Ecology" (B. C. Patten, ed.), Vol. 2, pp. 213-240. Academic Press, New York.

Williams, R. B. (1973). Nutrient levels and phytoplankton productivity in the estuary. *In* "Proceedings of the Second Coastal Marsh and Estuary Management Symposium" (R. H. Chabreck, ed.), pp. 59-89. Division of Continuing Education, Louisiana State University, Baton Rouge.

Williams, R. B., and Murdoch, M. B. (1966). Phytoplankton production and chlorophyll concentration in the Beaufort Channel, North Carolina. *Limnol. Oceanogr.* **11,** 73-82.

Williams, R. B., and Murdoch, M. B. (1969). The potential importance of *Spartina alterniflora* in conveying zinc, manganese and iron into estuarine food chains. *Radioecol., Proc. Natl. Symp., 2nd, 1967* U.S. A.E.C., CONF-670503, pp. 431-439.

Williams, R. B., and Murdoch, M. B. (1972). Compartmental analysis of the production of *Juncus roemerianus* in a North Carolina salt marsh. *Chesapeake Sci.* **13,** 69-79.

Williams, W. T., and Barber, D. A. (1961). The functional significance of aerenchyma in plants *Symp. Soc. Exp. Biol.* **15,** 132-144.

Wood, E. J. F., and Zieman, J. C. (1969). The effects of temperature on estuarine plant communities. *Chesapeake Sci.* **10,** 172-174.

Wood, E. J. F., Odum, W. E., and Zieman, J. C. (1969). Influence of sea grasses on the productivity of coastal lagoons. *In* "Coastal Lagoons." (A. Ayala Castanares and F. B. Phleger, eds.), pp. 495-502. Universidad Nacional Autonoma de Mexico, Mexico, D.F.

Woodhouse, W. W., Jr., Seneca, E. D., and Broome, S. W. (1974). "Propagation of *Spartina alterniflora* for Substrate Stabilization and Salt Marsh Development,'' Coastal Eng. Res. Cent., Tech. Memo. No. 46. U.S. Army Corps of Engineers, Washington, D.C.

Woodwell, G. M., Rich, P. H., and Hall, C. A. S. (1973). Carbon in estuaries. *In* ''Carbon and the Biosphere'' (G. M. Woodwell and E. V. Pecan, eds.), *AEC Symp. Ser.* **30,** CONF-720510, 221-240.

Worthington, E. B. (1943). ''Eleventh Annual Report of the Director, Year Ending March 31, 1943. ''Freshwater Biological Association of the British Empire, Ambleside, England.

Wurster, C. F., Jr. (1968). DDT reduces photosynthesis by marine phytoplankton. *Science* **159,** 1474-1475.

Zieman, J. C. (1970). The effects of a thermal effluent stress on the seagrasses and macroalgae in the vicinity of Turkey Point, Biscayne Bay, Florida. Ph.D. Thesis, University of Miami, Coral Gables, Florida.

Zieman, J. C. (1975a). Quantitative and dynamic aspects of the ecology of turtle grass, *Thalassia testudinum*. *In* Estuarine Research (L. E. Cronin, ed.), Vol. 1, pp. 541-562. Academic Press, New York.

Zieman, J. C. (1975b). Tropical sea grass ecosystems and pollution. *In* ''Tropical Marine Pollution'' (E. J. Ferguson Wood and R. E. Johannes, eds.), pp. 63-74. Am. Elsevier, New York.

Zieman, J. C., Thayer, G. W., Robblee, M. B., and Zieman, R. T. (1979). Production and export of seagrasses from a tropical bay. *In* ''Ecological Processes in Coastal and Marine Systems'' (R. J. Livingston, ed.), pp. 21-33. Plenum Press, New York.

Zobell, C. E. (1964). The occurrence, effects and fate of oil polluting the sea. *Adv. Water Pollut. Res. Proc. Int. Conf. 1st, 1962* pp. 85-118.

Marine Bacteria

3

L. Harold Stevenson and Thomas H. Chrzanowski

Biologists have come to expect the following things of bacteria; they divide very rapidly, they have a vigorous rate of metabolism, they can adapt to almost any natural environment, they can degrade any biologically produced organic compounds (and many man-made organics), they are very important in biogeochemical conversions of the elements, and they act as universal decomposers. Consequently, when one asks fundamental questions concerning the physiological adaptation of bacteria to the oceanic environments, one expects to find answers that are relative to the properties mentioned above. Many attempts to demonstrate the above physiological properties in deep-sea bacteria under *in situ* conditions have resulted in frustration. If one looks for the characteristics mentioned above, heterotrophic bacteria appear to be poorly adapted to life in the deep ocean.

The field of microbial ecology has expanded rapidly in recent years and microbial ecologists have been able to elucidate some of the factors governing the activities of bacteria. No longer can bacteria be viewed strictly as ''active decomposers'', but rather must be viewed as organisms having other properties and functions. It is unfortunate that contributions made by microbial ecologists have not as yet made their way into main-line sources of ecological education but are instead relegated to small sections of general microbiology texts. Consequently, the contents of this chapter will be directed, not toward the attention of the small group of marine microbial ecologists, but rather to students of marine ecology in general.

Bacteria are ubiquitously distributed in the oceans; however, neither their distribution nor the physiological activity of these organisms is uniform. A myriad of factors can potentially influence both the presence and functioning of oceanic bacteria, including temperature, pressure, anions, cations, organic carbon, nitrogen, inorganic nutrients, detritus, primary producers, and predators (Colwell and Morita, 1974). A consideration of the adaptations of marine bacteria to these factors must be undertaken with an awareness of the limitation that one faces when discussing the topic. One factor is the relative dearth of information that is specifically applicable; the second is the uncertainty of the information that is available. The situation was basically described by Campbell in the ''Conclusions'' chapter in his recent book on the ecology of microorganisms

FUNCTIONAL ADAPTATIONS OF MARINE ORGANISMS

ISBN 0-12-718280-2

when he wrote "Microbial ecologists are in the position that general ecologists were half a century ago . . ." (Campbell, 1977). A significantly large body of information is available on the physiology of bacteria; indeed, several text books have been devoted to the subject. Likewise, the area is reviewed annually in Advances in Microbial Physiology, which has been issued since 1967. A perusal of the indexes of the fifteen volumes of that publication reveals less than a half-dozen citations concerning marine bacteria. Only one article has been devoted to marine bacteria, that dealt with growth at high pressure. The issues of Annual Review of Microbiology published since 1967 contain a few more index citations of marine bacteria; but the information is miniscule in comparison to the information available on *Escherichia coli*.

The second limitation to a discussion of physiological adaptations might be called a "Compounded Bacterial Uncertainty Principle" (Doetsch and Cook, 1973). The properties of a population of cultured bacteria can be determined with some precision; however, it is impossible to quantitatively specify characteristics of individual organisms. This general uncertainty is compounded by both the nature of the marine bacterial community and by the methods available for study of that community. Many physiological experiments require the isolation and cultivation of bacteria in the laboratory. Laboratory experiments can, and often do, contribute useful information about the myriad of physiological factors regulating the ability of microorganisms to grow and survive in the sea. However, as Alexander (1971) so poignantly put it, "The point of focus . . . is nature and not the laboratory, the characteristic environments where the species grows and affects its surroundings and not the artificial conditions imposed on the isolates, which are pampered or mishandled—depending on the investigators propensities—to perform this or that reaction." Thus, although microorganisms can be isolated from the sea, cultivated, and manipulated in the laboratory, we must remain cognizant of the fact that in nature the activity of the organism may not be as we estimated it to be on the basis of laboratory findings.

This general uncertainty is amply illustrated in a discussion of a topic as basic as the distribution of bacteria in the marine environment. The vast expanse and depth of the oceans, combined with the large investments of time and capital required to obtain distribution data, have resulted in only a limited number of recent studies. In addition, there are significant scientific limitations concerning interpretation of much of the "older" data on bacterial distribution in the oceans. The greater part of that data was generated using various plating and most probable number procedures. These methods probably recover fewer than 6% of the actual number present depending on sampling locations (Sieburth, 1979). Data collected using the newer techniques of epifluorescence microscopy (see Daley and Hobbie, 1975), lipopolysaccharide quantification with *Limulus* amebocyte lysate (Watson *et al.*, 1977), and possibly various biochemical assays (White *et al.*, 1978; Fazio *et al.*, 1979) have not been employed extensively enough to generate a critical mass of published information.

The data available from earlier studies have indicated that bacterial numbers are usually greater near land and tend to decrease with increasing distance from shore (see Rheinheimber, 1974). This distribution has been accredited to increased availability of nutrients along coastal areas. The vertical distribution of bacteria in the deep ocean appears to be influenced by the degree of light penetration, density differences, and the distribution of organic matter. The upper 150 μm of the ocean surface, the microneuston, has been found to be enriched in monosaccharides and polysaccharides as well as in ATP extractable from material less than 3 μm in diameter (presumably bacterial) (Sieburth *et al.,* 1976). Dietz *et al.* (1976) have shown bacterial counts in the neuston to be higher than those in the plankton. In the zone 10–50 m below the neuston, bacteria have generally been found to be abundant, usually a few hundred cells per milliliter (ZoBell and Anderson, 1936). ZoBell (1946) reported that maximum numbers were found below the zone of maximum phytoplankton. He also reported that few cells can be found below 200 m. Sieburth (1971) presented evidence that seemed to collaborate ZoBell's work. Sieburth reported that during the Pacific leg of the *R/V Trident* cruise "Debut," 59% of the samples contained fewer than 5 colony-forming bacteria per milliliter and on the Atlantic leg, 63% contained fewer than 5 colony-forming bacteria per milliliter. Most of the higher count samples were in the upper 100 m. Frequently bacterial numbers have been found to increase just above the bottom (Rheinheimer, 1974).

Much of the additional information on the vertical and geographic distributional patterns has been collected by Kriss and his co-workers (Kriss, 1960, 1963, 1970, 1972, 1973, 1976; Kriss *et al.,* 1960a,b, 1961, 1971, 1972; Kriss and Mitskevitch, 1970; Kriss and Novozhilova, 1971; Kriss and Stupakova, 1972). However, considerable criticism has been levied against their work (Bogoyavlenskii, 1962; Sieburth, 1971; Lewin, 1974). Consequently, much of the detail must be overlooked and the work interpreted on a general basis. From the series of papers dealing with the enumeration of bacteria in the Indian, Pacific, and Atlantic Oceans, it would appear the bacterial numbers tend to increase as latitude decreases, the greatest numbers being found in the equatorial zones.

The development of the epifluorescent microscopic techniques for use in observing bacteria, the use of fluorescent dyes such as acridine orange, and the availability of perforated polycarbonate membranes with well-controlled pore sizes (see Hobbie *et al.,* 1977) have added a new dimension to the study of marine bacteriology. Although the use of direct observation techniques has been relatively limited, results to date indicate that the very nature of the marine bacterial community may be quite different from that which conventional wisdom would dictate. The change has been mildly revolutionary: the bacterial population appears to be much higher than previously suspected, there is only a relatively small decrease in the population from the photic to the aphotic zone, and a relatively small decrease is observed in the density from near-shore to the open ocean environments (Sieburth, 1979). These observations are well illustrated by the report of Ferguson and Palumbo (1979). Through the use of epifluorescence microscopy, they demonstrated that the total number observed per ml at a 500 m-deep station varied from a high of 0.9 million to a low of only 0.2 million per milliliter. Similarly, they reported populations varied from 2 million per milliliter near shore to 1.9 million at midshelf to 0.6 million at shelf break. To move from a position of being able to quantify 5 or 10 or even 500 bacteria per milliliter of seawater to a position where we can now quantify about a million cells per milliliter represents a substantial alteration in the status of marine bacteriology.

Direct observation of bacteria in samples of marine waters by epifluorescence microscopy has also resulted in a growing awareness that the suspended microbial community is not homogenous with respect to the nature of the organisms present. Sieburth (1979) has described two populations of osmotrophic heterotrophs; epibacteria and planktobacteria. The epibacteria are the larger bacteria that are frequently associated with seston and shallow coastal waters. They have organelles of attachment and locomotion as well as the ability to move across solid surfaces. The epibacteria are apparently those bacteria that are easily cultured on nutrient agar surfaces. If so, these forms would represent the bacteria that have most frequently been described and examined by individuals engaged in the study of marine bacteria. This group of bacteria represents those organisms that bacteriologists are best equipped to study; however, they represent the minor component of the bacterial community in terms of density. The planktobacteria represent the bacterial population of greatest density. They are small cells of differing morphologies that are gram negative, have little affinity for attaching to surfaces, and do not grow on conventional agar media. The epibacteria probably spend most of their existence attached to organic debris and depend largely on the particulate material for sources of carbon; on the other hand, the planktobacteria are suspended in the water column and depend on the dissolved organic carbon. It is the planktobacteria that account for the great increase in the bacterial populations observed by direct microscopic examination over that observed by culture techniques. In addition, it is apparently this population of very small heterotrophic bacteria that is responsible for the continuous breakdown of organic matter in offshore regions of the sea (Hoppe, 1976).

A second technical advance that has also allowed the demonstration of the presence of significant bacterial populations in the deep sea is the determination of adenosine triphosphate and related nucleotides as indicators of microbial populations. The measurement of adenosine triphosphate originally evolved as a measure of total microbial biomass; however, the basic limitation of the method—that all living cells contain adenosine triphosphate (Holm-Hansen and Booth, 1966; Holm-Hansen, 1973)—is of importance. Since adenosine triphosphate is found within all living cells, the measure does not discriminate between procaryotic or eucaryotic cells. However, a study in which the aquatic microbial community was fractionated into its procaryotic and eucaryotic components has shown that there is sufficient bacterial adenosine triphosphate in coastal waters to account for 10^8 cells per liter (Azam and Hodson, 1977). Sieburth (1979) also demonstrated the presence of procaryotic microbes in water samples taken from depth profiles by separating the bacterial community from the eucaryotic community by filtration with polycarbonate membranes. There was little variation in the amount of bacterial ATP with depth. Additional evidence for the existence of viable cell populations comes from several investigations reporting detectable adenosine triphosphate levels from depths where microbial populations are supposedly procaryotic (1000–3000m) (Devol *et al.*, 1976; Karl *et al.*, 1976, 1977; Karl 1978). In some of these studies, the density of bacteria, as indicated by ATP levels, did not decrease markedly with depth.

The direct observation of bacteria in samples collected from the deep ocean and the extractable adenosine triphosphate indicates that viable bacteria are present in that environment; however, this evidence should not be taken to mean that those bacteria are metabolically active in terms of the criteria that we have come to expect. The more recent evidence strongly suggests that those bacteria observed in the deep ocean are indeed not active.

An excellent overview of the bacteriology of the deep ocean was given by Baross *et al*. (1974). Direct electron microscopic evidence showing the presence of bacteria in the deep-sea was published by Carlucci *et al.* (1976). In that study, the bacterial cells collected and fixed at *in situ* pressures appeared much like organisms collected from surface water samples. The presence in the deep ocean of bacteria that are specifically adapted to function under high barometric pressure has been indicated by the work of ZoBell and Morita (1957). They obtained higher population densities (by a factor of about 10–1000×) when sediment samples were incubated under high pressure as compared to replicate samples incubated at 1 atm. Quantative recovery of organisms capable of starch hydrolysis and nitrate reduction was also enhanced by incubation at 700 atm. Schwarz *et al.* (1976) were able to demonstrate that a mixed flora obtained from amphipod gut had growth rates and substrate conversion rates that were equal at 750 atm and 1 atm. However, these organisms lost their barophilic properties upon laboratory cultivation (Pope *et al.*, 1976). The best example of a truely barophilic bacterium described to date was derived from an "amphipod lysate" held under high pressure (Yayanos *et al.*, 1979). The lysate yielded a spiral shaped bacterium that was capable of maximum growth when incubated at 250–675 atm. Biochemical characterization of this organism may provide new insights into the relationship between microbes and pressure.

Extreme barometric pressure appears to be the most important factor limiting the metabolic activity of the planktobacteria in the deep ocean. Much of what is known about the responses of bacteria to pressure is deductive in nature. Many attempts to demonstrate unique physiological adaptations by bacteria to provide for vigorous metabolism and growth at high pressure have resulted in frustration. There is even some question regarding the possibility of any bacterium adapting to life at high pressure (Marquis, 1976). An extensive consideration of the physical–chemical factors governing biochemistry under high pressure has been presented by Marquis (1976).

High pressure injures bacteria in a variety of ways and different bacteria show differing responses; in addition, all alterations do not occur simultaneously (Baross *et al.,* 1974; Arcuri and Ehrlich, 1977). Pressure seems to stabilize native states of DNA and does not tend to promote phase transitions in lipid bilayers (Marquis, 1976). Protein systems appear to be very sensitive, i.e., flagellar and ribosomal subunits along with other functional proteins are affected. Landau, Pope, and Smith have been able to show that the degree to which protein synthesis is inhibited by pressure parallels the degree to which growth of the organism is inhibited (Pope *et al.,* 1976; Smith *et al.,* 1975). Paul and Morita (1971) demonstrated that transport proteins in bacterial cell membranes were inhibited, resulting in starvation of the cells under pressure. Salts appear to protect bacteria against death under high pressure (Morita and Becker, 1970).

There is growing evidence to indicate that the metabolism of heterotrophic bacteria in the water column and at the sediment interface in the deep-sea is markedly inhibited by high pressure. This has been shown directly and indirectly. The direct demonstrations of low activity of native populations at *in situ* conditions have involved a lack of uptake of single amino acids (Deming *et al.,* 1980), lowered metabolism of an amino acid mixture (Jannasch and Wirsen, 1977a), and a lack of growth response upon enrichment with polysaccharides (Schwarz *et al.,* 1976). Insights into the activity of deep sea populations have also been obtained by determination of the adenylate energy charge (Chapman *et al.,* 1971; Karl and Holm-Hansen, 1978). Laboratory studies indicate that the adenylate energy charge is a measure of the metabolically available energy stored in a cell (Chapman *et al.,* 1971). While additional work in both methodology and data interpretation is needed, current data would suggest that the microbial community at varying depths within the ocean has an activity level corresponding to the activity level of stationary phase cultures (Karl *et al.,* 1977; Karl 1978). The above studies rely on our ability to study the properties of a mixed community and are subject to the basic uncertainty mentioned earlier. A direct method that looks at individual cells is micro-autoradiography. Using this technique, Peroni and Lavarello (1975) demonstrated very low ^{32}P-uptake activity below 150–200 m. Furthermore, when samples collected at 30 m depth were incubated at 320 m. a 93% decrease in phosphorous uptake was noted. Thus, normally active planktobacteria were inhibited by the increase in hydrostatic pressure.

There are several indirect indicators of low or nonexistent microbial activity in the deep ocean. Williams and Carlucci (1976) were able to show that "normal" bacterial activity would result in the development of anoxic conditions in the deep ocean. By careful analysis of the distribution of inorganic nutrients, Menzel (1970) reported that microbial decomposition does not seem to take place in the deep ocean. A third, rather unsuspected indication—that bacteria in the deep sea are not active—was provided by the work of Barber (1968). He was able to show that dissolved organic carbon derived from the deep ocean was not degraded by natural bacteria even if the carbon was concentrated. It is apparent that not only are the organisms in the deep ocean inhibited by the extreme temperature and pressure but they are also limited because of the refractory nature of their energy supply (Morita, 1979).

Despite these observations, it is apparent that decomposition and recycling of macro and micro particulate material must take place in the world's oceans. The observations of Jannasch and Wirsen (1977b) and Sieburth and Dietz (1974) amply demonstrate the importance of marine fauna in this process. The amphipods appear to be specifically adapted to function at abyssal pressures (Marquis, 1976), and the microbial flora of the gut of these eucaryotes appears to include bacteria adapted to that existence (Schwarz *et al.*, 1976). In addition, the study of barophilic bacteria has been limited because of the significant technical problems associated with collecting and maintaining water samples at high *in situ* pressures. Two laboratories in the United States have recently described the development of samplers designed to obtain, retrieve, and incubate deep-sea samples at *in situ* pressures (Jannasch and Wirsen, 1977a,b; Deming *et al.*, 1980). The new technology, together with the description of a true bacophilic microbe (Yayanos *et al.*, 1979), may represent significant advances in this area.

Oceanic waters are generally characterized as having low levels of both suspended particulate (POC) and dissolved organic carbon (DOC) (Menzel and Ryther, 1970). Where the conditions of pressure and nature of the carbon permit (see Morita, 1979), bacteria seem adapted to respond efficiently to these low concentrations (Stevenson and Erkenbrecher, 1976). Some of the adaptations to carbon limitation appear to include a very efficient mechanism of nutrient uptake, possible attachment, and the activation of chemotaxis. The ability of bacteria to compete effectively for dissolved nutrients has been accepted for some time (Wright and Hobbie, 1966). Jannasch (1967, 1969) was able to demonstrate growth of marine bacteria at very low carbon levels; however, he was not able to demonstrate growth at the limiting levels measured *in situ*. Button and his co-workers observed a similar pattern when a single substrate (i.e., glucose) was employed to support growth (Law and Button, 1977; Button, 1978; Robertson and Button, 1979). The threshold concentration for growth of a marine bacterium in continuous culture was 0.21 mg/liter; however, the addition of amino acids reduced the threshold to only 0.008 mg/liter. Thus, they pointed out that bacteria in the marine environment are able to utilize the low DOC levels in the ocean since this carbon is undoubtedly a mixture of various compounds. Marine bacteria appear to be adapted to metabolize at a maximum rate when a variety of dissolved organic compounds are available. Terresterial bacteria, on the other hand, appear to be better adapted to growth on single substrates (see references of Button and co-workers above).

Morita (1979) has suggested that the free bacteria (planktobacteria) in the deep ocean are in effect starved for a utilizable energy supply and that this limits growth. A possible relaxation of this inhibition in oceanic environments has been demonstrated in the Galapagos hot spring area and the oxygen-minimum zones in the Black Sea and the Cariaco Trench (Karl, 1978). The energy source in the oxygen-minimum areas associated with the Black Sea is apparently sulfide (see Karl, 1978, for a complete discussion). In fact, when the data from the Galapagos Trench becomes available, it will probably revolutionize our understanding of deep-ocean microbiology.

Two other adaptations contribute to the functioning of the heterotrophic epibacteria where conditions of pressure permit. The first, attachment, has been observed since the early days of marine bacteriology (see Corpe, 1974). Various conditions have been demonstrated to influence the attachment of bacteria to surfaces, and microbial populations have been recovered from and observed on the surface of a variety of materials (see Stevenson and Erkenbrecher, 1976). The ability of epibacteria to attach to and colonize particulate organic material may provide them with a microenvironment higher in nutrient concentration than the surrounding water. Although the data are preliminary and derived from studies of nonmarine bacteria, there is some indication that bacteria in the natural environment produce an exocellular polysaccharide polymer that is unusually efficient in allowing bacteria to attach to substrates. This polymer, termed glycocalyx, has been shown to be of importance in attachment of bacteria to surfaces in streams (Geesey *et al.*, 1977; Costerton *et al.*, 1978), and there is a possibility that marine bacteria have a similar adaptation. Glycocalyx is produced when cells are subject to environments that are low in dissolved nutrients; consequently, one might expect the particulate detritus in the oceanic waters to be heavily colonized by bacteria. Indeed, this has been part of the "conventional wisdom" among biologists for some time. Wiebe and Pomeroy (1972) demonstrated that this is apparently not the case. In a direct microscopic examination of particles from oceanic water, they were able to show directly that much of the material is not heavily colonized.

A second adaptation to the low concentrations of organic nutrients that bacteria encounter in the oceanic waters is chemotaxis. The activation of chemotaxis by a marine bacterium has been noted in response to low carbon concentrations by Torella and Morita (Morita, 1979). Starvation of the organism for 48 hours stimulated the synthesis of flagella so that a chemotaxis response was possible by that organism.

Even though vigorous metabolic activity of bacteria is yet to be demonstrated at *in situ* temperature and pressures encountered in the deep ocean, one should not infer that bacteria are not adapted to that environment. Bacteria are apparently well adapted to survive long exposures to the combined stresses of nutrient limitation and high pressures. ZoBell and Morita (1957) and Morita (1979) have suggested that the cells in the water column of the deep ocean are in a dormant state. Stevenson (1978) further proposed that dormancy is a general survival trait of marine bacteria. The first testing of these suggestions was provided by Wright (1978) who combined observations made by direct epifluorescence microscopy with heterotrophic uptake and was able to show that "bacteria are adapted to conditions of nutrient starvation by becoming 'dormant,' existing for an unknown period of time in a reversible physiological state that reflects the availability of organic nutrients."

The dormant state of bacteria in the marine environment has also been demonstrated by Novitsky and Morita using another approach. The term "growth" in bacteriology generally applies to an increase in the population density of microbes; e.g., from 5×10^6 to 8×10^7 per milliliter. Generally, this increase in numbers is accompanied by an increase in cell mass at the expense of exogenous nutrients. A somewhat different process has been described by Novitsky and Morita (1976, 1978a,b, 1979). When a psychrophilic marine vibrio (Ant-300) was subjected to an environment deficient in organic carbon, such as may be encountered in the ocean, the bacterium apparently underwent an increase in population density (growth) without a concurrent increase in mass (Novitsky and Morita, 1978b). Prior to starvation, the cells were relatively large, rod or vibrio-shaped. After starvation, the resulting cells were much smaller and coccus-shaped (Novitsky and Morita, 1976). The cells produced in response to starvation appear to be small dormant versions of the vegetative or active cells. In the small cells, the endogenous respiration decreased markedly without loss in viability and the intracellular pool of low molecular metabolites was completely eliminated.

Initially, the increase in cell number appeared to result from a fragmentation of "healthy" active cells in response to starvation conditions. The number of nuclei in the cells was directly related to the increased yield upon fragmentation (Novitsky and Morita, 1978b). For example, when cells with four nuclei per cell were starved, there was a 400% increase in the population. DNA synthesis was apparently not required for fragmentation. However, the increase in numbers appears to be more complex than a simple fragmentation. A later observation by the same authors (Novitsky and Morita, 1979) reported a 92,000% increase in cell numbers when the organism was starved. It therefore appears that at least three mechanisms may be involved in the increase in the population of Ant-300 when subjected to low nutrient levels; (1) the larger, rod-shaped cells may fragment, (2) there may be an efficient utilization of reserve material and/or cellular constituents, and (3) there may be a possible utilization of trace nutrients in the environment. Novitsky and Morita (1979) pointed out, however, that the addition of trace levels of amino acids to the starvation medium did not enhance the increase in cell density. The ''growth'' or multiplication of marine bacteria at very low nutrient concentrations has been reported by Carlucci and Shimp (1974).

The increased dispersion of the gene pool of the organism by fragmentation and the conversion to the dormant state would obviously enhance the long-term survival of the organisms in nutrient-poor oceanic waters. However, there is yet another fortuitous consequence of these processes—the small cells produced in response to starvation are resistant to inactivation by elevated barometric pressure (Novitsky and Morita, 1978a). The larger ''active'' cells are killed readily by elevated pressure (250 atm), whereas a short period of starvation prior to the application of pressure increased barotolerance. The active cells would be undergoing a variety of metabolic activities, such as protein synthesis, which could be interrupted by hydrostatic pressure. The small ''inactive'' cells would not be involved in similar activities and would consequently be spared.

Thus, bacteria are not well adapted for metabolic activity in the deep ocean, but they are well adapted for survival in that environment.

References

Alexander, M. (1971). Biochemical ecology of microorganisms. *Annu. Rev. Microbiol.* **25,** 361–392.

Arcuri, E. J., and Ehrlich, H. L. (1977). Influence of hydrostatic pressure on the effects of the heavy metal cations of manganese, copper, cobalt, and nickel on the growth of three deep-sea bacterial isolates. *Appl. Environ. Microbiol.* **33,** 282–288.

Azam, F., and Hodson, R. E. (1977). Size distribution and activity of marine microheterotrophs. *Limnol. Oceanogr.* **22,** 492–501.

Barber, R. T. (1968). Dissolved organic carbon from deep waters resists microbial oxidation. *Nature (London)* **220,** 274–275.

Baross, J. A., Hanus, F. J., and Morita, R. Y. (1974). Effects of hydrostatic pressure on uracil uptake, ribonucleic acid synthesis, and the growth of three obligately psychrophilic marine vibrios, *Vibrio alginolyticus,* and *Escherichia coli. In* ''Effect of the Ocean Environment on Microbiol Activities'' (R. R. Colwell and R. Y. Morita, eds.), pp. 180–202. University Park Press, Baltimore, Maryland.

Bogoyavlenskii, A. N. (1962). On the distribution of heterotrophic micro-organisms in the Indian Ocean and in Antartic waters. *Okeanologiya* **2,** 293–297.

Button, D. K. (1978). On the theory of control of microbial growth kinetics by limiting nutrient concentrations. *Deep-Sea Res.* **25,** 1163–1177.

Campbell, R. (1977). ''Microbial Ecology.'' Wiley, New York.

Carlucci, A. F., and Shimp, S. L. (1974). Isolation and growth of a marine bacterium in low concentrations of substrate. *In* ''Effect of the Ocean Environment on Microbial Activities'' (R. R. Colwell and R. Y. Morita, eds.), pp. 363–367. University Park Press, Baltimore, Maryland.

Carlucci, A. F., Shimp, S. L., Jumars, P. A., and Paerl, H. W. (1976). *In situ* morphologies of deep-sea and sediment bacteria. *Can. J. Microbiol.* **22,** 1667–1671.

Chapman, A. G., Fall, L., and Atkinson, D. E. (1971). Adenylate energy charge in *Escherichia coli* during growth and starvation. *J. Bacteriol.* **108,** 1072–1086.

Colwell, R. R., and Morita, R. Y., eds. (1974). "Effects of the Ocean Environment on Microbial Activities." University Park Press, Baltimore, Maryland.

Corpe, W. A. (1974). Periphytic marine bacteria and the formation of microbial films on solid surfaces. *In* "Effect of the Ocean Environment on Microbial Activities" (R. R. Colwell and R. Y. Morita, eds.), pp. 397–417. University Park Press, Baltimore, Maryland.

Costerton, J. W., Gessey, G. G., and Cheng, K.-J. (1978). How bacteria stick. *Sci. Am.* **238,** 86–95.

Daley, R. J., and Hobbie, J. E. (1975). Direct counts of aquatic bacteria by a modified epi-fluorescent technique. *Limnol. Oceanogr.* **20,** 875–882.

Deming, J. W., Eabor, P. S., and Colwell, R. R. (1980). Deep ocean microbiology. *In* "Advanced Concepts in Ocean Measurements for Marine Biology" (F. P. Diemer, F. J. Vernberg, and D. Z. Mirkes, eds.). Univ. of South Carolina Press, Columbia (in press).

Devol, A. H., Packard, T. T., and Holm-Hansen, O. (1976). Respiratory electron transport activity and adenosine triphosphate in the oxygen minimum of the eastern tropical North Pacific. *Deep-Sea Res.* **23,** 963–973.

Dietz, A. S., Albright, L. J., and Tuominen, T. (1976). Heterotrophic activities of bacterioneuston and bacterioplankton. *Can. J. Microbiol.* **22,** 1699–1709.

Doetsch, R. N., and Cook, T. M. (1973). " Introduction to Bacteria and their Ecobiology." University Park Press, Baltimore, Maryland.

Fazio, S. D., Mayberry, W. R., and White, D. C. (1979). Muramic acid assay in sediments. *Appl. Environ. Microbiol.* **38,** 349–350.

Ferguson, R. L., and Palumbo, A. V. (1979). Distribution of suspended bacteria in neritic waters south of Long Island during stratified conditions. *Limnol. Oceanogr.* **24,** 697–705.

Geesey, G. G., Richardson, W. T., Yeomans, H. G., Irvin, R. T., and Costerton, J. W. (1977). Microscopic examination of natural sessile bacterial populations from an alpine stream. *Can. J. Microbiol.* **23,** 1733–1736.

Hobbie, J. E., Daley, J. R., and Jasper, S. (1977). Use of nucleopore filters for counting bacteria by fluorescence microscopy. *Appl. Environ. Microbiol.* **3,** 1225–1228.

Holm-Hansen, O. (1973). Determination of total microbial biomass by measurement of adenosine triphosphate. *In* "Estuarine Microbial Ecology" (L. H. Stevenson and R. R. Colwell, eds.), pp. 73–89. Univ. of South Carolina Press, Columbia.

Holm-Hansen, O., and Booth, C. R. (1966). The measurement of adenosine triophosphate in the ocean and its ecological significance. *Limnol. Oceanogr.* **11,** 510–519.

Hoppe, H. -G. (1976). Determination and properties of actively metabolizing heterotrophic bacteria in the sea, investigated by means of microautoradiography. *Mar. Biol.* **36,** 291–302.

Jannasch, H .W. (1967). Growth of marine bacteria at limiting concentrations of organic carbon in seawater. *Limnol. Oceanogr.* **12,** 264–271.

Jannasch, H. W. (1969). Estimation of bacterial growth rates in natural waters. *J. Bacteriol.* **99,** 156–160.

Jannasch, H. W., and Wirsen, C. O. (1977a). Retrieval of concentrated and undecompressed microbial populations from the deep sea. *Appl. Environ. Microbiol.* **33,** 642–646.

Jannasch, H. W., and Wirsen, C. O. (1977b). Microbial life in the deep sea. *Sci. Am.* **236,** 42–52.

Karl, D. M. (1978). Distribution, abundance, and metabolic states of microorganisms in the water column and sediments of the Black Sea. *Limnol. Oceanogr.* **23,** 936–949.

Karl, D. M., and Holm-Hansen, O. (1978). Methodology and measurement of adenylate energy charge ratios in environmental samples. *Mar. Biol.* **48,** 185–197.

Karl, D. M., LaRock, P. A., Morse, J. W., and Sturges, W. (1976). Adenosine triphosphate in the North Atlantic Ocean and its relationship to the oxygen minimum. *Deep-Sea Res.* **23,** 81–83.

Karl, D. M., LaRock, P. A., and Shultz, D. J. (1977). Adenosine triphosphate and organic carbon in the Cariaco Trench. *Deep-Sea Res.* **24,** 105–113.

Kriss, A. E. (1960). Micro-organisms as indicators of hydrological phenomena in seas and oceans. I. Methods. *Deep-Sea Res.* **6,** 88–94.

Kriss, A. E. (1963). "Marine Microbiology. Deep Sea" (transl. by J. M. Shewan and Z. Kabata). Wiley (Interscience), New York.

Kriss, A. E. (1970). Ecological-geographical patterns in the distribution of heterotrophic bacteria in the Atlantic Ocean. *Mikrobiologiya* **39,** 362–371.

Kriss, E. A. (1972). Characterization and distribution of the heterotrophic microbial population in different regions of the Atlantic Ocean. *Mikrobiologiya* **41,** 1091–1098.

Kriss, A. E. (1973). Quantitative distribution of heterotrophic bacteria in the southern ocean between New Zeland and Antartica. *Mikrobiologiya* **42,** 913–917.

Kriss, A. E. (1976). The vertical distribution patterns of heterotrophic bacteria in the depths of the world ocean and the importance of their regularity to oceanography. *Int. Rev. Gesamten Hydrobiol.* **61,** 417–438.

Kriss, A. E., and Mitskevitch, I. N. (1970). Quantitative distribution of heterotrophic bacteria in the equatorial-tropical zone of the Pacific Ocean near South America. *Mikrobiologiya* **39,** 1087–1094.

Kriss, A. E., and Novozhilova, M. I. (1971). Quantitative distribution of heterotrophic bacteria in the southeastern Atlantic Ocean. *Mikrobiologiya* **39,** 892–897.

Kriss, A. E., and Stupakova, T. P. (1972). Ecological-geographical regularities of heterotrophic bacteria distribution in the west and central Pacific. *Int. Rev. Gesamten Hydrobiol.* **57,** 497–506.

Kriss, A. E., Lebedeva, M. N., and Mitzekevich, I. N. (1960a). Micro-organisms as indicators of hydrological phenomena in the seas and oceans. II. Investigation of deep circulation of the Indian Ocean using microbiological methods. *Deep-Sea Res.* **6,** 173–183.

Kriss, A. E., Abyzov, S. S., and Mitzkevich, I. N. (1960b). Micro-organisms as indicators of hydrological phenomena in the seas and oceans. III. Distribution of water masses in the central part of the Pacific Ocean (according to microbiological data). *Deep-Sea Res.* **6,** 335–345.

Kriss, A. E., Mitzkevitch, I. N., Mishustina, I. E., and Abyzov, S. S. (1961). Micro-organisms as hydrological indicators in seas and oceans. IV. The hydrological structure of the Atlantic Ocean, including the Norwegian and Greenland Seas, based on microbiological data. *Deep-Sea Res.* **7,** 225–236.

Kriss, A. E., Mishustina, I. E., Mitskevitch, I. N., Novzhilova, M. I., and Stupakova, T. P. (1971). Ecological-geographical regularities of heterotrophic bacteria distribution in the equatorial-trophical zone of the world ocean. *Int. Rev. Gesamten Hydrobiol.* **56,** 689–730.

Kriss, A. E., Stupakova, T. P., and Tsyban, A. V. (1972). Ecological-geographical pattern of distribution of the heterotrophic microbial population in the central part of the Pacific Ocean. *Mikrobiologiya* **41,** 542–549.

Law, A. T., and Button, D. K. (1977). Multiple-carbon-limited growth kinetics of a marine coryneform bacterium. *J. Bacteriol.* **129,** 115–123.

Lewin, R. A. (1974). Enumeration of bacteria in sea water. *Int. Rev. Gesamten Hydrobiol.* **59,** 611–619.

Marquis, R. E. (1976). High-pressure microbiol physiology. *Adv. Microb. Physiol.* **14,** 159–241.

Menzel, D. W. (1970). The role of *in situ* decomposition of organic matter on the concentration of nonconservative properties in the sea. *Deep-Sea Res.* **17,** 751–764.

Menzel, D. W., and Ryther, J. H. (1970). Distribution and cycling of organic matter in the oceans. *Sym. Org. Matter Nat. Waters, 1968* pp. 31–54.

Morita, R. Y. (1979). Deep-sea microbial energetics. *Sarsia* **64,** 9–12.

Morita, R. Y., and Becker, R. R. (1970). Hydrostatic pressure effects on selected biological systems.

In "High Pressure Effects on Cellular Systems" (A. M. Zimmerman, ed.), pp. 71–83. Academic Press, New York.

Novitsky, J. A., and Morita, R. Y. (1976). Morphological characterization of small cells resulting from nutrient starvation of a psychophilic marine vibrio. *Appl. Environ. Microbiol.* **32,** 617–622.

Novitsky, J. A., and Morita, R. Y. (1978a). Starvation-induced barotolerance as a survival mechanism of a psychophilic marine vibrio in the waters of the antarctic convergence. *Mar. Biol.* **47,** 7–10.

Notivsky, J. A., and Morita, R. Y. (1978b). Possible strategy for the survival of marine bacteria under starvation conditions. *Mar. Biol.* **48,** 289–295.

Novitsky, J. A., and Morita, R. Y. (1979). Survival of a psychophilic marine vibro under long-term nutrient starvation. *Appl. Environ. Microbiol.* **33,** 635–641.

Paul, K. L., and Morita, R. Y. (1971). Effects of hydrostatic pressure and temperature on the uptake and respiration of amino acids by a facultatively psychophilic marine bacterium. *J. Bacteriol.* **108,** 835–843.

Peroni, C., and Lavarello, O. (1975). Microbial activities as a function of water depth in the Ligurian Sea: An autoradiographic study. *Mar. Biol.* **30,** 37–50.

Pope, D. H., Orgrinc, W. P., and Landau, J. V. (1976). Protein synthesis at 680 atm: Is it related to environmental origin, physiological type, or taxonomic group? *Appl. Environ. Microbiol.* **31,** 1001–1002.

Rheinheimer, G. (1974). "Aquatic Microbiology." Wiley (Interscience), New York.

Robertson, B. R., and Button, D. K. (1979). Phosphate-limited continuous culture of *Rhodotorula rubra:* Kinetics of transport, leakage, and growth. *J. Bacteriol.* **138,** 884–895.

Schwarz, J. R., Yayanos, A. A., and Colwell, R. R. (1976). Metabolic activities of the intestinal microflora of a deep-sea invertebrate. *Appl. Environ. Microbiol.* **31,** 46–48.

Sieburth, J. McN. (1971). Distribution and activity of oceanic bacteria. *Deep-Sea Res.* **18,** 1111–1121.

Sieburth, J. McN. (1979). "Sea Microbes." Oxford Univ. Press, London and New York.

Sieburth, J. McN., and Dietz, A. S. (1974). Biodeterioration in the sea and its inhibition. *In* "Effect of the Ocean Environment on Microbial Activities" (R. R. Colwell and R. Y. Morita, eds.), pp. 318–326. University Park Press, Baltimore, Maryland.

Sieburth, J. McN., Willis, P. J., Johnson, K. M., Burney, C. M., Lavoic, D. M., Hinga, K. R., Caron, D. A., French, F. W., Johnson, P. W., and Davis, P. B. (1976). Dissolved organic matter and heterotrophic microneuston in surface microlayers of the North Atlantic. *Science* **194,** 1415–1418.

Smith, W., Pope, D., and Landau, J. V. (1975). Role of bacteria ribosome subunits in barotolerance. *J. Bacteriol.* **124,** 582–584.

Stevenson, L. H. (1978). A case for bacterial dormancy. *Microb. Ecol.* **4,** 127–133.

Stevenson, L. H., and Erkenbrecher, C. W. (1976). Activity of bacteria in the estuarine environment. *In* "Estuarine Processes" (M. Wiley, ed.), Vol. 1, pp. 318–394. Academic Press, New York.

Watson, S. W., Novitsky, T. J., Quinby, H. L., and Valois, F. W. (1977). Determination of bacterial number and biomass in the marine environment. *Appl. Environ. Microbiol.* **33,** 940–946.

White, D. C., Bobbie, R. J., Herron, J. S., King, J. D., and Morrison, S. J. (1978). Biochemical measurements on microbial mass and activity from environmental samplings. *In* "Native Aquatic Bacteria, Enumeration, Activity and Ecology" (R. R. Colwell and W. Costerton, eds.), pp. 69–81. Am. Soc. Test. Mater., Philadelphia, Pennsylvania.

Wiebe, W. J., and Pomeroy, L. R. (1972). Microorganisms and their association with aggregates and detritus in the sea: A microspic study. *Mem. Ist. Ital. Idrobiol.* **29,** 325–352.

Williams, P. M., and Carlucci, A. F. (1976). Bacterial utilization of organic matter in the deep sea. *Nature (London)* **262,** 810–811.
Wright, R. T. (1978). Measurement and significance of specific activity in the heterotrophic bacteria of natural waters. *Appl. Environ. Microbiol.* **36,** 297–305.
Wright, R. T., and Hobbie, J. E. (1966). Use of glucose and acetate by bacteria and algae in aquatic ecosystems. *Ecology* **47,** 447–464.
Yayanos, A. A., Dietz, A. S., and Boxtel, R. V. (1979). Isolation of a deep-sea barophilic bacterium and some of its growth characteristics. *Science* **205,** 808–809.
ZoBell, C. E. (1946). "Marine Microbiology." Chronica Botanica, Waltham, Massachusetts.
ZoBell, C. E., and Anderson, D. Q. (1936). Vertical distribution of bacteria in marine sediments. *Bull. Am. Assoc. Pet. Geol.* **20,** 258–269.
ZoBell, C. E., and Morita, R. Y. (1957). Barophilic bacteria in some deep sea sediments. *J. Bacteriol.* **73,** 563–568.

Zooplankton

4

Donald R. Heinle

I. Introduction

Many of the functional adaptations of marine zooplankton are in response to stresses or selective forces in the environment that can be readily perceived. The breadth of tolerance or degree of adjustment afforded by the adaptation thus can be determined experimentally. While experiments have helped us gain insight into adaptations to some singly varying stresses, other phenomena have been observed that are considered to be adaptations because of their energetic cost to the organisms, but adaptations to what? The principal example is the extensive vertical migration by some marine zooplankton which probably represents a balanced response to two or more selective forces.

Among the most discernible biological features of the marine environment to which zooplankton must adapt are scarcity of food, predation, and competition. While the reality of competition is debated by some, it must occur to some degree when food is scarce. Physical variables that may stress zooplankton include temperature, salinity, mass movements of water, and—under special coastal circumstances—low concentrations of dissolved oxygen. The relative importance of the seven selective forces mentioned above varies geographically. For example, temperature and salinity are relatively constant in mid-oceanic areas

FUNCTIONAL ADAPTATIONS OF MARINE ORGANISMS

ISBN 0-12-718280-2

where food may be scarce. Conversely, in areas of coastal upwelling or temperate estuaries, food may be abundant while temperature and salinity vary greatly, and predation is intense. In boreal seas, food and predators may be seasonally abundant, but temperatures are so low that some species require more than a year to complete their life cycles.

Discussions of adaptations of marine zooplankton are necessarily dominated by work on copepods, since they have by far received the most attention.

II. Scarcity of Food

A. DIMENSIONS OF THE PROBLEM

Conover (1968) succinctly stated a major problem faced by zooplankton in general but particularly by marine zooplankton; i.e., their food resources are widely dispersed in a three-dimensional space. Thus they must be able to find, collect, and ingest it. We will assume that selective forces lead to strategies that maximize energy gain per unit food eaten and minimize energy cost per unit food. Under special circumstances, as in estuaries, food may be relatively abundant and other selective forces such as predation much more important.

In addition to absolute scarcity of food, there are temporal changes in abundance associated with seasonal cycles that may include low temperatures. Two adaptations have evolved that allow planktonic animals to compensate for temporal changes in abundance of food; storage of energy reserves, principally as lipids, and arrested development at varying states of maturity. Many animals demonstrate both strategies simultaneously, sometimes coupled with lowered rates of metabolism during periods of food scarcity. The following discussion is intended to compliment rather than repeat the review by Marshall (1973) on feeding by zooplankton.

B. STORAGE OF LIPIDS AND REDUCED METABOLISM

1. Life Cycles in Relation to Food and Body Lipids

Although some investigators have concluded that lipids do not serve principally as storage reserves in certain copepods (Marshall and Orr, 1955a; Linford, 1965), there is substantial evidence that other species use stored lipids as an energy source. Conover (1962) observed that between November and March, females of *Calanus hyperboreus,* a cold water form, became ripe and stage V copepodids molted rapidly into adults. The females began laying eggs within a few weeks. Approximately two months were required for growth to stage III copepodid. Egg laying ceased by the end of March and by May most of the population had grown to stage IV or V copepodid. Conover found no feeding in natural populations during the warm months, and metabolic rates (respiration) were greatly reduced during that period. Known losses of weight were compared with theoretical losses based on measured respiration and assumed metabolism of lipds [respiratory quotient (R.Q.) assumed to be 0.7] and assumed metabolism of carbohydrates (R.Q. = 1.0). The observed losses of weight were generally consistent with those calculated for metabolism of lipids during the nonfeeding period. Conover calculated that a copepod metabolizing stored lipids would decrease dry weight by about 50%, while an animal using carbohydrate would consume itself in 3 months. In a later paper, Conover and Corner (1968) showed that both the dry weight and the lipid content of *Calanus hyperboreus,* expressed as percent of dry weight, declined gradually during the nonfeeding period and then fell rapidly during the reproductive and early feeding period. The loss in average dry weight of individuals was at least partly explained by the gradual replacement of fat, older individuals by thinner individuals of the new generation. The decrease in the lipid percentage of body weight indicated that lipids were metabolized to a greater extent than protein during the reproductive period. Dry weight and percent lipids increased again during the spring bloom of phytoplankton. During Conover and Corner's (1968) studies, the nitrogen content varied seasonally with dry weight, but nitrogen was relatively constant when expressed as percent of body weight (6–9%).

In a review of life histories of copepods, Heinrich (1962) related the storage of lipids to the way in which reproduction and growth were linked to the seasonal production of phytoplankton. Three types of life histories were identified. In the first, breeding and development of young coincided with the spring bloom of phytoplankton and only the molt from copepodid V to the adult occurred outside the spring bloom. The life cycle of *Calanus finmarchicus* generally falls within this pattern. The second type of life history may include some breeding and considerable growth of juveniles outside the annual period of phytoplankton abundance. Heinrich placed *Calanus cristatus* and *Calanus plumchrus* in this group. Based on structural changes in the mandibles, these two species may not feed as adults (Campbell, 1934; Beklemishev, 1954, cited by Heinrich, 1962) and were assumed to subsist and reproduce with stored lipids as their major source of energy (Heinrich, 1962). Several other species, capable of feeding as adults, were included in the second group, among them *Calanus hyperboreus,* the species studied by Conover (1962) and Conover and Corner (1968). Heinrich (1962) noted that in at least some species of this group, i.e., *Calanus hyperboreus, Metrida longa, Rhincalanus gigas,* and *Microcalanus pygmaeus,* the adults were heavier than stage V. He concluded that feeding on something other than diatoms might account for the increases in weight. Conover (1962) commented that *Calanus hyperboreus* does not feed during the period of egg laying, but also noted (Conover and Corner, 1968) the consistent difference in weight between stage V and adult females. Conover (1967) experimentally fed ripe female *Calanus hyperboreus*. He found little effect of feeding on production of eggs, but did note a decrease in viability and an increase in buoyancy of eggs from starved copepods (Fig. 1). Thus, feeding by adults may (Conover, 1967) or may not (Heinrich, 1962) occur among those copepods that have the second type of life history described by Heinrich (1962).

Species that reproduce throughout year or for extended periods, so that the size

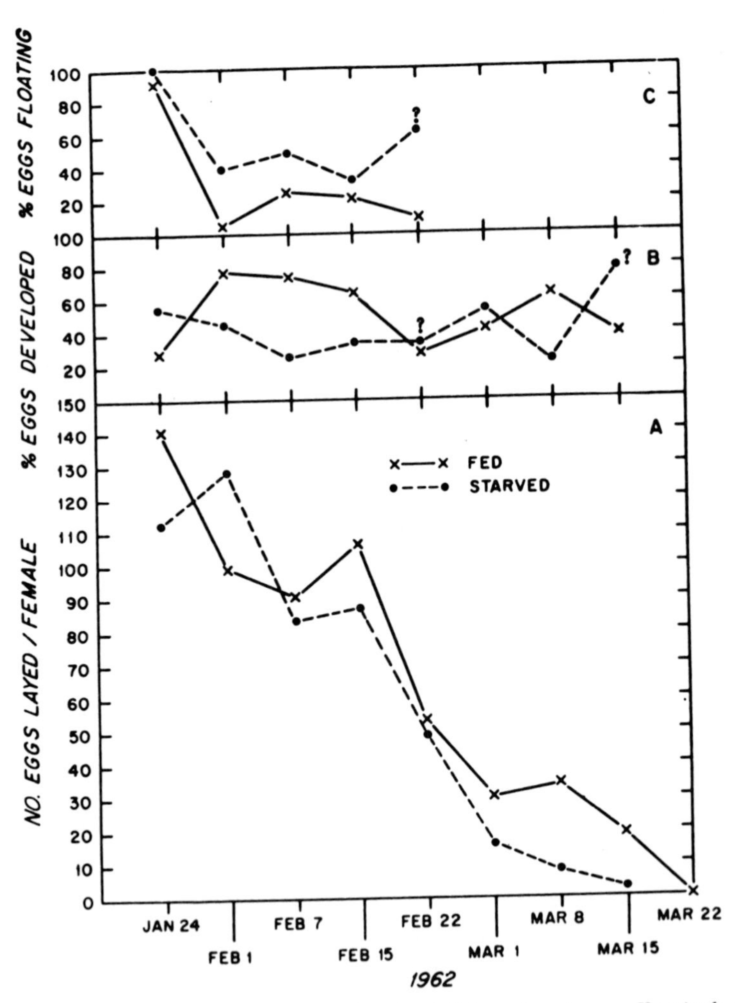

Fig. 1. Egg production by fed and starved females of *Calanus hyperboreus* (Kryer) taken from over the continental slope in a gravid condition. A, weekly egg production; B, percentage eggs developing to hatching; C, percentage eggs which were less dense than sea water. The question mark indicates sample too small to be reliable. (From Conover, 1967, by permission of E. J. Brill.)

of broods are affected by the supply of food, have the third type of life history proposed by Heinrich (1962). Many tropical species and coastal and estuarine temperate and boreal species exhibit the third type of life history. Members of this group cease egg laying and may die more quickly when starved. Heinle *et al.* (1977) found that the estuarine copepod, *Eurytemora affinis,* which normally lays from 10 to 14 broods of 20–40 eggs each when cultured at 20°C, produced a second brood only 25% of the time when placed in filtered estuary water after producing a normal first brood. Survival of starved individuals was much shorter than fed copepods. Heinle *et al.* (1977) also found that an estuarine harpacticoid, *Scottolana canadensis,* was sometimes completely prevented from reproducing by a period of starvation shortly after the terminal molt, even when later provided with abundant food. Species with the third type of life history grow rapidly and reproduce continuously for a least part of the year (Heinrich, 1962). Heinrich suggested that up to 10 generations per year occur. More recent studies suggest that generation times are even shorter than previously thought, i.e., as little as 7 days (Heinle, 1966, 1969a); and population turnover times are even shorter, i.e., as little as 1.5–2.0 days (Heinle, 1966, 1969a, 1974; Petipa *et al.*, 1970). The short turnover times found by Heinle (1966, 1969a, 1974) were for *Acartia tonsa* in a temperate estuary; those of Petipa *et al.*, were for a mixture of herbiverous and omniverous zooplankton in the Black Sea. Heinrich (1962) suggested that zooplankton with the third type of life history generally reached their maximum biomass at the times of phytoplankton maxima. Although that may generally be true, maximum production and biomass of zooplankton do not necessarily coincide, particularly when predation is intense (Heinle, 1969a, 1974; Kremer, 1975). This pattern is similar to that found in the Baltic Sea (Heinrich, 1962). Stored lipids are insufficient to maintain zooplankton in this group for extended periods.

The amount of stored lipids in relation to body size was dramatically demonstrated by Ikeda (1974) (Fig. 2). Large lipid droplets were observed in *Calanus cristatus* and *Calanus plumchrus,* two species that rely solely on stored lipids for reproduction (Heinrich, 1962). Ikeda (1974) observed no lipid droplets in tropical copepods. Lee *et al.* (1971a) found that, in general, the amount of stored lipids was greatest in boreal and deep-water zooplankton, and least in tropical and surface-dwelling species. The results of Ikeda (1974) support those conclusions; however, exceptions have been noted, particularly similarities among family groups regardless of habitat (Lee *et al.*, 1971a) and seasonal variations within species, perhaps reflecting collecting locations (Ikeda, 1974).

2. The Oxygen to Nitrogen Ratio

Experimental evidence for the use of stored lipids has been obtained by measuring the ratio of oxygen consumed to nitrogen excreted by zooplankton under various trophic conditions. Since carbohydrate comprises a very small

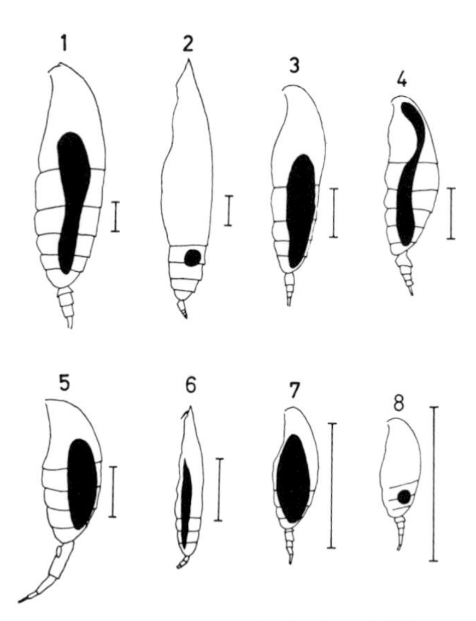

Fig. 2. Various sizes and shapes of oil-sac in the bodies of different copepods (lateral view). Vertical bars alongside of animals show a relative length of 1 mm. *Calanus cristatus* V; (1) *C. plumchrus* V (3), *C. glacialis* (4), and *Eucalanus bungii bungii* (2) were collected from the Bering Sea (June–August 1968); *Metridia okhotensis* (5) and *Pseudocalanus minutus* V (7) off Kitami (September 1970); *Rhincalanus nasutus* (6) off Cape Garnet, Spanish Sahara, Africa (January 1972); and *Paracalanus parvus* (8) from Oshoro Bay (June–July 1970). (From Ikeda, 1974, by permission of the Faculty of Fisheries, Hokkaido University.)

fraction of the body weight of copepods (Raymont and Krishnaswamy, 1960; Raymont and Conover, 1961; Beers, 1966; Ikeda, 1972), protein and lipids are the only materials available for energy reserves. When ammonia is the principal nitrogenous excretion product as some studies suggest (Corner *et al.*, 1965; Corner and Newell, 1967: Butler *et al.*, 1969, 1970), the atomic ratio (atoms of oxygen consumed per atoms of nitrogen excreted) of the metabolism of protein alone is about eight. When roughly equal amounts of protein and lipids are metabolized, the ratio is near 24. Higher values indicate metabolism of predominantly lipids, and since no nitrogenous waste would be produced by metabolism of lipids alone, the ratio in theory could approach infinity (Ikeda, 1974).

Some studies have suggested that amino acids account for a substantial part of the nitrogen excreted by marine zooplankton (Johannes and Webb, 1965; Webb and Johannes, 1967). Corner and Newell (1967) were able to measure amino acids only when their experimental animals (*Calanus helgolandicus*) were extremely crowded. Webb and Johannes (1969) have argued that rapid uptake of amino acids by bacteria caused the results obtained by Corner and Newell (1967). Butler *et al.* (1969) measured excretion of *Calanus finmarchicus* treated with antibiotics over short periods of time. They found that ammonia accounted for 78% of the excreted nitrogen, a value similar to that reported by Corner and Newell (1967). Harris (1973) found that ammonia always comprised over 75% of the total nitrogen excreted by *Tigriopus brevicornis,* a littoral harpacticoid. The criticism by Corner and Cowey (1968) of the results of Johannes and Webb (1965) based on the high densities of zooplankton employed and their use of membrane filters may be well founded. I attempted to collect *Acartia tonsa* on Millipore HA filters for subsequent clearing and counting but found that at 0.7 Kg/cm^2 filtration pressure, the body walls of adult *Acartia* were frequently broken, and body fluids (presumably amino acids) were sucked through the filters. While there is some evidence for excretion of urea and small amounts of amino acids by marine zooplankton (Jawed, 1969), most of the nitrogenous excretion is probably in the form of ammonia. Excretion of incompletely oxidized amino acids would lower the O:N ratio. The approximate value of 24 or greater, which suggests metabolism of lipids, might thus be somewhat lower at times.

Conover and Corner (1968) found O:N ratios of 6–10 for stages IV and V of *Calanus hyperboreus,* presumably from actively feeding animals in April. The ratio then rose to about 75 for stage V copepodids and 180 for adults in May, about the time when feeding ceases (Conover, 1962). In October, November, and January, Conover and Corner (1968) found mean O:N ratios ranging from 20 to 40 with considerable scatter. The ratios found during May for animals of the same stage from the continental slope were higher than those from the Gulf of Maine, but during April similar values were observed from both locations. Conover and Corner (1968) concluded that the O:N ratio fell in late winter because stored lipids were used up and protein metabolism was increased.

Harris (1959) found an O:N ratio of 7.7 for *Acartia clausi,* a coastal species from Long Island Sound. The quantities of nitrogen excreted amounted daily to 48% of the body nitrogen. Harris felt that the excretion of ammonia may have been overestimated in his experiments. However, Corner and Davies (1971) have concluded that the levels of excretion measured by Harris (1959) were probably accurate, since the higher levels of excretion per unit body weight (relative to results with adult *Calanus finmarchicus*) were the result of the smaller body size of *A. clausi*. Corner *et al.* (1965), working with copepodids II, III, and IV of *C. finmarchicus,* found levels of nitrogen excretion between those for adult *A. clausi* (Harris, 1959) and adult *C. finmarchicus* (Cowey and Corner, 1963). Nauplii of *C. finmarchicus,* similar in size to adult *A. clausi,* excreted comparable amounts of ammonia (Corner *et al.*, 1967). Corner and Cowey (1968), citing the population studies of the copepods of Long Island Sound by Conover (1956), suggest that at the time of Harris' (1959) measurements competition was occurring between *A. clausi* and *A. tonsa,* so that *A. clausi* may have been catabolizing protein. Corner and Cowey (1968) also noted that O:N ratios observed for *C. finmarchicus* and *C. helgolandicus* (Corner al., 1965) approached the theoretical ratio of 17 suggested by Redfield *et al.* (1963) for the oxidation of typical particulate organic matter in the sea.

Ikeda (1974) has extensively documented the inverse relationship between body size and weight specific excretion rates (Fig. 3). In addition, Ikeda observed significant latitudinal differences within a size range. Rates of excretion increased from boreal to temperate to subtropical to tropical species, as shown in Fig. 3. A similar relationship was found for weight specific respiration (Ikeda, 1974). The data of Ikeda show that for both respiration and excretion, the latitudinal differences were greatest for smaller animals and least for larger animals.

The O:N ratio was determined for a large number of species of marine zooplankton by Ikeda (1974) during starvation. The frequency distributions shown by Ikeda are reproduced in Fig. 4. The O:N ratio was commonly in excess of 24 for boreal species. Ikeda reported that no nitrogen was detectable in some experiments with boreal species, i.e., that the O:N ratio was infinity. The O:N ratio was below 24 for all temperate and most subtropical and tropical species.

Two O:N ratios in excess of 100 were observed by Ikeda (1974) for tropical species during starvation, both for decapod megalopa. Sulkin (1975) has suggested that lipids may be an essential component of the diet of larvae of the decapod *Callinectes sapidus*. Frank *et al.* (1975) found that the lipid content of another decapod, *Rithropanopeus harrisii,* increased from 3 to 9% of fresh weight during zoeal development. Protein similarly increased, particularly be-

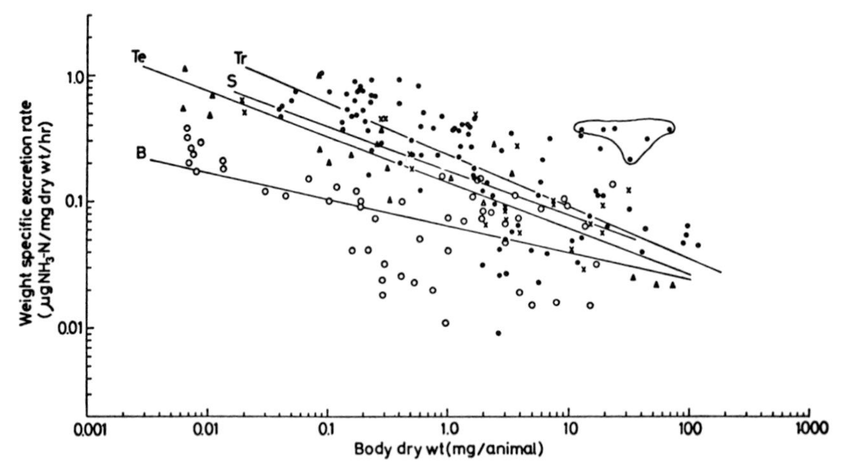

Fig. 3. Relationship between weight specific excretion rate of ammonia-nitrogen and body dry weight from tropical, subtropical, temperate, and boreal zooplankton. Points encircled are data for tropical fish larvae collected with fish larvae net. Symbols are as follows: ●, tropical species; x, subtropical species; Δ, temperate species; ○, boreal species. (From Ikeda, 1974, by permission of the Faculty of Fisheries, Hokkaido University.)

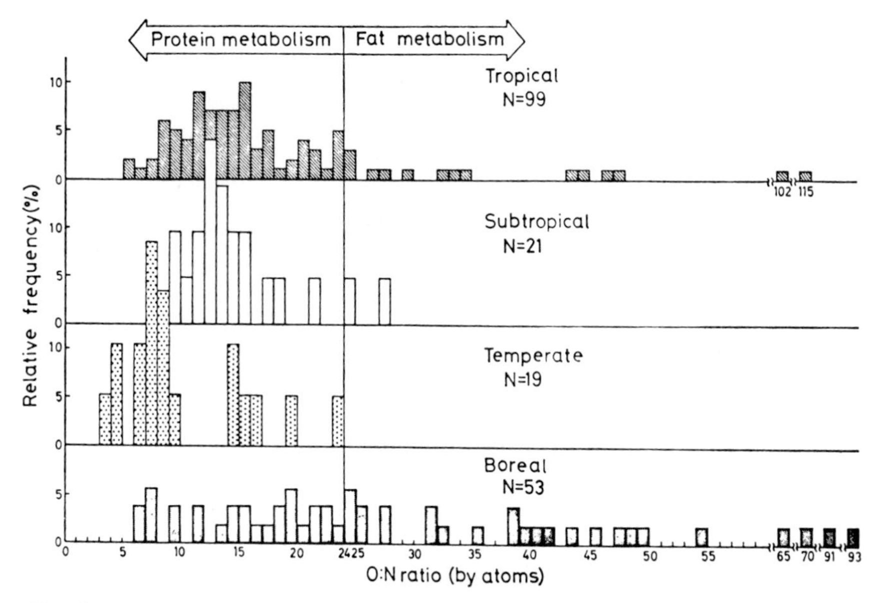

Fig. 4. Relative frequency (%) of the O:N ratio by atoms, calculated from respiration and ammonia-nitrogen excretion, for tropical, subtropical, temperate, and boreal zooplankton. N = number of experiments from each area. (From Ikeda, 1974, by permission of the Faculty of Fisheries, Hokkaido University.)

tween the second and third zoeal molt. Lipid appears to serve as an important organic reserve for decapoda in general (O'Conner and Gilbert, 1968).

Extremely low O:N ratios (7–10) were found by Ikeda (1974) for some boreal species, particularly chaetognaths and pteropods, but also for some copepods, e.g., *Acartia longiremis* (two of three determinations), and surprisingly, *Calanus plumchrus* (one of ten experiments). The experiment which gave the low O:N ratio for *C. plumchrus* was done with five copepodid IV's during May and thus may have involved recently molted animals with low fat reserves. All three experiments done with stage IV gave O:N ratios of between 8 and 15, while those done during the same period with copepodid V gave ratios ranging from 25 to 91, which indicated catabolism of lipids in agreement with the observations of Heinrich (1962). Experiments done by Ikeda (1974) with *C. plumchrus* stage V during June at 13° to 16°C gave O:N ratios of 8–17. However, these animals were captured in temperate waters. Ikeda found low O:N ratios for *Acartia clausi* that were nearly identical to those reported by Harris (1959). The only O:N ratios in excess of 24 for subtropical species were found in experiments with the amphipod *Scina conigera* (Ikeda, 1974). Among the tropical species, O:N ratios in excess of 24 were observed by Ikeda only for decapods, tunicates, and one chaetognath, *Sagita bipuncta*. All of the tropical chaetognaths examined by

Ikeda had higher O:N ratios than Reeve *et al.* (1970) found for *Sagita hispida*. Ikeda (1974) concluded that the O:N ratio is affected most by the temperature of the habitat from which the animals were collected. Herbivorous species in boreal seas had more variable O:N ratios than did the carnivores, including some lower than 24. Ikeda also concluded that herbivores and carnivores from temperate, subtropical, and tropical seas are less distinguishable by their O:N ratios than are boreal species. The morphological and chemical evidence thus supports the conclusions of Heinrich (1962); that stored lipids are important where food is seasonally scarce.

3. The Nature of Stored Lipids and the Role of Wax Esters

Nevenzel (1970) reviewed the status of knowledge of the occurrence of wax esters in marine animals and first noted the large quantities that occur in marine copepods. Benson and Lee (1975) succinctly describe the distinction between triglycerides and wax esters and summarize more recent knowledge about the distribution of wax esters in marine animals. Early interest in these compounds arose for two major reasons. Wax esters still are not known to occur in marine algae but are common among terrestrial plants. Thus, the biochemical synthesis of these compounds by copepods was of considerable interest, since terrestrial insects apparently acquire them only through their diet. Also, catabolism of waxes by animals is not common as triglycerides are the commonly used fatty reserve, so the possible role of the wax esters as energy reserves was intriguing. Nevenzel (1970) suggested that wax esters might provide buoyancy, thermal insulation, and might serve as energy reserves. He cited unpublished work of Lee's that demonstrated the use of wax esters for energy reserves by *Calanus helgolandicus*.

Lee *et al.* (1971a) determined the amounts of triglycerides, wax esters, phospholipids, and total lipids in a large number of marine species from several phyla. They found that not only were lipids more abundant in boreal and deep-water forms, but that the percentage of total lipids comprising wax esters was higher in these groups. The pattern was most striking among the copepods. The same trends with latitude and depth have been observed for individual species, e.g., *Gaussia princeps* (Lee and Barnes, 1974) (Tables I and II). Lee *et al.* (1971a) found that some copepods collected from tropical latitudes contained substantial amounts of wax esters as percentage of total lipids (Table I). These were all either boreal species as described by Brodskii (1950) and Rose (1933) or were of the genus *Eucheata*. No copepods collected from tropical latitudes had more than 37% of their dry weight as lipids, or more than 46% of their lipids as wax esters (Table I). Two of three temperate copepods had high lipid contents and high percentage of wax esters. Adult female *Calanus plumchrus* were known to be nonfeeding, thereby relying on lipids as a store of energy for reproduction. The other copepod, *Euchaeta japonica,* deposits lipid reserves in the eggs suffi-

TABLE I

The Lipid Composition and Weight of Calanoid Copepods from Subtropical, Temperate, and Arctic Regions[a]

	Lipid per animal (mg)	Lipid (% of dry weight)	Triglyceride (% of lipid)	Wax ester (% of lipid)
I. Subtropical latitudes				
La Jolla, California (33°N, 117°W; upper 50 m)				
Valero Basin (31°N, 119°W; upper 250 m)				
Calanus gracilis (♀)	0.08	26	17	21
Calanus helgolandicus (♀)[b]	0.04	14	4	33
Calanus robustior (♀)	0.03	8	3	21
Eucalanus sp. 1 (♀,♂)	0.08	31	42	1
Rhincalanus nasutus (♀)[b]	0.06	37	1	46
Euchirella rostrata (♀)	0.14	21	37	
Euchirella (♀)	0.13	15	28	0
Euchaeta acuta (♀)	—[c]	—	8	37
Euchaeta media (♀)	—	—	6	42
Candacia curta (♀)	0.01	3	10	4
Candacia aethiopica (♀)	0.02	9	11	1
Labidocera trispinosa (♀)	0.03	14	17	2
II. Temperate latitudes				
Vancouver, British Columbia (50°N, 123°W; upper 200 m)				
Calanus plumchrus (♀)	0.41	59	7	86
Gaetanus intermedium (♀)	0.06	25	21	5
Euchaeta japonica (♀)	0.44	41	18	54
III. Polar latitudes				
Arctic Station T-3 (84°N, 106°W, upper 500 m)				
Calanus finmarchicus (♀)	0.22	50	11	63
Calanus hyperboreus (♀)	0.60[d]	73	2	92
Euchaeta sp. (C-V)	0.16	31	4	61
Metridia longa (♀)	0.06	34	2	62

[a] From Lee *et al.* (1971a) by permission of Pergamon Press.
[b] Copepods described as boreal species by Brodskii (1950) and Rose (1933).
[c] Not analyzed.
[d] Misprinted as 5.90 in the original table of Lee *et al.* (1971a).

TABLE II

The Medians and Ranges of the Lipid, Triglycerides, and Wax Esters in Copepods from Different Depth Intervals[a]

Group		Lipid (% of dry weight)	Triglyceride (% of lipid)	Wax ester (% of lipid)	Genera examined	Family
I. Near-surface copepods	median	18	14	2·5	*Calanus*	Calanidae
0–250 m, day–night;					*Eucalanus*	Eucalanidae
8 groups analyzed	range	3–42	3–42	0–69	*Rhincalanus*	Eucalanidae
					Euchirella	Aetideidae
					Candacia	Candaciidae
II. Migrating copepods	median	18	29	2·5	*Euchirella*	Aetideidae
0–250 m, night only;					*Chirundina*	Aetideidae
25 groups analyzed	range	4–53	6–55	0–72	*Undeuchaeta*	Aetideidae
					Pseudochirella	Aetideidae
					Scottocalanus	Scolecithricidae
					Pleuromamma	Metridiidae
					Gaussia	Metridiidae
III. Deep-living copepods	median	40	4	63	*Megacalanus*	Calanidae
absent in 0–250 m;					*Bathycalanus*	Calandiae
40 groups analyzed	range	10–61	1–29	11–82	*Gaetanus*	Aetideidae
					Paraeuchaeta	Euchaetidae
					Amallothrix	Scolecithricidae
					Metridia	Metridiidae
					Lucicutia	Lucicutidae
					Disseta	Heterorhabdidae
					Heterorhabdus	Heterorhabdidae
					Euaugaptilus	Augaptilidae

[a] From Lee *et al.*, 1971a, by permission of Pergamon Press.

cient to sustain early nonfeeding naupliar stages (Campbell, 1934; Lewis and Ramnarine, 1969; Lewis *et al.*, 1971; Lee *et al.*, 1974). Four polar species of copepods were found to have 31–73% of their dry weight as lipids (Lee *et al.*, 1971a; Table I). Wax esters comprised 61% or more of the total lipids in these species. *Calanus hyperboreus* was found to have 92% of its lipids as wax esters.

Lee *et al.* (1971a) measured the loss of lipids during starvation of *Gaussia princeps, Megacalanus longicornis,* and *Gaetanus brevicornis,* mesopelagic or bathypelagic carnivorous copepods. At 5°C in filtered surface water, the starved copepods lost 0.09–1.04% of their lipids per hr during experiments up to 198 hr long. The changes in lipid composition of *G. princeps* were also measured and only the triglycerides decreased during 120 hr of starvation.

Lee *et al.* (1971a) noted that Mattson *et al.* (1970) had observed the inhibition of pancreatic lipase by long chain alcohols, which are breakdown products of wax esters, and they suggested a hypothesis concerning the role of wax esters. They proposed that triglycerides might serve as the lipid reserve used first, with wax esters being used only when triglycerides were depleted. The catabolism of wax esters would be self-inhibiting and thus force reduced metabolism on the copepods. The large amounts of wax esters found in species known to rely on lipid reserves, e.g., *Calanus hyperboreus, C. plumchrus,* and *Euchaeta japonica,* lend credence to this hypothesis. Lee and co-workers have since demonstrated that triglycerides are used first and that wax esters are later catabolized at a slower rate (Lee, 1974a,b; Lee and Barnes, 1974; Lee *et al.*, 1974). The deep-water carnivorous copepod, *Gaussia princeps,* withstood starvation for up to 90 days in good condition but generally consumed their triglycerides within 1 week (Lee and Barnes, 1974). *Calanus hyperboreus* from an arctic population never had more than 8% of their lipids as triglycerides, while wax esters varied seasonally from 34 to 91% of total lipids (Lee, 1974a). Lipid content as percent of dry weight was lowest in May and June, rising rapidly to highest levels in August (Table III). During February, March, and June, Lee found that *C. hyperboreus* from deep water had slightly higher total lipids and wax esters than copepods from near the surface (upper 100m). The seasonal changes in percent lipids were as expected from herbiverous feeding and the seasonality of production of phytoplankton (English, 1961). During starvation, arctic *C. hyperboreus* used all of their triglycerides within 2 weeks. Wax esters declined from 90 to 32% of total lipids in 90 days, while lipids decreased from 68 to 22% of dry weight (Lee, 1974a). Lee observed that the decrease in lipid content of the wild copepods between September and December was similar to those used in the starvation experiments. After December however, the lipid content of the wild *C. hyperboreus* decreased more slowly than predicted by the decreases during starvation. Lee suggested that either carnivorous feeding by the overwintering copepods (Heinrich, 1962) or reduced metabolism as observed by Conover (1962) could account for the slower rates of decrease in lipid content after

TABLE III

Seasonal Changes in Lipid Content of *Calanus hyperboreas*[a]

Collection date	Lipid per individual (mg)	Lipid content (% dry weight)	Wax ester (% lipid)	Triglyceride (% lipid)
February	0.8	42	71	1
March	0.8	44	67	2
April	0.7	38	60	1
May	0.6	37	45	0
June	0.4	29	34	0
July	0 9	52	82	2
August	1.7	74	86	7
September	2.1	66	91	4
October	1.3	64	82	3
November	1.2	62	89	1
December	0.9	51	77	1

[a] From Lee, 1974a, by permission of Springer-Verlag.

December. Changes in the fatty acid composition of the wax esters of *C. hyperboreus* were thought by Lee (1974a) to reflect the shift from active feeding and storage of lipids to catabolism of lipids during starvation. Copepods from one sample collected in March may have been feeding carnivorously.

In some species, the wax esters appear to be consumed simultaneously with the triglycerides (Lee *et al.*, 1970). In *Calanus helgolandicus* the wax esters are apparently used for short-term metabolic requirements. Lee *et al.* (1970, 1971b, 1972) demonstrated not only that *C. helgolandicus* could synthesize wax esters from algae that contained none, but also that the rates of decrease of labled pools of wax esters and triglycerides were equal. Structural lipids such as phospholipids and cholesterol were used more slowly. While Lee *et al.* (1970) do not say so, they probably used adult *C. helgolandicus* in their experiments. Since *C. helgolandicus* relies on the spring production of phytoplankton for reproduction, long-term reserves of lipids may be of lesser importance than in species such as *Calanus hyperboreus* and *Calanus plumchrus,* which use stored lipids for reproduction.

Some species, like *Euchaeta japonica,* that pass lipid reserves on to their offspring via the eggs may also rely on wax esters as long-term reserves. Lee *et al.* (1974) observed a gradual decrease in wax esters from 58% of total lipids in laboratory cultured *E. japonica* eggs to 9% in copepodid I followed by an increase to 60% in mature females (Table IV). Triglycerides were absent in nauplius VI and did not reappear until copepodid IV. Triglycerides never comprised more than 19% of total lipids. There was no change in the percentage composition of wax esters and triglycerides through nauplius II, although lipids as percent of dry weight decreased. Both reserve fractions were thus used non-selectively until nauplius II, after which the triglycerides were consumed rapidly and the wax esters more slowly. Animals collected from the field (copepodid I) contained more total lipids and had a higher percentage of wax esters than the same instar in laboratory cultures, possibly a consequence of the diet provided for the cultured copepods.

Lipids might also be used for short-term storage of energy for reproduction. Corkett and McLaren (1969) observed that individual females of *Pseudocalanus minutus* produced relatively constant volumes of eggs in successive clutches. At reduced food levels, the time interval between clutches lengthened while the constancy of volumes of eggs was maintained. Females that produced clutches of eggs at greater intervals had reduced oil sacs. Unlike *Euchaeta japonica,* the early larval stages of *P. minutus* appear to require food to survive (Corkett and McLaren, 1970). Ikeda (1974) observed O:N ratios below 24 for *Pseudocalanus*

TABLE IV

Lipid Changes During Development of *Euchaeta japonica*[a,b]

Stage	Lipid/individual (mg)	Lipid (% dry weight)	Wax ester (% of lipid)	Triglyceride (% of lipid)
Eggs	0.59[c,d]	64.4	58	19
Eggs (late stage)	0.39[c]	58.1	50	17
Nauplius 2	0.02	43.8	61	17
Nauplius 3	0.02	30.8	56	5
Nauplius 4	0.02	25.0	20	3
Nauplius 5	0.04	21.2	15	1
Nauplius 6	0.04	17.0	12	absent
Copepodid I	0.03	14.2	9	absent
Copepodid I (field collected)	0.05	23.6	29	absent
Copepodid II	0.03	11.6	12	absent
Copepodid IV	0.20	31.2	40	3
Copepodid V	0.52	50.1	81	2
Adult ♀ (immature)	0.44	41.3	54	18
Adult ♀ (mature)	0.60	52.2	60	17
Adult ♂ (mature)	0.58	49.2	78	9

[a] From Lee *et al.*, 1974, by permission.

[b] Egg sacs were removed from females and placed in filtered sea water. Naupliar stages 2–4 were reared in filtered sea water at 10°C without feeding. Naupliar stages 5–copepodid II were raised in the laboratory by feeding them a mixture of phytoplankton. The remaining stages were field collected.

[c] Per egg cluster.

[d] Average 0.04 mg lipid/egg.

elongatus during May but obtained a ratio of 42 during a single experiment in June. As implied by Corkett and McLaren (1969), the taxonomy of species within the genus *Pseudocalanus* may be somewhat uncertain, making direct comparison of experiments difficult.

Two of the boreal and temperate taxa that Ikeda (1974) found to have low O:N ratios, the chaetognaths and ctenophores, were similarly found by Lee (1974b) to contain small amounts of lipid (less than 14% of dry weight) with little wax ester (less than 12% of total lipids). Lee (1974b) observed chaetognaths feeding on copepods that contained oil droplets. The oil droplets disappeared during digestion and Lee concluded that the lipids were converted to protein. Starving chaetognaths shrink in size (Reeve *et al.*, 1970; Ikeda, 1974) as do pteropods (Lalli, 1970). The single decapod from Bute Inlet, British Columbia, that Lee (1974b) analyzed had a moderate amount of lipid composed largely of triglycerides. In light of the observations by Ikeda (1974) and Sulkin (1975) that lipids are important metabolic and dietary constituents, lipids may serve mainly as a short-term reserve of energy for decapods.

Ackman *et al.* (1974) have suggested that the wax esters in copepods and other marine animals may have a second function. They suggested that since the wax esters are deposited during periods of heavy feeding and accumulation of protein, the synthesis by the copepods of low-density monoethylenic fatty acids from shorter chain lengths might be a way of compensating for the negative buoyancy of the proteins. Gatten and Sargent (1973) observed lower lipid contents and greater activity of synthesis of wax esters by *Calanus finmarchicus* near the surface than at depth.

It would be of adaptive value for marine zooplankton to use compounds like wax esters for more than one purpose. The hypothesis of Lee *et al.* (1971a) that the adaptive significance of the wax esters is the self-inhibition that occurs during catabolism is appealing because it suggests a means of slowing catabolism during extended shortages of food. It is possible that as lipid metabolism is studied in light of life histories, a variety of adaptive mechanisms will emerge. For at least one species, *Calanus hyperboreus,* the experimental observations regarding the role of wax esters (Lee, 1974a) and metabolism (Conover and Corner, 1968), and also the observations of the life history (Conover, 1962), agree with the hypothesis of Lee *et al.* (1971a) about the self-inhibition of catabolism of wax esters.

C. ARRESTED DEVELOPMENT

1. Arrested Development at Immature Stages

Marshall and Orr (1955a) reviewed the life cycle of *Calanus finmarchicus* at the extremes of its distribution in the Atlantic Ocean. Over most of its range, particularly the southern part, most *C. finmarchicus* overwinter as copepodid V.

At the northern extremes of its range much of the population is copepodid III and IV during the winter. Overwintering populations are frequently found in greater abundance in deeper waters. The molting to adults, or in northern populations, to later copepodids and then adults, begins in December or January. Marshall and Orr (1955a) noted the important fact that the onset of resumed development does not vary latitudinally. Southern populations may then produce two or three distinct generations in surface waters, while northern populations generally produce only one. Copulation usually occurs shortly after molting to adulthood, while the production of eggs does not occur until sufficient food is present. During August and September, most of the final generation has reached copepodid V (Marshall and Orr, 1955a, Fig. 29). Much higher percentages are found in copepodid III and IV east of Greenland (Marshall and Orr, 1955a, Fig. 30). Marshall and Orr suggest that *Calanus helgolandicus* has essentially the same life cycle as *C. finmarchicus*. Although it might be possible to ascribe the cessation of growth of *Calanus finmarchicus* in the fall to a decline in phytoplankton or falling temperatures, resumption of molting during December and January must depend on an environmental cue other than food or rising temperature. More recent studies (Butler *et al.*, 1970) have shown that overwintering *Calanus* have gradually decreasing rates of excretion of nitrogen and phosphorus between October and February. At least part of the decreased metabolism was due to the decrease in dry weight observed by Butler *et al.* (1970) during that period, since the weight-specific rates of excretion decreased much less than rates per animal. The life history events that these authors observed indicated that egg laying began *in advance* of major increases in chlorophyll *a*. The excretion of nitrogen and phosphorus observed were greater than could be accounted for by loss of weight by the copepods. The experiments of Corner *et al.* (1974) indicate that carnivorous feeding by overwintering copepodid V or adult female *Calanus helgolandicus* could account for the results of Butler *et al.* (1970). Corner *et al.* (1974) observed that either live or dead nauplii of the barnacle *Elminius modestus* could be captured and eaten by *Calanus*. In some areas at least, food is thus available to overwintering *Calanus* in the form of nauplii of other species.

A more restricted life cycle is shown by *Calanus hyperboreus* (Conover, 1962, 1967), and particularly *Calanus plumchrus* (Fulton, 1973). As mentioned previously (Section B, 1), adult *C. hyperboreus* are capable of feeding (Heinrich, 1962; Conover, 1967) and are present in natural populations during the entire winter in the Gulf of Maine (Conover, 1962) and throughout the year in the arctic (Lee, 1974a). Lee observed copepodids IV and V only during the summer months. The observation by Lee that adults were common throughout the year suggests that the life cycles of the arctic and the Gulf of Maine populations may differ. Conover (1967) observed that nauplius I and II *C. hyperboreus* do not feed. He isolated females as copepodids and observed them in the laboratory. Most breeding occurred during the winter months, but some females produced eggs all year. *Calanus hyperboreus* thus is not totally restricted to its described life cycle (Conover, 1962).

The life cycle of *Calanus plumchrus* observed by Fulton (1973) appears to be much more restricted. His schematic summary is reproduced in Fig. 5. Reproduction and growth occur during the months of January through May, accompanied by extensive upward migration of the growing instars. The remainder of the year is spent as copepodid V with the bulk of the population retreating gradually to deeper water during May to December. Fulton observed that in addition to the lack of feeding by adults reported previously by Campbell (1934) and Heinrich (1962), copepodid V *C. plumchrus* did not feed while in deep waters. The nauplii of *C. plumchrus* also do not feed, according to Benson and Lee (1975). Fulton (1973) commented that *C. plumchrus* was present in surface waters only during the period of maximum production of phytoplankton. According to Heinrich (1962), *Calanus cristatus* has a similar life cycle. Growth in these species is suspended in a specific instar, copepodid V, a pattern identical to that observed from some freshwater cyclopoid copepods (Smyly, 1961).

Species that occur sympatrically with *Calanus plumchrus* do not have such a restricted life cycle. LeBrasseur *et al.* (1969) found *Calanus helgolandicus* (*pacificus*), *Pseudocalanus minutus*, *Microcalanus* sp., *Metridia* sp., and *Acartia* sp. with *C. plumchrus* in the Strait of Georgia. The advantage gained by *C.*

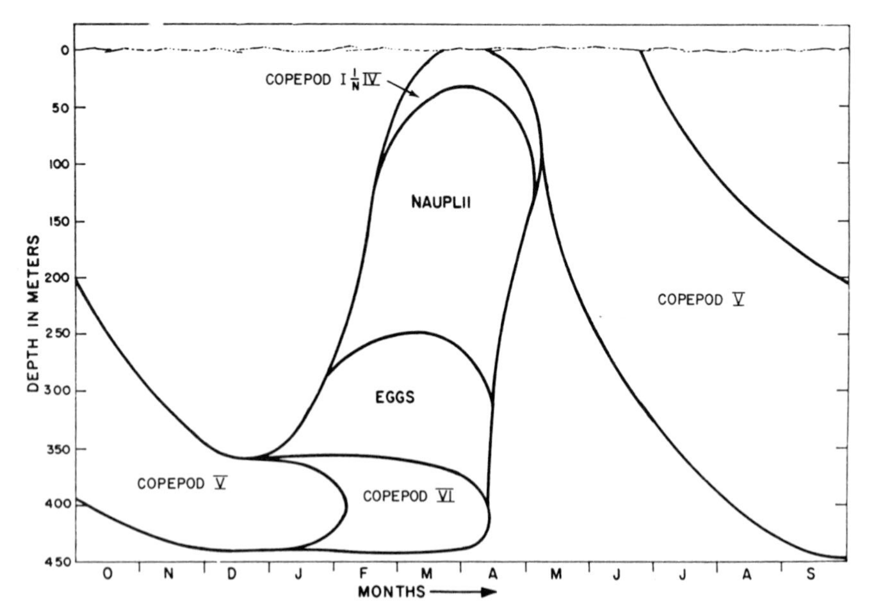

Fig. 5. Schematic diagram of *Calanus plumchrus* life cycle in the Strait of Georgia. (From Fulton, 1973, by permission of the Fisheries Research Board of Canada.)

plumchrus by its restricted life cycle in the Strait of Georgia is not readily apparent, but LeBrasseur *et al.* (1969) and Fulton (1973) have suggested that predators might be avoided, since the copepods are in deep water for much of the year where juvenile fish are less abundant.

Some species that have arrested growth appear to stop in any copepodid stage. McLaren (1969) observed that both *Pseudocalanus minutus* and *Oithona similis* overwintered in all copepodid stages, including adults, but had an annual (one generation per year) life cycle. He concluded that nauplii produced late in the summer perished before the following spring, although it is not apparent from his figures that the nauplii were absent during the winter. Carter (1965) observed a similar life cycle for *P. minutus* in another fjord in northern Canada. Despite the apparent flexibility of their life cycle, McLaren (1969) concluded that the secondary production (size of a generation) was controlled more by physical factors in Ogac Lake on Baffin Island, Canada, than by the amount of primary production. The chaetognath *Sagita elegans* was also annual in Ogac Lake with its life cycle closely keyed to the timing of the annual production of copepods. Cairns (1967) observed that a high latitudes *P. minutus* may take 2 years to complete its life cycle. Two age groups were abundant in his samples from May and August; i.e., early copepodids and adult females (Cairns, 1967). It appears to this author that the arrested development of *P. minutus* and *O. similis,* as described by Carter (1965), McLaren (1969), and Cairns (1967), might have been simply the cessation of growth because temperatures were low and the necessary biochemical reactions proceeded slowly or stopped. McLaren *et al.* (1969) found, however, that biological zero for *P. minutus* was −13.4°C.

2. *Resting Eggs*

Many temperate species appear not to employ arrested development as nauplii or copepodids but instead disappear from the plankton during unfavorable conditions. One of the best known examples is the seasonal alternation of *Acartia tonsa* and *Acartia clausi* on the Atlantic coast of North America (Deevey, 1948; Bousfield, 1955; Conover, 1956; Jeffries, 1962; Martin, 1965; Herman *et al.*, 1968; Heinle, 1969a). *Acartia tonsa* is the dominant summer copepod in Bras d'Or Lake, Nova Scotia (Geen and Hargrave, 1965). While *A. clausi* was not reported from Bras d'Or Lake, it may occur as a winter form as its distribution extends well to the north of Nova Scotia (Pinhey, 1923). In the Miramichi estuary, New Brunswick, *A. tonsa* is abundant at lower salinities in the summer, while *A. clausi* persists during the summer at higher salinities (Bousfield, 1955). Although Bousfield did not sample during the winter, one can probably assume that *A. tonsa* disappears as it does farther south (Conover, 1956), whereas *A. clausi* is present throughout the year.

At the southern extreme of the distribution of *Acartia clausi* in Chesapeake Bay, that species is abundant only during late winter and early spring, whereas

A. tonsa is predominant from May through December (Herman *et al.*, 1968; Heinle, 1969b) and is present all year (Heinle, 1969b). The seasonal succession of the two species of *Acartia* appears to be partly explainable on the basis of their differing responses to the effects of temperature and salinity (Conover, 1956; Jeffries, 1962). One of the more puzzling features of the succession has been the suddent reappearance of *A. tonsa* in northern waters during the summer following a period of total absence. The discovery of resting eggs of *Acartia tonsa* by Zillioux and Gonzalez (1972) suggests the probable cause for the sudden reappearance of that species in the spring. They found that a portion of the eggs of *A. tonsa* produced at temperatures below 14.5°C in laboratory cultures were true resting eggs, i.e., they failed to hatch when held past the time required for normal hatching. At temperatures between 9.0° and 14.5°C, a mixture of active and resting eggs were produced. All of the eggs were found to be resting eggs in experiments done at 5°C. Zillioux and Gonzales fitted Bělehrádek's (1935) equation to the hatching times of *A. tonsa* at temperatures as described by McLaren *et al.* (1969). Normal hatching times at lower temperatures were then estimated by extrapolation of the equation that was derived. While there may still be some doubt concerning the use of Bělehrádek's equation to predict development rates of copepods (e.g., Bernard, 1970), Zillioux and Gonzalez (1972) kept eggs up to three times longer than their predicted normal hatching time, sufficiently long to suggest that they were observing true dormancy. They caused the eggs to hatch by raising the temperature to 20°–30°C. Hatching was often observed in less than 24 hr after the increase in temperature. Forty-three percent of one group of eggs held for 135 days at 5°C hatched within 24 hr after the temperature was raised to 20°C. Short periods of freezing were sufficient to kill the resting eggs. *Acartia tonsa* is absent from Narragansett Bay, Rhode Island, for about 3 months, February through April (Frolander, 1955; Martin, 1965). The dormancy of 135 days that Zillioux and Gonzalez (1972) observed for eggs of *A. tonsa* from that location is thus sufficient to carry the species through the unfavorable period. These authors were able to demonstrate that nauplii of *Acartia* could be hatched from eggs collected with sediments from Narrganset Bay during March of 1967. As there was a possibility that the nauplii were *A. clausi* rather than *A. tonsa,* they repeated their experiment 3 years later. A single *A. tonsa* female was produced and reared to adulthood within 8 days at 21°C.

Zillioux and Gonzalez (1972) reviewed the earlier literature concerning the possibility of resting eggs for a number neritic copepod species. The seasonal alternate of *Acartia tonsa, A. clausi,* was among those suggested, although direct evidence was lacking. Kasahara *et al.* (1974) have identified resting eggs of *A. clausi* and five other species from samples of sediment from the Seto Inland Sea, Japan. They observed up to 3.4 million eggs per square meter in one sample, with those of *A. clausi* comprising 96% of the total. That sample was collected in the early summer and, if representative, would lead to a substantial population of *A. clausi* upon hatching. Other species observed by Kasahara *et al.* were *Acartia erythraea, Tortanus forcipatus, Calanopia thompsoni, Centropages abdominales,* and *Centropages yamadai*. There is a seasonal succession of species in the Inland Sea (Hirota, 1962, 1964). The abundance of the eggs that were identified by Kasahara *et al.* (1974) in the sediments was greatest just before the disappearance of adults of the respective species, suggesting clearly that these were resting eggs (Kasahara *et al.*, 1975a).

Kasahara *et al.* (1975b) described the conditions that promoted hatching of only the resting eggs of *Tortanus forcipatus*. They found that eggs of *T. forcipatus* collected during November and December hatched when exposed to temperatures above 13°C. No eggs hatched at 10°C. Since *Acartia clausi* is a winter and spring form in the Seto Inland Sea (Kasahara *et al.*, 1975a), high temperatures are probably not the proper stimulus for hatching of that species. *Acartia clausi* attained its maximum density in the plankton during May (about 3×10^4 copepodids and adults per m^3) and nearly disappeared in July. None were collected between mid-August and mid-October (Fig. 6). During the period that *A. clausi* was absent from the plankton, the eggs of that species were abundant in the sediments (Fig. 6). Temperatures at the station sampled by Kasahara *et al.* (1975a) were in excess of 20°C during that period. Landry (1975a) observed that hatching of the eggs of *A. clausi* could be inhibited for a short period of darkness. As the eggs used in Landry's experiments were held in the dark for only up

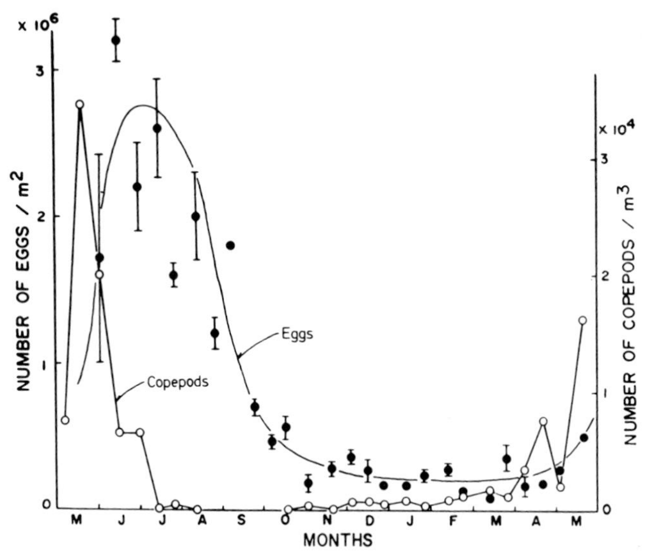

Fig. 6. *Acartia clausi*. Seasonal cycles of abundance of populations of copepods and eggs in sea-bottom mud. (From Kasahara *et al.*, 1975b, by permission of Springer-Verlag.)

to 42 hr, it is not yet known if darkness alone causes inhibition of hatching for the much longer times needed to survive the summer months in temperate estuaries. J. K. Johnson (personal communication) has observed that nauplii of *A. tonsa* can be hatched from sediments of Yaquina Bay, Oregon at almost any time. Grice and Gibson (1975) demonstrated experimentally that resting eggs were the probable mechanism by which *Labidocera aestiva* survives winters in neritic temperate waters. It is not possible to tell with certainty from these experiments whether the overwintering eggs of *L. aestiva* are truly dormant or greatly slowed in their development because of low temperatures. The very high percentages of eggs that hatched after long times (up to 180 days at 5°–6°C and 150 days at 2°–3°C) suggest true dormancy (Table V). The eggs were stimulated to hatch by slowly warming them to 18°C. The viability of *L. aestiva* eggs appeared to vary somewhat between females. There was no apparent decrease in viability of the eggs of one female over 5 months of incubation at natural winter temperatures. It is thus possible that some of the variation in viability observed in other experiments of Grice and Gibson (1975) was due to differences between females. Adult *L. aestiva* were absent from the plankton during the winter but were observed over sediment collected in April and incubated for 20 days.

TABLE V

Hatching of Resting Eggs of *Labidocera aestiva* Incubated at 2°–3°C and 5°–6°C in the Laboratory[a,b]

Incubation (days)	Number of eggs	Condition[c]	Percent hatched
2°–3°C			
30	6	E	100
60	11	G	82
90	10	G	80
120	57	G	72
150	53	G	64
180	21	F–G	0
188	13	F	15
5°–6°C			
30	8	E	50
60	31	G	47
90	19	G	89
120	17	G	94
150	15	G	87
180	66	G	83

[a] From Grice and Gibson, 1975, by permission of Springer-Verlag.
[b] Hatching was caused by warming the eggs to 18°C over a period of 10 days.
[c] Based on appearance at beginning of experiment: E, excellent; G, good, F, fair.

The observation of resting eggs from eight species in five genera of temperate neritic species (Zillioux and Gonzalez, 1972; Kasahara *et al.*, 1974; Grice and Gibson, 1975) suggests that the resting phenomenon may represent a widespread adaptation for surviving unfavorable conditions. Zillioux and Gonzalez (1972) list a number of species whose seasonal occurrence suggests the probability of resting eggs. It is possible that dormancy during periods of low temperature may represent greatly retarded normal development, but the experiment of Zillioux and Gonzalez (1972) suggests otherwise. Delayed hatching during exposure to high temperatures, as described by Kasahara *et al.* (1975a) for *Acartia clausi,* and possibly *Centropages abdominales,* clearly must represent true dormancy.

D. EFFICIENCIES OF ASSIMILATION OF FOOD

1. Measurement of Assimilation Efficiency

The requirements for food of an animal (R) can be expressed as the amount used for growth and reproduction (G), the amount used for metabolism (T), and the amount excreted and egested (E) (Richman, 1958).

$$R = G + T + E. \tag{1}$$

It is apparent that if any three parts of the equation above can be measured, the other can be determined by subtraction and the efficiency of assimilation (A) can be determined from

$$A = (R - E)/R. \tag{2}$$

To express assimilation efficiency as a percent, equation (2) is multiplied by 100. Because of the small size and fragility of many species of zooplankton, the measurement of three of the four parts of equation (1) has frequently not been possible. Conover (1966a) therefore devised an indirect method for estimating assimilation efficiency. Conover's method is based on the assumption that the ash present in the food of zooplankton is not assimilated. The percent ash content of the food and percent ash content of the feces were measured and the following relationship was derived:

$$A = \frac{(F^1 - E^1)}{(1 - E^1)\,(F^1)} \times 100 \tag{3}$$

where A is assimilation efficiency as percent, and F^1 and E^1 are the ratios of ash-free dry weight to dry weight of the food and feces, respectively. The principal utility of Conover's method is that it does not require quantitative measurement of any of the elements of equation (1). Conover compared estimates of assimilations based on equations (1) and (3) using dinoflagellate *Exuviella* sp. as food for 12 individual female *Calanus hyperboreus*. There was no significant difference between these methods (Conover, 1966a). Conover (1966b) found that the assimilation efficiency of *C. hyperboreus* was relatively unaffected by temperature, age of food culture, concentration of food, or the amount of food ingested. Foods with low ash contents were assimilated more efficiently than those with high ash contents, however. The constancy of assimilation efficiency over a range of cell concentrations (Fig. 7B) was particularly interesting, since the number of cells ingested was highest at intermediate concentrations in that experiment (Fig. 7A), an observation similar to those of Mullin (1963) and Haq (1967). The slight negative slope found by Conover (1966b) in the regression of percent assimilation versus cells ingested (Fig. 7C) was not significantly different from zero. Corner *et al.* (1967) calculated assimilation efficiencies of nitrogenous compounds of algae by *Calanus finmarchicus* and *Calanus helgolandicus* and compared them with efficiencies calculated by the method of Conover (1966a). They found good agreement when *Skeletonema costatum* was used as food, but often could not calculate efficiencies based on changes in nitrogen when *Brachiomonas submarina* and *Cricosphaera elongata* were used as food, because the nitrogen concentration of the particulate fraction of the experimental medium (food) frequently increased in spite of grazing by the copepods.

Parsons and Takahashi (1973) have suggested that assimilation of carbon can be accurately estimated by the method of Conover (1966a) but that estimates of assimilation of nitrogen and phosphorus might be less accurate. Butler *et al.* (1969, 1970) have shown that the percentage assimilation of nitrogen and phosphorus does differ as phosphorus turns over more rapidly in copepods. A method that employs the elemental ratios of nitrogen and phosphorus in the food excretory products, new animal tissue, and fecal pellets was used by Butler *et al.* (1970) to calculate assimilation efficiencies. Their method requires that accurate accounting be made of the nitrogen and phosphorus content of the experimental animals, their daily growth increments, and their daily excretion. In addition, the nitrogen and phosphorus content of the phytoplankton and feces and the net efficiency of growth in terms of nitrogen and phosphorus must be determined. Table VI shows the data used by Butler *et al.* (1970) to calculate assimilation efficiencies of nitrogen and phosphorus. The nitrogen to phosphorus ratios that are coded by letters in parentheses, i.e., a_1, a_2, a_3, a_4, and m, were used to calculate the efficiencies as follows (see Butler *et al.*, 1970, for details). An additional term,

$$k = \frac{1}{\mathrm{m}} \left(\frac{\mathrm{a}_3}{\mathrm{a}_1} \right) \tag{4}$$

is needed. If D_{p} is the assimilation efficiency for phosphorus in percent, then

$$D_{\mathrm{p}} = \frac{\mathrm{a}_1 - \mathrm{a}_4}{k\mathrm{a}_1 - \mathrm{a}_4} \times 100 \tag{5}$$

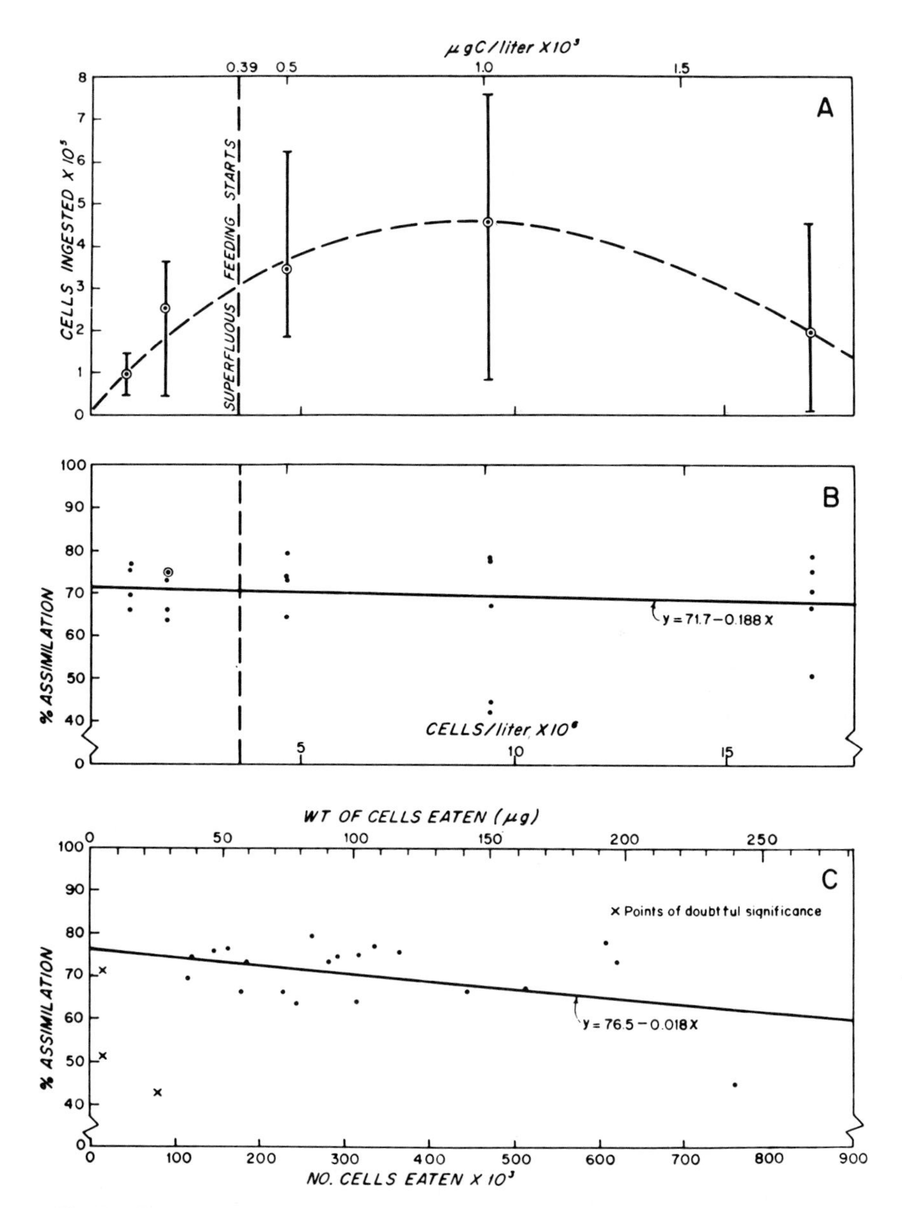

Fig. 7. Relationship between food concentration, total food ingested, and percentage of assimilation in *Calanus hyperboreus* feeding on *Thalassiosira fluviatilis;* (A) cells ingested at different concentrations, (B) percentage of assimilation in relation to cell concentration, (C) percentage of assimilation in relation to cells ingested. Points of doubtful significance were not included in statistical analysis. (From Conover, 1966b, by permission of the American Society of Limnology and Oceanography.)

TABLE VI

Data for Estimating Assimilation Efficiency of *Calanus* During Spring 1969[a]

	N	P	N:P ratio[b]
Average body content (μg)	23·8	2·25	10·60
Amount added daily (μg)	0·85	0·068	12·50 (a_3)
Amount excreted daily (μg)	1·14	0·238	4·78 (a_2)
Net growth efficiency (%)	42·7	22·3	1·91 (m)
Percentage body content used daily for growth and metabolism	8·4	13·6	—
Phytoplankton			8·05 (a_1)
Fecal pellets			13·19 (a_4)

[a] From Butler *et al.*, 1970, by permission.

[b] The following are coded ratios used to calculate the assimilation efficiencies of phosphorus and nitrogen (Eqs. 4, 5, and 6): a_1, the N:P ratio of the food; a_2, the N:P ratio of animal execratory products; a_3, the N:P ratio of animal growth; a_4, the N:P ratio of fecal pellets; and m, the net growth efficiency.

and the assimilation efficiency for nitrogen, D_n, is found by the following:

$$D_n = kD_p \tag{6}$$

The method by Butler *et al.* (1970) for calculating assimilation requires considerably more effort than that of Conover (1966a) but the added information gained is at times desirable.

Assimilation has been measured by feeding radioactively labeled foods and subsequently measuring the ratioactivity of zooplankton, their eggs, and fecal pellets (Marshall and Orr, 1955b, 1961; Lasker, 1960; Sorokin, 1966). Conover and Francis (1973) have noted a number of ways that the use of tracers can lead to incorrect conclusions. Their paper should be consulted for details, but it appears that more than one pool of an element can exist in an animal, each with different turnover rates, thus causing the outcome of labeling experiments to vary with the length of time that the animals are fed.

2. *Reported Efficiencies of Assimilation*

Beklemishev (1962) suggested that when food was abundant, copepods fed superfluously, i.e. either cell contents were spilled or assimilation efficiences decreased. Assimilation efficiencies calculated by him were in the range of 25–33%. Corner and Cowey (1968) reviewed much of the literature on assimilation efficiency through 1967. Their summary is presented in Table VII. While the observed values vary considerably, most were well over 50% and some as high as 99%. Conover (1964) reported similarly high values for marine zooplankton and somewhat lower ones for freshwater zooplankton. Conover's

TABLE VII

Assimilation of Phytoplankton by Various Zooplankton[a]

Species	Food	Method	Assimilation (%)	Reference
Calanus finmarchicus	Various diatoms and flagellates	tracer-isotope [^{32}P]	15–99	Marshall and Orr (1961)
	Skeletonema costatum	tracer-isotope [^{14}C]	60–78	Marshall and Orr (1955b)
Temora longicornis	*S. costatum*	tracer-isotope [^{32}P]	50–98	Berner (1962)
Euphausia pacifica	*Dunaniella primolecta*	tracer-isotope [^{14}C]	85–99	Lasker (1960)
Ostrea edulis (larvae)	*Isochrysis galbana*	tracer-isotope [^{32}P]	13–50	Walne (1965)
Calanus helgolandicus	natural particulate matter	chemical analyses	74–91	Corner (1961)
Metridia lucens	*T. nordensköldii*	"ratio" method	50–84	
	Ditylum spp.		35	Haq (1967)
	Artemia nauplii		59	
Calanus hyperboreus	*Exuviella* spp.	"ratio method"	39·0–85·6	
		chemical analyses	54·6–84·6	
Natural zooplankton	natural particulate material	"ratio" method	32·5–92·1	Conover (1966a)
Calanus finmarchicus	*Skeletonema costatum*	"ratio" method	53·8–64·4	Corner *et al.* (1967)
	S. costatum	chemical analyses	57·5–67·5	

[a] From Corner and Cowey, 1968, by permission.

(1966b) experiments, and those of Haq (1967), were done over a range of food concentrations that Beklemishev (1962) suggested should cause superfluous feeding. Conover and Haq both found that ingestion rates were highest at intermediate concentrations of food, as observed also by Mullin (1963), but that assimilation efficiencies were relatively constant and high (Conover, 1966b) (Fig. 7). The view of Beklemishev (1962) was partly supported by the fact that the ingestion rates upon which his calculations were based (30–50% of the copepod's body weight per day) seemed much higher than necessary to support the growth and metabolism of marine copepods. More recent studies have shown that the daily requirements for food by marine zooplankton range from a few percent to nearly 100% of their body weight (Mullin, 1963; Bell and Ward, 1970; Parsons and LeBrasseur, 1970; Sushchenya, 1970). Some neritic species may eat in excess of 100% of their body weight daily (Petipa, 1966; 1967; Heinle, 1974;

Heinle *et al.*, 1977). Even oceanic forms such as *Rhincalanus nasutus* and *Calanus helgolandicus* have growth rates that suggest daily rations well in excess of 50% (Mullin and Brooks, 1967, 1970a) or over 100% for certain instars (Petipa, 1967). For example, Mullin and Brooks (1970a) reported that *C. helgolandicus* copepodid IV to adults feeding on *Thalasiosira* had a daily growth coefficient of 37%. Assuming that assimilation efficiency was about 75%, the daily ration for growth alone was 49%. If metabolic requirements were in excess of 50% of the ration (Corner, 1961; Petipa, 1966; Sushchenya, 1970), the daily ration must have exceeded 100%. Corner (1972) calculated daily rations from the data of Mullin and Brooks (1970a) and found that for *R. nasutus* they were as high as 164% of body weight daily.

Conover (1966b) noted that assimilation efficiencie of *Calanus hyperboreus* was higher with some species of algae than with others. Conover found that those species with a lower ash content were assimilated with greater efficiency. Corner *et al.* (1972) found that the assimilation efficiency of *Calanus helgolandicus* was lowered when a large spiny diatom, *Biddulphia sinensis,* was used as food. Corner *et al.* (1972) contrasted the assimilation values that they observed of 34.1% for nitrogen and 40.4% for phosphorus with those of Marshall and Orr (1955b) of 71.2–94% for phosphorus and Corner *et al.* (1967) of 61.7% for nitrogen. Butler *et al.* (1970) found that assimilation of nitrogen averaged 62.5% and phosphorus averaged 77%. The reduced assimilation of *Biddulphia* by *Calanus* that Corner *et al.* (1972) observed was not caused by spilling of cell contents prior to ingestion since only ammonia-nitrogen could be detected in the experimental water. Butler *et al.* (1970) attributed the differing assimilation efficiencies for nitrogen and phosphorus to different rates of turnover of the two elements in the animal. It is relatively easy to see that if a copepod were actively storing lipid at the expense of catabolized plant protein, then the estimated assimilation efficiencies for carbon and nitrogen might differ greatly because of disproportionally high levels of excretion of nitrogen when comparing the method of Butler *et al.* (1970) with that of Conover (1966a).

Assimilation efficiencies of small coastal and estuarine planktonic species have not been measured directly. Petipa (1966) assumed that the assimilation efficiency of *Acartia tonsa* was 80% based on the work of Winberg (1956) and Marshall and Orr (1955b). Efficiency of assimilation of nitrogen by *Tigriopus brevicornis,* a benthic harpacticoid copepod, has been measured by Harris (1973). He found values of assimilation efficiencies that averaged 75.4%, quite similar to those observed for marine forms. Heinle *et al.* (1977) measured gross efficiencies of growth (as reproduction) (K_1 of Ivlev) of adult female *Eurytemora affinis,* a brackish and freshwater copepod from the Patuxent River estuary. They found K_1 to be 9–10% when the daily ration was over 200% of body weight. A K_1 of 17% was observed when the daily ration was reduced to 100% of body weight. These measurements can be compared with those reported from the

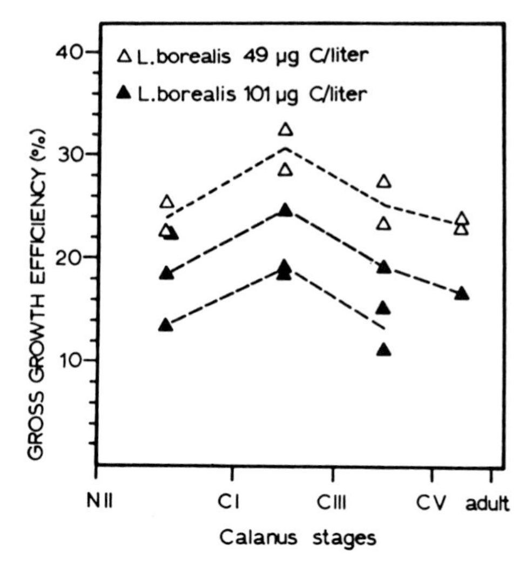

Fig. 8. Gross growth efficiency of *Calanus helgolandicus* feeding for various periods of its juvenile life on *Lauderia borealis* at different food concentrations. The data are based on μg C/liter organic carbon. The two experiments with *L. borealis* at 49 μg C/liter were run simultaneously and the results are combined. The data represented by the lower line at 101 μg C/liter are from one experiment that was split into two when the copepods attained CI. N = nauplii; C = Copepod. (From Paffenhöfer, 1976, by permission of the American Society of Limnology and Oceanography.)

literature by sushchenya (1970) for a number of freshwater and marine zooplankton. Marine forms, such as *Calanus helgolandicus* (Corner, 1961) and *Euphausia pacifica* (Lasker, 1960), had values of K_1 ranging from 37.2 to 48.9% and from 7.1 to 31.8%, respectively. Paffenhöfer (1976) found that K_1 varied between 11 and 31% for *C. helgolandicus* and was clearly dependent on the concentration and species of food and on the stage of development (Fig. 8). Thus, while the data vary considerably, it appears that coastal and estuarine zooplankton might be less efficient than marine forms (Conover, 1956; Petipa, 1966; Heinle *et al.*, 1977).

III. Predation and Competition

A. VERTICAL MIGRATION AND LIFE HISTORY

1. Energetic Gains from Vertical Migration Patterns

Predation and competition have been placed together here for two reasons. The adaptations to both are part physiological, behavioral, and morphological, and at times it may be difficult to distinguish which of these two selective forces is responsible for an adaptation.

The environmental stimulus for vertical migration, particularly diurnal migration, is frequently light (Cushing, 1951), sometimes modified by thermal or density gradients (Harder, 1968; Boyd, 1973). McLaren (1963) proposed a theory for the adaptive value of vertical migration that suggested an energetic advantage to migrants. He has more recently (1974) explicitly included predation in his calculations and expressed the advantages of migrating copepods in demographic terms. McLaren's papers should be consulted for details because his arguments are extensive, as is his review (1963) of the various theories of the adaptive value of vertical migration. An essential feature of McLaren's (1963) hypothesis is that the coefficients of weight-specific anabolism and catabolism are different functions of temperature. This hypothesis followed from the fact that the final size of copepods and other planktonic animals has frequently been observed to be a function of temperature (Deevey, 1960). If the coefficients of anabolism and catabolism do indeed have different temperature relationships, a net effect of temperature on a measure of efficiency of growth, K_1 or K_2, should be apparent. Data are available that suggest that efficiencies of growth may either vary (Conover, 1968) or remain relatively constant over a wide range of concentrations of food (Shushkina, 1968). However, most experiments appear to have been done under presumed optimum temperatures [see Corner and Davies (1971) and Parsons and Takahashi (1973) for a review of the literature on efficiencies of growth]. Mullin and Brooks (1970a) calculated gross efficiencies of growth (K_1) for *Rhincalanus nasutus* and *Calanus helgolandicus* at 10° and 15°C. The differences that they found for *R. nasutus* were not statistically significant. The values of K_1 for *C. helgolandicus* were nearly identical at the two temperatures. The most striking demonstration of an effect of temperature on K_1 is that of Kersting (1973), who worked with *Daphnia*. He observed maximum values of K_1 (near 40%) at about 8°C with decreasing values at higher or lower temperatures. If marine copepods can indeed acquire their daily ration by intermittent feeding [e.g., *Metridia lucens* feeding carnivorously (Haq, 1967), *Acartia tonsa* migrating diurnally (Petipa, 1966), *A. tonsa* and *A. clausi* migrating slightly (Conover, 1956), or *A. tonsa* not migrating (Heinle, 1974)], a demonstration of a temperature effect on K_1, like that found by Kersting (1973) for *Daphnia* would support McLaren's (1963) hypothesis that metabolizing food at temperatures different from that at which food is collected would enhance feeding efficiencies. The changes in feeding rates and respiration that Anraku (1964a) found for several marine copepods suggest that K_1 might indeed vary with temperature.

There is some evidence that vertical migration may not require large amounts of energy (Hutchinson, 1967; Vlymen, 1970), as was previously assumed (e.g., see Petipa, 1966). If the hypothesis of Ackman *et al.* (1974) that wax esters provide buoyancy and food reserves is correct, then the energy requirements for vertical migration might be negligible since Vlymen's (1970) calculations for horizontally swimming *Labidocera trispanosa* were based on neutral buoyancy.

The energetics of copepod swimming is presently the center of an unresolved controversy (Enright, 1977a,b; Vlymen, 1977; Strickler, 1977; Lehman, 1977).

Kerfoot (1970) has suggested that vertical migration is the inevitable outcome of the use of light as a frame of reference (Harris, 1953). Kerfoot's (1970) calculations also indicate that, for the natural population of *Calanus finmarchicus* cited, the use of light and the resulting vertical distributions place the copepods in depth strata where the amount of food available (as primary production) is maximum. He suggested that the adaptive value of the use of light as a means of orientation is particularly high at higher latitudes because noctural feeders have longer daily periods of feeding during the winter when food is less abundant. As McLaren (1974) noted, the hypothesis of Kerfoot (1970) does not fit all of the patterns that have been observed, particularly the seasonal variations in vertical migration (Marshall and Orr, 1955a). Exceptions are also found among species with restricted life cycles, like *Calanus plumchrus* (Fulton, 1973).

2. Avoidance of Predators

LeBrasseur *et al.* (1969) suggested that the restricted life cycle of *Calanus plumchrus* (Fulton, 1973), including an extended absence from surface waters, might protect that species from size-selective predation by juvenile fish. LeBrasseur *et al.* (1969) discussed the possibility that the presence of other species that were preferred prey for juvenile fish may have actually provided some protection for the developing *C. plumchrus*. The inclusion of differential mortalities in McLaren's (1974) hypothesis adds strength to his arguments, since absence from surface waters during the day helps avoid visual predation as suggested by Hutchinson (1967).

B. REPRODUCTIVE STRATEGIES

1. Intrinsic Rates of Increase

Allan (1976) reviewed the life histories of the major groups of freshwater zooplankton, with some examples of marine species, in light of the relative magnitudes of competition among and predation upon these groups. He observed that the life history strategies of the groups (rotifers, cladocerans, and copepods) often involve differing balances of reproductive potential, competitive ability, and predator evasion. While cladocerans and rotifers are not notably successful marine groups, some species are abundant in estuaries (Krazhan, 1971; Rodriguez, 1973; Bosch and Taylor, 1973). Allan (1976) observed that the reproductive potential, defined is the maximum intrinsic rate of increase (r_{max}) (Birch, 1948), was highest among the rotifers and least among the copepods at any temperature. It appears likely that variation within groups is large also. For example, Allan (citing Edmondson *et al.*, 1962) reported instantaneous rates of birth (in theory, nearly the same as r_{max}) of 0.25–0.4 per day for three species of freshwater copepods. Heinle (1969b) observed productivities (instantaneous rates of increase of biomass) of up to 0.4 per day for *Eurytemora affinis* and up to 0.8 per day for *Acartia tonsa*. Since productivities might also be similar to r_{max}, those relatively high rates can be contrasted to the range of 0.5–0.15 per day for *Pseudocalanus minutus* that Allan (1976) calculated from the data of McLaren (1974). Allan (1976) noted that r_{max} for any species was relatively more sensitive to changes in the time to first reproduction than to changes in duration of reproductive period or to the total numbers of offspring produced (see also Birch, 1948; Leslie, 1945). Highest rates of reproduction are thus achieved by animals that require the shortest time to reach sexual maturity. Coastal and estuarine species in general mature more rapidly than oceanic species. Coastal species mature in days to weeks (Conover, 1956; Jacobs, 1961; Zillioux and Wilson, 1966; Heinle, 1966, 1969a; Petipa, 1966, 1967; Katona, 1970; Heinle and Flemer, 1975) whereas oceanic species mature in weeks to months (Marshall and Orr, 1955a; Conover, 1962; Mullin and Brooks, 1967, 1970a,b; Katona and Moodie, 1969; Corkett, 1970; Nassogne, 1970; Paffenhöfer, 1970). Omori (1973) should be consulted for a review of the literature on culture of copepods. One can thus generalize that reproductive rates are higher and efficiencies of assimilation lower (Section II,D,2) in coastal and estuarine species than in oceanic species.

2. *Changes in Sex Ratio*

Allan (1976) noted that in addition to short generation times, cladocerans and rotifers achieve relatively higher rates of reproduction than copepods by parthenogenetic reproduction. Alteration of sex ratios may be a way in which copepods gain added reproductive potential during critical periods (Heinle, 1970). Heinle found that the proportion of females in laboratory and wild populations of *Acartia* (Fig. 9) was higher when the densities of the populations were lower. Eggs from the same population raised in isolation on natural food resulted in a sex ratio of unity over a wide range of densities (Fig. 10). Heinle proposed that the genetic sex ratio was indeed unity, and that a portion of the potential males developed into functional females in response to low population densities in the wild populations and populations maintained in the laboratory. There was no indication of early death or retarded development of males as found for other species (Campbell, 1934; Bogorov, 1939). Heinle's (1970) data suggested that the density of the older instars (copepodid I–VI) had the greatest influence on the sex ratio. Some effect on younger instars was proposed because *Acartia tonsa* from the wild population reared as cohorts had sex ratios near unity at most densities (Fig. 10). A high proportion of females would be highly adaptive, since the females of that species can lay eggs for up to 2 weeks after a copulation

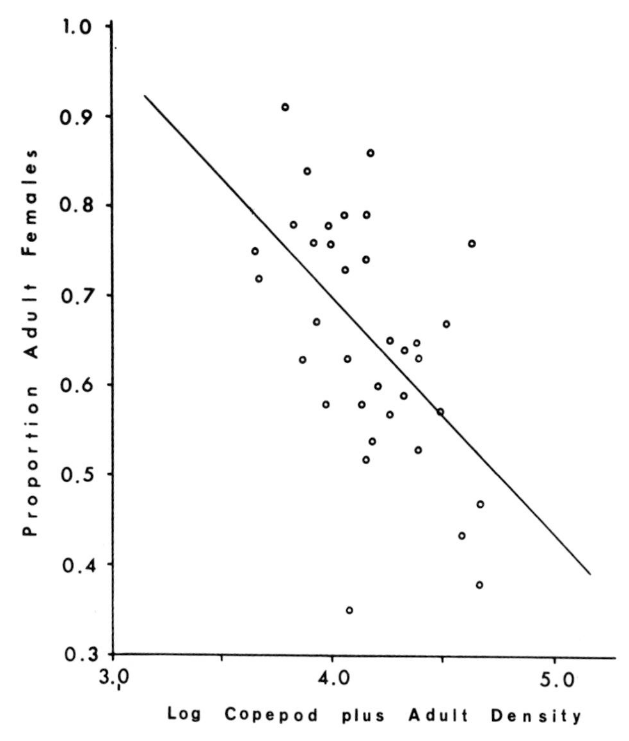

Fig. 9. Proportion of adult female *Acartia tonsa* versus logarithm of copepodid plus adult density (numbers/m^3) in samples taken from the Patuxent estuary during June–September. Proportion of females = (1.754−0.264) × (log density of adults plus copepodids). (From Heinle, 1970, by permission of Boyens and Co.)

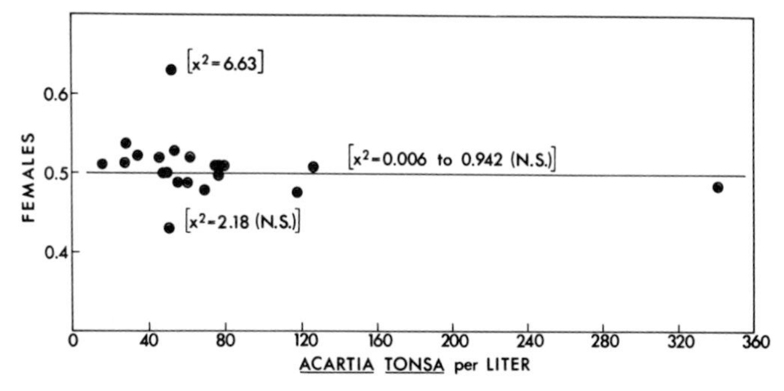

Fig. 10. Sex ratios of *Acartia tonsa* raised from eggs collected in the Patuxent estuary. Deviations from 0.5 were tested by chi-square (N.S. indicates a nonsignificant value of chi-square). (From Table I of Heinle, 1970.)

(Wilson and Parrish, 1971). Mednikov (1961) first suggested that increased proportions of females among calanoid copepods was a general adaptation for survival at low population densities. Bayly (1965) studied three sympatric species in an estuary and found that only the species that were scarce at a particular time had a higher proportion of females. Heinle (1970) also suggested that the sex ratio of *Acartia tonsa* was altered in response to the density of that species (but not in response to the density of others).

Several environmental stimuli have been shown to alter sex ratios of copepods in the laboratory. Katona (1970) found that temperatures affected sex ratios of *Eurytemora affinis* and *Eurytemora herdmani*. The sex ratio in these species are frequently weighed in favor of males, and, since copulation is required before production of each egg sac (Heinle, 1969a; Heinle and Flemer, 1975), Katona's (1970) suggestion that production of surplus males assures copulation during unfavorable conditions may be correct. Paffenhöfer (1970) found an effect of food on the sex ratio of *Calanus helgolandicus,* but mortalities were about 50% among growing instars in his cultures, as in those of Mullin and Brooks (1970a), who noted only mature females in their cultures of *C. helgolandicus*. Hydrostatic pressure has also been shown to affect changes in sex ratios of *Tigriopus californicus* by Vacquier and Belser (1965). Their results were particularly interesting as they showed that sex is determined prior to the copepodid stages in *Tigriopus*.

Although a number of stimuli have been shown to alter sex ratios in the laboratory, one should realize that the physical stimuli, temperature and pressure, may act directly on the physiological mechanisms for sex determination, possibly including a hormonal system (Carlisle, 1965). These stimuli may thus give little insight concerning the adaptive significance of altered sex ratios. The biological stimuli, poor food quality, and low population densities may occur together in nature (Mednikov, 1961), but the evidence for *Acartia tonsa* (Heinle, 1970) and the three species studied by Bayly (1965) suggests that the true adaptive value lies in the increased reproductive potential of a population with proportionally more females.

IV. Temperature

A. EFFECTS ON RATES OF PROCESSES

1. Descriptive Equations

McLaren (1963) proposed that Bělehrádek's (1935) equation

$$V = a\,(T + \alpha)^{b} \tag{7}$$

could be used to describe the relationship between the velocity of a process V and temperature T. This equation plots as a straight line on a log-log scale.

The constant a is a scale correction related to the units of V and describes the Y-intercept on a log-log plot; b is the slope of a log-log plot and describes the curvature of an arithmetic plot; and α describes a biological zero, the temperature at which V becomes zero. McLaren (1963) showed that the time required for development, length of life, and adult size of various species of zooplankton could all be described adequately in Bělehrádek's equation. He later showed (McLaren, 1966) that the development times of several species of copepods were very accurately described by equation (7) (Fig. 11). McLaren (1966) further suggested that the constant b in the equation might be truly constant with a value of about -1.68. When b was assumed to be -1.68, the fitted values of α were

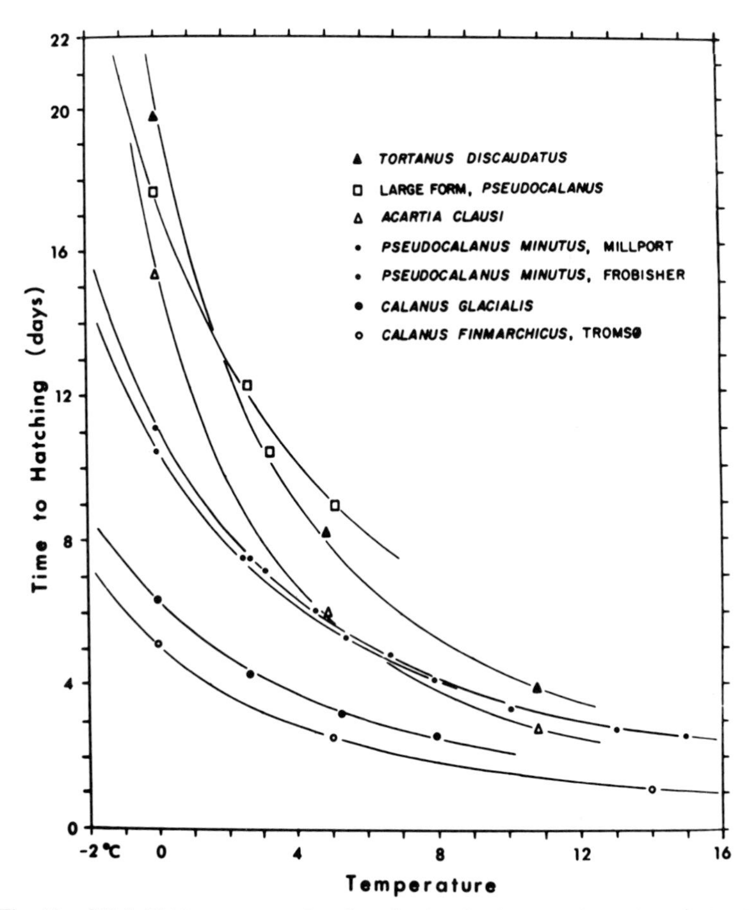

Fig. 11. Bělehrádek's temperature functions fitted to development times of copepod embryos. The parameter b is taken as -1.68 for all curves. (From McLaren, 1966, by permission of the Marine Biological Laboratory, Woods Hole.)

more interpretable, cold-water species having lower values than species living in warmer waters. McLaren also observed a relationship between the constant a and the diameter of eggs within a species or between closely related species.

Bernard (1970) noted that plots of copepod egg-hatching rates versus temperature frequently display asymptotes or inflections (Fig. 12). She found that data of this sort, which are common for tropical or subtropical species, were best described by an elliptical function shown in Fig. 13(3), described by the formula

$$(X_1 - X_0)^2 \left(\frac{1}{a^2} + \frac{K^2}{b^2} \right) + 2(X - X_0)(Y - Y_0)\, K \left(\frac{1}{a^2} + \frac{1}{b^2} \right)$$

$$+ (Y - Y_0) \left(\frac{K}{a^2} + \frac{1}{b^2} \right) - K^2 = 1 \qquad (8)$$

In formula (8), Y equals the time required for a process, X equals temperature, and K is an angular coefficient ($\tan \alpha$). Parts 1 and 2 of Fig. 13 show graphically the alteration of the basic formula of an ellipse to that of equation (8). The intercept of the vertical tagent X_1, in part 3 of Fig. 13, represents biological zero, analogous to α in Bělehrádek's formula (7), and the temperature range X_1 to X_2 defines the vital limits of a species. The temperature at which the rate of a

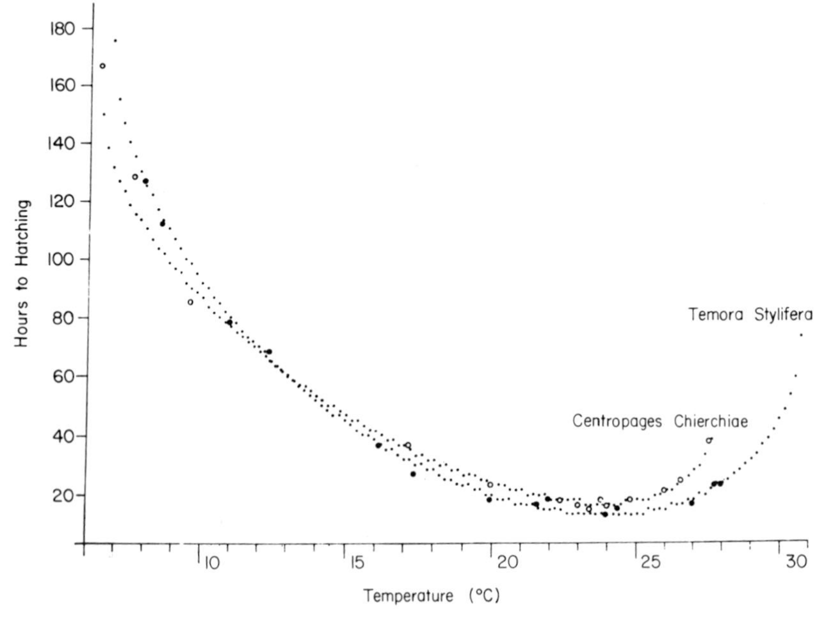

Fig. 12. Theoretical curves of the elliptical function fitted to hatching times (hr) versus temperature (°C) of *Temora stylifera* (●) and *Centropages chierchioe* (○). Calculated (fitted) curves are shown by small dots. (From Bernard, 1970, by permission of le Institut Océanographique.)

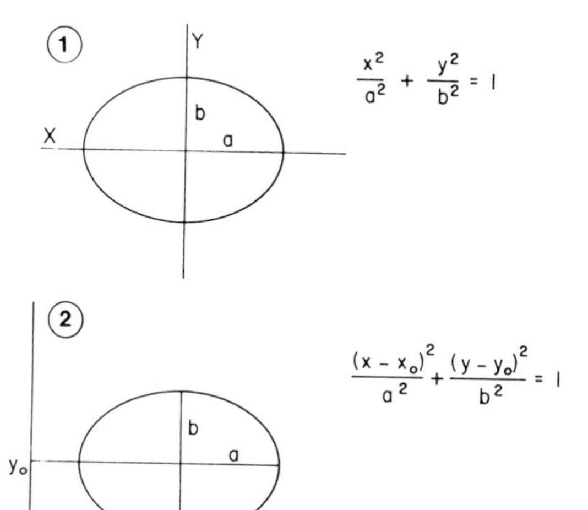

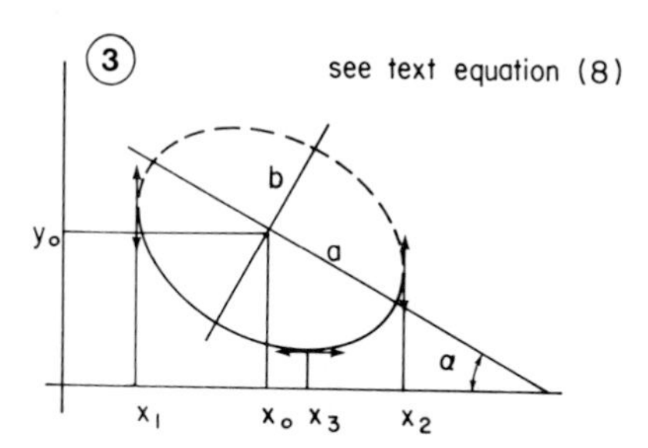

Fig. 13. Diagramatic illustration of the modification of the basic equation for an ellipse showing the variables in text equation (8). (From Bernard, 1970, by permission of l'Institut Océanographique.)

process is maximum (X_3) is the theoretical optimum for a species. Bernard suggested that the coefficient of the long axis K was a measure of thermal sensitivity, somewhat analogous to a in Bělehrädek's equation. It is possible that K is a quantitative measure of acclimation.

When data on rates of biological processes or their reciprocals are plotted as natural logarithms versus temperature (an Arrhenius plot), inflections in the curves are sometimes noted at about 15° and 30°C (Figs. 14 and 15). Growth rates and frequency of molting of two copepods plotted against temperature show inflections or changes in slope at about 15° and 30°C (Fig. 14). Vernberg and Vernberg (1972) have shown that plots of respiration rates of some tissues excised from *Uca pugnax* acclimated to cold and warm temperatures had

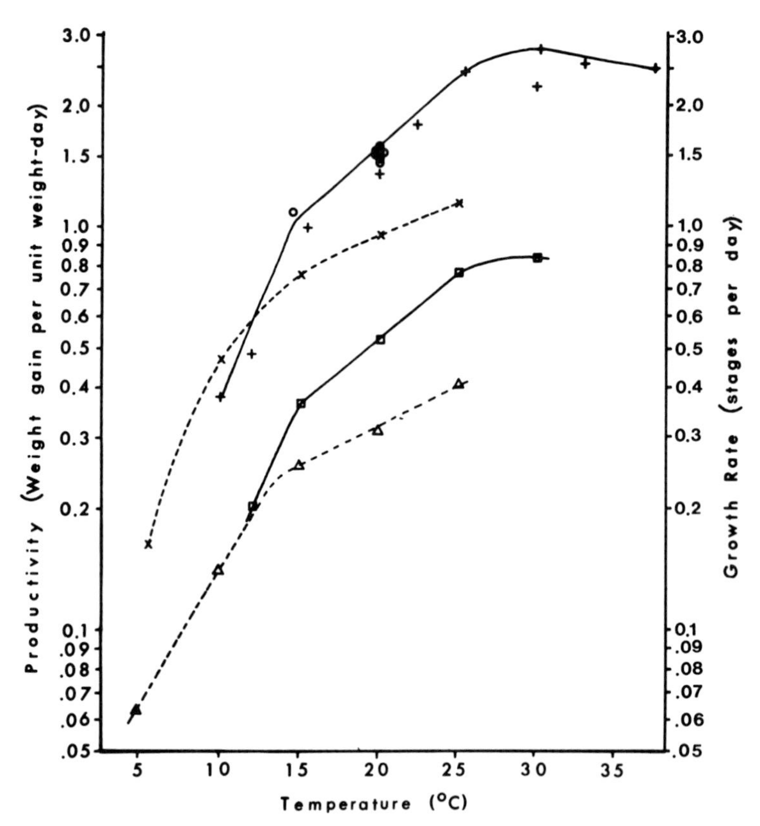

Fig. 14. Growth rates (upper two curves) and productivity (lower two curves) of *Acartia tonsa* (solid curves) and *Eurytemora affinis* (broken curves) plotted against temperature. The logarithmic scale is the same for growth rate and productivity. The curves are fitted to the data points by eye. (From Heinle, 1969b, by permission of the University of Maryland, Chesapeake Biological Laboratory.)

changes in direction of curvature at about 15° and 30°C. Not all tissues responded to changes in temperature in the same way, however, and the ability to acclimate to cold and warm temperatures varied among tissues. An important point made by Vernberg and Vernberg (1972) was that the integrated (whole body) response of the organism does not explain the effects of temperature at the tissue or cellular level. In a not widely known paper, Drost-Hansen (1969) has suggested that such inflections may be due to abrupt changes that occur in the physical properties of water and aqueous solutions near interfaces. Drost-Hansen suggested that the inflection points might represent thermal boundaries defining the limits of adaptation for some organisms. Thermal boundaries might also be analogous to resistance and capacity adaptation of Precht (1958). The semi-log plots of rate versus temperature have a value in that the slopes of the linear portions represent changes of rate with temperature (e.g., Q_{10}) and are thus useful for purposes of comparison.

All three methods of describing changes in rate with temperature are fitted well by certain experimental data. The definition of a biological zero, α in equation

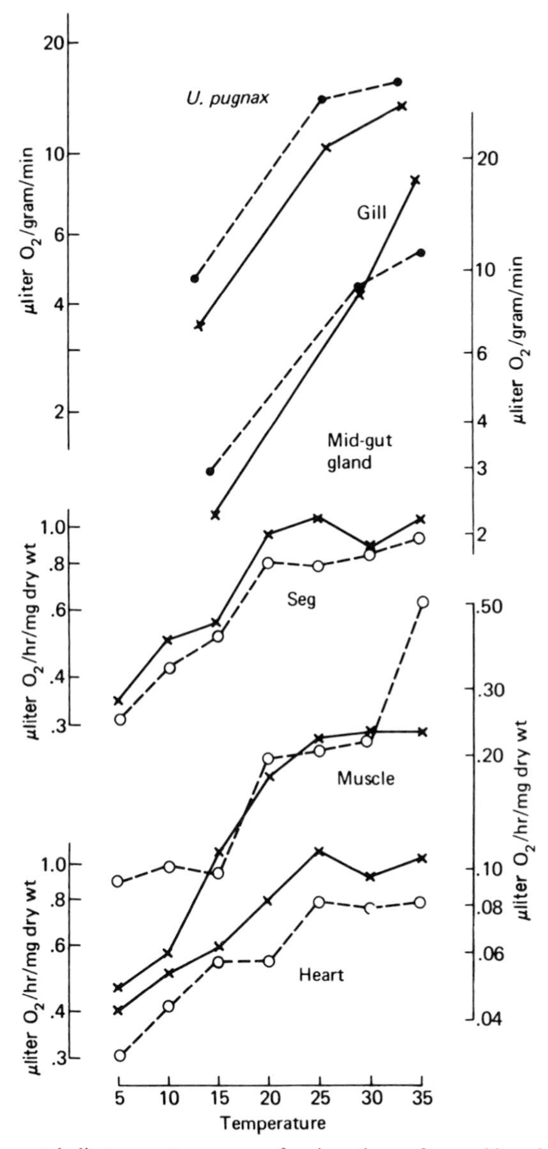

Fig. 15. The metabolic-temperature curves of various tissues from cold- and warm-acclimated *Uca pugnax*. Metabolism measured as μliter O_2/hr/mg dry wt. (From Vernberg and Vernberg, 1972, by permission of Springer-Verlag).

(7) and X_1 in equation (8) has considerable utility, as does the definition of vital thermal range by the elliptical expression of Bernard (1970). McLaren's (1963, 1966) and Bernard's (1970) papers should be consulted for review of the considerable physiological basis for equations (7) and (8). The observations of Drost-Hansen (1969) suggest that information might be lost by masking the inflection points of a semi-log representation (Figs. 14 and 15) by fitting data to other equations.

Bottrell (1975) has proposed a fourth method for describing the effects of temperature, involving a reduced quadratic equation. While an excellent data fit was achieved, the biological meaning of the parameters of the quadratic are obscure.

B. THERMAL ADAPTATIONS

1. Genetic Adaptation

It has been suggested that the main mechanism used by aquatic invertebrates for compensating for changes in temperature is metabolic adaptation (non-genetic) (Kinne, 1964a, 1967; Gilles, 1975a). There is however some evidence for genetic adaptation as well (Bradley, 1975a,b, 1976, 1978a,b). Bradley (1975b, 1976) used a shock-recovery technique that allowed repeated testing of individual *Eurytemora affinis,* and more importantly, the breeding of copepods with known thermal tolerances. The evidence for genetic adaptation is not yet complete, but Bradley (1975a,b) found that the thermal tolerance of laboratory populations of *E. affinis* collected from the Patuxent Estuary at two different times (March and August) differed significantly. The "August" population had a higher thermal tolerance at each of the salinities used in the assay even though the two populations were grown under identical conditions in the laboratory. Bradley (1975a) found the correlation of thermal tolerance of full siblings to be 0.5 when 12 broods were raised separately and five individuals from each brood tested. Bradley concluded that the heritability estimate of 50% indicated considerable genetic variance. In more extensive studies, Bradley (1978b) found heritability of tolerance of elevated temperatures to be 0.11 among females and 0.89 among males. Females, however, displayed greater physiological adaptation. Bradley (1978a) suggested that the high amount of genetic adaptation was somewhat surprising in light of the previously known range of physiological adaptation and current genetic theory. Bradley (1978a) also demonstrated an interaction between temperature and salinity tolerance in *E. affinis*. Bradley (1975a) was also able to demonstrate an increase in thermal tolerance following one generation of parental selection. A fourth indication of genetic adaptation described by Bradley was the observation that the progeny of *E. affinis* collected from the effluent of a power plant had higher thermal tolerances than the progeny of copepods collected from the intake on the same day. Animals with lower thermal tolerances experienced higher mortalities in passing through the cooling condenser.

Another indication of genetic adaptation can be found in studies of *Acartia tonsa,* a species that, like *Eurytemora affinis,* occurs over a wide range of temperatures. The upper thermal limit of *A. tonsa* from Biscayne Bay, Florida, was in excess of 35°C (Reeve and Cosper, 1970), whereas *A. tonsa* from Chesapeake Bay cultured at 25°C survived exposure to 30°C but not 35°C (Heinle, 1969b). Heinle (1969b) and Reeve and Cosper (1970) found that acclimation to higher temperature increased the thermal tolerance of *A. tonsa*. Gonzalez (1974) compared the critical thermal maxima (temperatures at which the locomotion of animals became disorganized) and upper lethal temperatures of *A. tonsa* from Mount Hope Bay, Rhode Island, Biscayne Bay, Florida, and Bahía Fosforescente, Puerto Rico. He found that acclimation (non-genetic adaptation) affected both variables. One-half of the *A. tonsa* from Puerto Rico acclimated (by culturing) at 27°C tolerated a temperature of 37°C for 4 hr. The regional differences in thermal tolerance led Gonzalez to the conclusion that *A. tonsa* was capable of genetic adaptation in addition to the non-genetic adaptation that he observed (changes in tolerance following acclimation).

The differences in thermal metabolic patterns observed by Moreira and Vernberg (1968) for dimorphic males of *Euterpina acutifrons* suggest genetical adaptation of unknown functional significance. They found that cold-acclimated small males had much lower metabolic rates than warm-acclimated small males at 15°C, although metabolism was similar at 25°C. In contrast, warm-acclimated large males had lower metabolic rates at 25°C than cold-acclimated large males did, but metabolism at 15°C was similar. These data suggest that one morph is genetically adapted to low temperatures while the other is genetically adapted to high temperatures. Since the ecological significance of low metabolic rates is unknown for this species we can not tell which morph is better adapted to a particular temperature.

The constant in Bělehrádek's equation that describes biological zero [α in equation (7)] has been found to vary with latitude (McLaren, 1966; McLaren *et al.*, 1969). McLaren *et al.* (1969) suggest that estimates of α are the simplest expressions of temperature adaptation. They found a nearly linear relationship between α and average environmental temperature, using data on hatching times of 11 species of copepods (Fig. 16). An important observation of McLaren *et al.* (1969) is that over a wide geographic (and temperature) range, α varied less than 1°C for separate populations of *Pseudocalanus minutus* and *Calanus finmarchicus*. They suggested that α was genetically adapted to conditions over the entire range of a species. These data indicate that non-genetic adaptation to low temperatures may not be possible.

2. *Non-Genetic Adaptation*

Landry (1975b) observed an effect of acclimation temperatures (warm, 18°–20°C; cold, 8°–10°C) of female *Acartia clausi* on the subsequent development rates of their eggs. At 10°C, development times of the eggs of warm- and

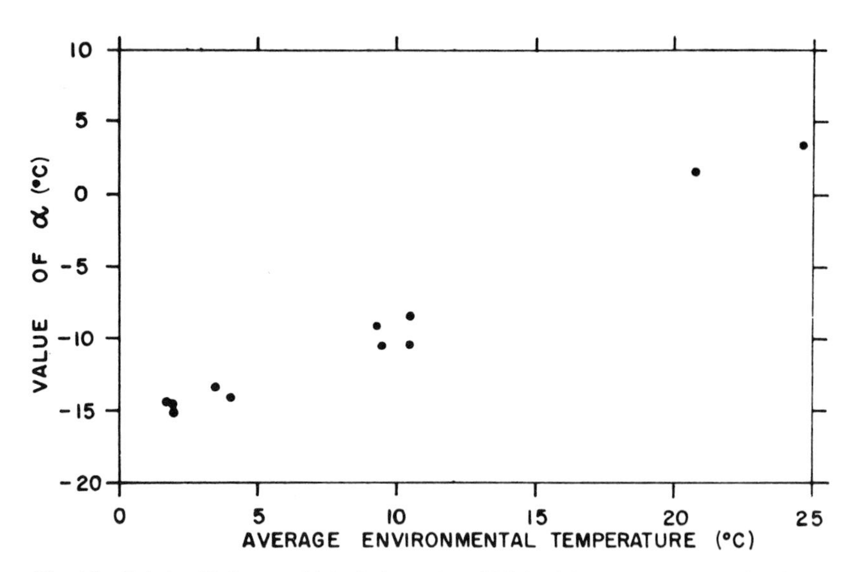

Fig. 16. Relationship between biological zero (α) of Bělehrádek's temperature function for eggs of 11 species of copepods and estimates of average temperature in their environmental ranges. (From McLaren *et al.*, 1969, by permission of the Marine Biological Laboratory, Woods Hole.)

cold-acclimated females were about equal. At higher temperatures however, eggs of cold-acclimated females developed more rapidly (Fig. 17). These results suggest the possibility of irreversible phenotypic adaptation, which could be easily confused with genetic adaptation. Bradley (1978a) observed a similar phenomenon in *E. affinis,* irreversible acclimation being acquired by growing juveniles. The results shown by Landry rather clearly suggest seasonal temperature compensation. It is tempting to speculate that the shift in the curves of Landry (Fig. 17) might be described by a change in the coefficient K in the elliptical function of Bernard (1970), equation (7). Landry (1975b) suggested that egg size did not vary in his experiments. It should be noted, however, that Corkett (1970) found that larger eggs of *Calanus* took longer to develop than small eggs. It has also been suggested that the lipid content of eggs might affect development rates (McLaren, 1966). Hart and McLaren (1978) more fully explored the effects of size of eggs, the length of females (and males since assortive mating occurred), and acclimation on the duration of embryonic development of *Pseudocalanus* sp. They suggested that factors other than acclimation were more important for temperature compensation in that species.

The studies cited above all involved the measurement of growth or developmental variables that are the net result of metabolic processes. Work at the molecular level with higher organisms indicates that the actual physiological processes of thermal regulation (non-genetic adaptation) are complex, involving

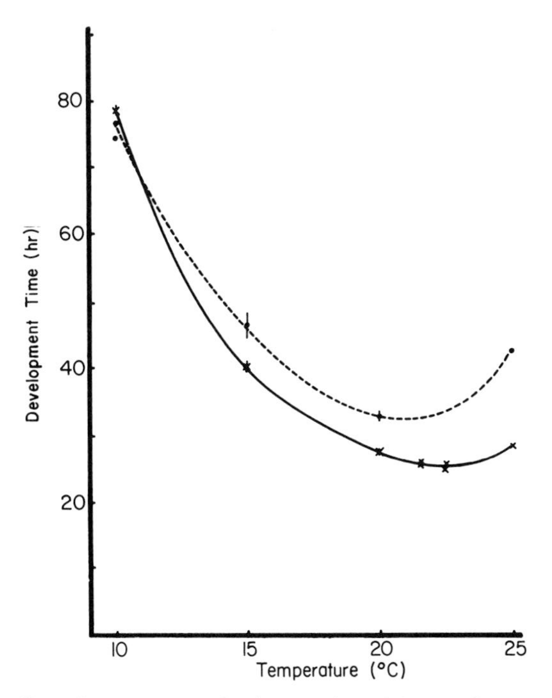

Fig. 17. The effect of temperature on development time of *Acartia clausi* eggs for two seasonal groups. Winter-collected animals, ——; summer-collected animals, - - - . Vertical lines denote 95% confidence intervals. (From Landry, 1975b, by permission of the American Society of Limnology and Oceanography.)

shifts in metabolic pathways (Gilles, 1975a). The metabolic changes with varying temperatures are not simply changes in rates of reactions, but involve activation and inhibition of metabolic pathways. In particular, changes have been noted in the relative activity of the glycolytic pathway and the pentose-phosphate shunt (Gilles, 1975a). An example of typical evidence for non-genetic adaptation is shown in Fig. 18 taken from Mihursky and Kennedy (1967). There is a clear effect of acclimation temperatures on the thermal tolerance of the crustacean, *Neomysis americana*. Animals acclimated to successively higher temperatures had respectively higher thermal tolerances. with the exception of those acclimated at 5° and 10°C, which had identical thermal tolerances. While several reviews are available that discuss thermal acclimation (e.g., Kinne, 1964a,b, 1967, 1970; Vernberg and Vernberg, 1972; Gilles, 1975a), little experimental work has been done on zooplankton beyond that already cited in this section and in Section IV,B,1. Seasonal changes in rates of respiration (e.g., Conover, 1959; Marshall and Orr, 1966) do not provide separate evidence for non-genetic adaptation. Kinne (1970) has cautioned that considerable care should be exercised in

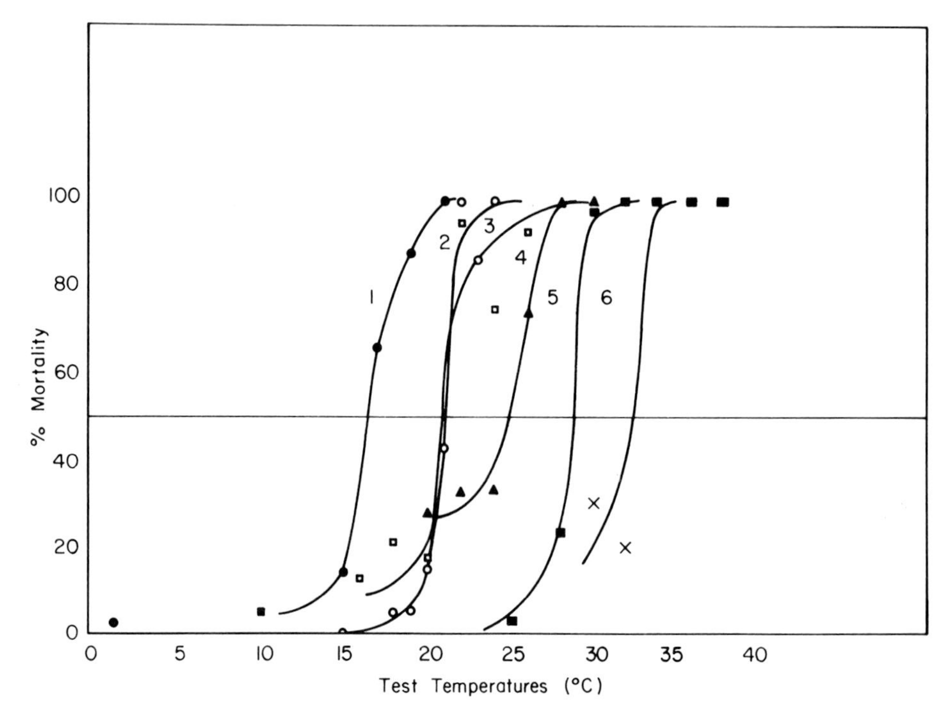

Fig. 18. Temperature mortality curves for *Neomysis americana*. (From Mihursky and Kennedy, 1967, by permission of the American Fisheries Society.)

the conduct and interpretation of experiments in which the consumption of oxygen is the dependent variable and temperature or salinity the independent variables. Kinne noted that respiration by some animals is affected by the partial pressure of oxygen in solution which is directly related to temperature and salinity. In addition, crowding of experimental animals may also affect oxygen concentrations. Some species show considerable tolerance to low concentrations of oxygen (Vargo, 1974), and Kinne (1970) has indicated that there are oxygen conformers, regulators, and partial regulators. Oxygen conformers reduce their consumption of oxygen in response to lowered concentrations whereas regulators do not. Partial regulators are intermediate between conformers and regulators.

Future studies of the effects of temperature on respiration of marine zooplankton should eventually provide insight to the mechanisms of non-genetic adaptation. The possibility that adaptation to extreme low temperatures may be primarily genetic, whereas adaptation to high temperatures may be both genetic and non-genetic, merits further study.

V. Salinity

A. EFFECTS ON RATES OF PROCESSES

1. Effects at the Molecular Level

Changes in salinity alter some metabolic pathways in much the same way that changes in temperature do, especially pathways of energy metabolism (Maetz, 1974; Gilles, 1975b). For that reason, much of the literature on effects of salinity on marine organisms reports combined effects of temperature and salinity (Ranade, 1957; Costlow *et al.*, 1960, 1962, 1966; Lance, 1963; Kinne, 1964a,b, 1967, 1970; Battaglia and Bryan, 1964; Gilfillan, 1972; Vernberg and Vernberg, 1972; Bradley, 1975a,b). Results of such studies can often be represented as iso-response lines plotted against temperature and salinity (e.g., Costlow *et al.*, 1960) or as three-dimensional response surfaces (e.g., McLeese, 1956). Reviews such as Kinne (1964a,b), Maetz (1974), and Gilles (1975b) should be consulted for details of what has been described at the metabolic level, based mainly on work with larger invertebrates and fish. Important points are (1) the relative roles of the glycolytic pathway and the pentose-phosphate shunt are possibly affected by salinity (Gilles, 1975b); (2) the rates of metabolism of amino acids and consequent excretion of ammonia are also altered (Gilles, 1975b), probably directly affecting the exchange of sodium in hypoosmotic environments (Maetz, 1974); and (3) amino acids are heavily involved in intracellular osmotic balance (Gilles, 1975b). The small size of most marine zooplankton will probably preclude direct measurement of the metabolism of specific organs or tissues for some time. The understanding of osmotic effects will therefore have to come from comparison of metabolic effects on whole organisms with the effects on larger invertebrates whose metabolism is more amenable to study.

2. Effects on Rates of Whole Animals

Most of the studies of salinity effects on marine zooplankton have involved neritic or estuarine species, since they are subjected to the greatest variations in salinity. It has been demonstrated that two euryhaline copepods, *Acartia tonsa* and *Centropages hamatus,* are classic osmotic conformers. Their hemolymph remains isomotic or slightly hyperosmotic over a wide range of salinities (Lance, 1965; Bayly, 1969). Bayly (1969) also found that a copepod from inland saline waters, *Calamoecia salina,* and a fresh-to-brackish water species, *Limnocalanus macrurus,* are also osmoconformers. Lance (1965) observed a linear relationship between the percent salts (NaC1) in the external medium and in the body fluids of *A. tonsa* (Fig. 19). The highest concentrations shown represent 100% seawater (salinity 36.4‰) while the lowest represent 15% seawater. Lance found that the body fluids reached stable osmotic concentrations within 1–2 hr after transfer from 100% seawater to 50% seawater. While it is not clear from her paper (Lance, 1965), the copepods were presumably acclimated stepwise to lower salinities as described in an earlier paper (Lance, 1963). Since only sodium chloride was measured, the precise tonicity of the hemolymph cannot be deduced from Fig. 19. Brand and Bayly (1971) have demonstrated that three other estuarine species, *Gladioferens pectinatus, G. spinosus,* and *Sulcanus conflictus,* are hypo- and hyperosmotic regulators. A freshwater species sometimes found in inland saline waters with high carbonate and biocarbonate concentrations (Bayly, 1969), *Boeckella triarticulata,* maintained its body fluid in a hyperosmotic state at low salinities and was isomotic at higher salinities. The concentration of the hemolymph and external medium, expressed as freezing point depression, for the four species studied by Brand and Bayly (1971) (Fig. 20) can be contrasted with the results of Lance (1965) (Fig. 19). Brand and Bayly (1971) found that the three estuarine species that were osmotic regulators became isomotic between 13 and 18‰. The freezing point depressions found for *L. macrurus* and *C. salina* in freshwater were also included in Fig. 21 (Bayly, 1969). Battaglia and Bryan (1964) found what appeared to be slight hypoosmotic regulation by Tisbe re-

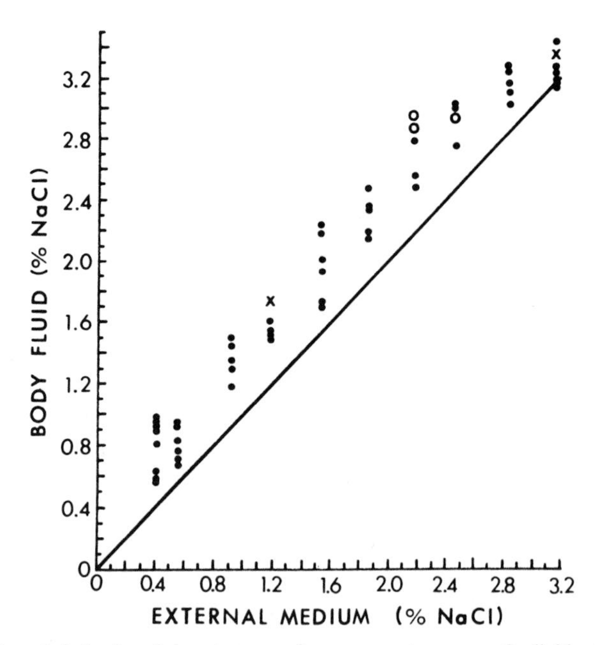

Fig. 19. Osmotic behavior of *Acartia tonsa* after exposure to a range of salinities for 12 hr. The straight line represents the relationship where the body fluid is isosmotic with the external medium; ●, single datum; ○, two identical data; x, three identical data. (From Lance, 1965, by permission of Pergamon Press.)

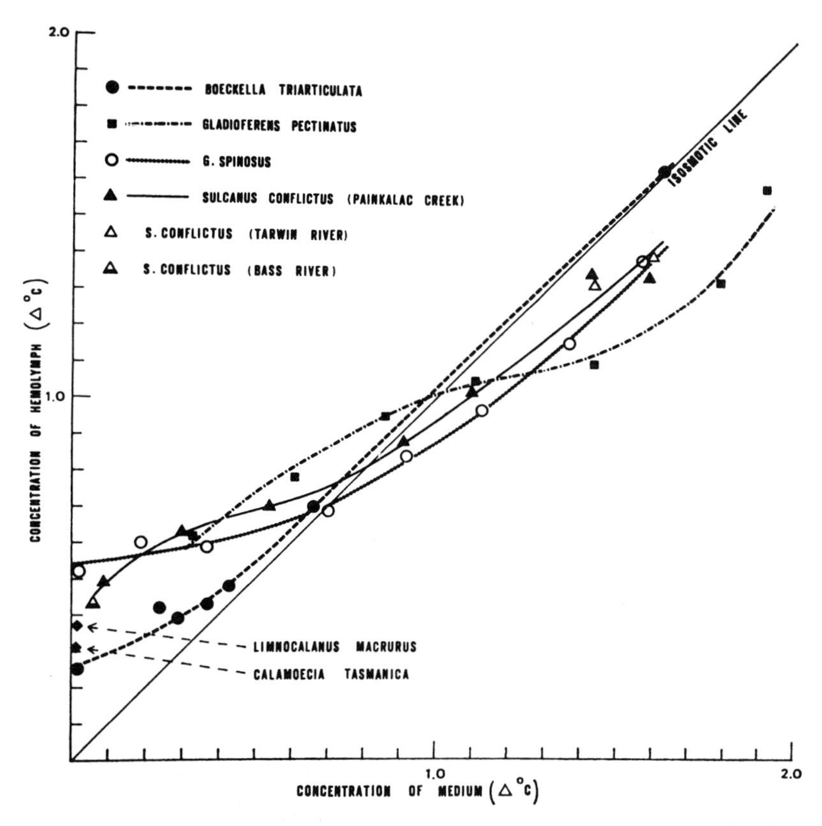

Fig. 20. Osmotic regulation of *Boeckella triarticulata, Gladioferens pectinatus, G. spinosus,* and *Sulcanus conflictus*. Hemolymph concentrations for *Limnocalanus macrurus* and *Calamoecia tasmanica* in fresh water are also indicated. The isosmotic line is shown passing through the origin of the axes. (From Brand and Bayly, 1971, by permission of Pergamon Press.)

ticulata, measured as sodium concentration. Their experiments were done at salinities only down to 33% seawater (about 11.5‰).

Osmoregulation presumably is accomplished at a metabolic cost, and although no measurements of metabolic rates have been made "chloride" secreting cells similar to those found in the gills of fish (*Fundulus*) have been observed in the mouthparts of the copepod, *Gladioferens pectinatus* (Ong, 1969). Secretion of chloride presumably requires energy. Unless intracellular osmoregulation occurs, the rates of metabolic processes of conforming species like the copepod, *Acartia tonsa,* might be directly affected by salinity.

Kinne (1966) suggested four types of changes in metabolic rates that might be observed with changes in salinity. In type 1, metabolic rates are highest at subnormal and lowest at supranormal salinities. Metabolic rates were lowest at

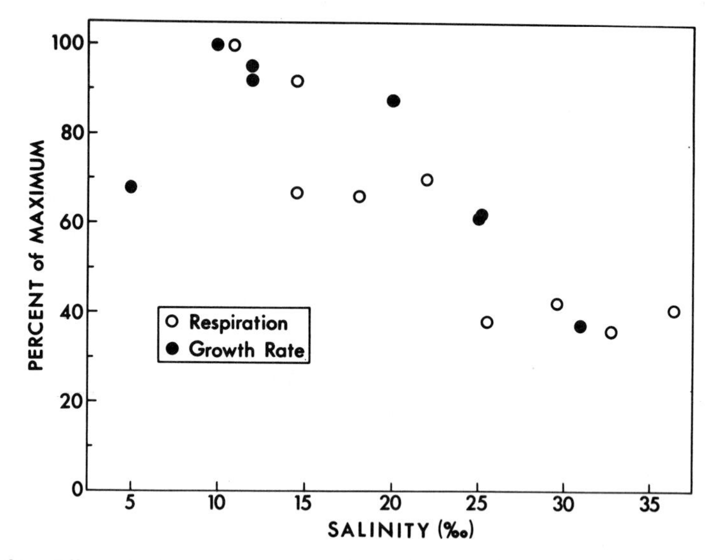

Fig. 21. Effects of salinity on respiration and growth rates of *Acartia tonsa*. Respiration data are from Lance (1965) and growth rates from various sources (Table VIII). All values have been expressed as percent of maximum.

normal salinities in type 2 and highest at normal salinities in type 3. Metabolic rates were unaffected by salinity in type 4. Gilfillan (1972) found that respiration of *Euphausia pacifica* was reduced by subnormal salinities (down to to 21‰), indicating metabolic type 3 of Kinne (1966). Exposure to lower salinities killed *E. pacifica*. Lance (1965) observed an increase in respiration of *Acartia tonsa* at salinities lower than that at which the copepods were captured, suggesting metabolic type 1 of Kinne (1966). Respiration was measured at salinities ranging from 36.4 to 10.9‰ and the increased respiration at lower salinities was interpreted as an indication of stress. Lance (1963) had previously noted that survival of *A. tonsa* was decreased only at salinities lower than the range over which respiration was measured. An alternate explanation of the changes in respiration measured by Lance (1965) is that they simply represent changes in metabolic rates between 10 and 36‰ salinity. Review of the literature on growth rate of *A. tonsa* (Zillioux and Wilson, 1966; Heinle, 1966, 1969a,b; Johnson, 1974) suggested that growth rates were highest at 10–12‰ salinity. Growth rates of *A. tonsa* over a range of salinities and three temperatures were corrected to 20°C using the change in rate per °C shown for that species by Heinle (1969a,b) as shown in Fig. 13 (Section IV,A,1). Growth is much more rapid at lower salinities (Table VIII).

In Fig. 21, the respiration data of Lance (1965) and growth rates calculated from Table VIII are plotted as percent of maximum values. All of the measurements made by Lance in 100% seawater were averaged to estimate respiration at that salinity. The measured rates of respiration were then divided by respiration in 30% seawater to obtain the percentages. It is apparent that over a salinity range of 10–31‰, changes in respiration and growth rate are proportional. The decreased growth rate observed by D. R. Heinle and M. A. Ross (unpublished) at 5‰ salinity was accompanied by high mortalities during growth to adulthood. It thus appears that the change in respiration with change in salinity observed for *Acartia tonsa* by Lance were the result of changes in rates of metabolism rather than stress. In contrast to the results above, McLaren *et al.* (1968) observed no effect of salinity on hatching times of eggs of the marine copepod, *Pseudocalanus minutus*. Eggs of some species of copepods may have relatively impermeable chorionic membranes and thus be less affected by changes in salinity (Bernard *et al.* 1967).

Rates of phosphorus excretion by *Acartia tonsa* were found to be lowest at low salinities and highest at high salinities (Hargrave and Geen, 1968). Temperature had an effect on phosphorus excretion similar to that of salinity. While these results may seem at first to conflict with the data on respiration and growth (Fig. 21), they may reflect changes in efficiency of assimilation since feeding was always possible, the turnover rates of body phosphorus were rapid according to Hargrave and Geen. An additional indication that excretion of phosphorus was linked to feeding and assimilation is provided by the observation that excretion rates were greatly reduced when densities of *Acartia* in the experimental containers was high. Anraku (1964b) found reduced feeding rates when adult *A. tonsa* were confined in less than 13 ml per animal.

Several studies suggest that the metabolic rates of larval decapods are rela-

TABLE VIII

Growth Rates of *Acartia tonsa* at Different Salinities[a]

Salinity (‰)	Experimental temperature (°C)	Days from egg to adult	Corrected to 20°C	Source
5	20.0	12.12	12.12	D. R. Heinle and M. A. Ross, unpublished
10	20.0	8.19	8.19	D. R. Heinle and M. A. Ross, unpublished
12	20.0	8.58	8.58	Heinle (1969b)
12	20.0	8.92	8.92	Heinle (1969b)
20	20.0	9.36	9.36	D. R. Heinle and M. A. Ross, unpublished
25	21.5	11.5	13.34	Johnson (1974)
25	21.5	11.6	13.45	Johnson (1974)
31	17.0	25.0	22.0	Zillioux and Wilson (1966)

[a] Data from Johnson (1974) and Zillioux and Wilson (1966) corrected to 20°C by using a ratio calculated from Heinle (1969a,b) (see Fig. 14).

tively unaffected by changes in salinity (type 4 of Kinne, 1966) or slightly reduced at sub- and supranormal salinities (type 3 of Kinne). Costlow *et al.* (1960) found that growth rates of *Sesarma cinereum* were relatively unaffected, even at salinities where survival was poor. Similar results were obtained with zoeae of *Pagurus longicarpus* by Roberts (1971) who observed that the growth rate was unaffected over a salinity range of 10–30‰ whereas mortalities were high at salinities below 18‰. By contrast, Costlow *et al.* (1966) found that growth rates of zoeae of *Rithropanopeus harrisii* were somewhat reduced above and below salinities of 15–25‰. The results of Sulkin and Minasian (1973) done over a narrower range of salinities appear to contradict those of Costlow *et al.* (1966), since growth rate of *R. harrisii* was relatively unaffected by salinity. Several species of marine decapoda thus appear to have constant growth rates over a range of salinities, including lethal ones. Unfortunately, data on respiration by those species does not seem to be available.

B. ADAPTATIONS TO ALTERED SALINITIES

1. Genetic Adaptation

Studies by Battaglia (1959, 1967) have suggested genetic adaptation to varying salinities by the estuarine harpacticoid copepod, *Tisbe reticulata*. Battaglia (1959) found that one homozygote of *T. reticulata* from Chioggia (lagoon of Venice) was more tolerant of salinities of 18–25‰, while another homozygote was more tolerant of salinities of 25–35‰. Heterozygotes were found to have higher rates of reproduction, suggesting a balanced polymorphism (Battaglia, 1958, 1962). Battaglia (1967) obtained further evidence of genetic adaptation during experiments with populations of *T. furcata* from three geographic areas. The copepods were all cultured in seawater with a salinity of 34‰ for over 2 years. Battaglia then measured the time required for recovery from shock produced by sudden transfer to a salinity of 18‰. The recovery times observed were longest for the population from the area with the highest mean salinity and shortest for the population from the area with lowest mean salinity.

The experiments of Bradley (1975b) were designed to test genetic and nongenetic adaptations to temperature by *Eurytemora affinis,* but they included salinity as a variable, and therefore were inconclusive with regard to genetic adaptation to salinity. Interestingly, Bradley's results show that the natural distribution of *E. affinis* is anomolous with the tolerances of that species to combinations of temperature and salinity and its seasonal alteration with *Acartia tonsa*. Relative growth rates of the two species (Fig. 14) and high rates of predation (Heinle, 1966, 1969a,b; Heinle and Flemer, 1975) may have a greater effect on seasonal succession than temperature and salinity. Work with geographic races of *E. affinis* should be rewarding since that species is known to vary morphologically and metabolically (Katona, 1970, 1971) and has recently invaded fresh waters in North America (Davis, 1969). Brand and Bayly (1971) have suggested that osmotic adaptations by the genus *Eurytemora* might be similar to those observed for several species of Centropagidae because of similarities of habitat.

2. Non-Genetic Adaptation

In general, non-genetic adaptations to salinity are similar to temperature adaptations, and reviews should be consulted (e.g., Kinne, 1963, 1964a,b). Evidence for non-genetic adaptation is provided by studies that show changes in salinity tolerance or metabolic rates following acclimation.

Lance (1963) demonstrated that tolerance of *Acartia bifilosa* to reduced salinities was increased by acclimation to lowered salinities. Lance noted that the ability of *A. bifilosa, A. tonsa,* and *A. discaudata* to survive reduced salinities was initially greatest at the temperatures to which they were exposed in the natural environment. Thermal acclimation of *A. tonsa* and *A. bifilosa* increased their tolerances of reduced salinities at the acclimation temperatures. Of the three species studied, *A. discaudata* was least resistant to reduced salinities. Lance exposed *A. bifilosa* for short (1–17 min) periods to lethal salinities (15 and 20% seawater). The percentage of the copepods that were alive after 20 hr in 100% seawater following exposure to 15 and 20% seawater decreased in a nearly linear fashion with increased time of exposure up to 17 min. No *A. bifilosa* survived 16 min exposure to 15% seawater and none survived 17 min exposure to 20% seawater. These results suggest that irreversible osmotic damage may have occurred during short exposure to lethal salinities.

Battaglia and Bryan (1964) noted without documentation that *Tisbe reticulata* could be acclimated to salinities as low as 50% seawater. They found that ionic regulation of potassium occurred in 75, 50, and 33% seawater. *Tisbe* did not regulate sodium ions in diluted seawater. Gilfillan (1972) similarly commented that *Euphausia pacifica* "were acclimated" but did not demonstrate changes associated with acclimation. The osmotic regulation by Centropagid copepods that Brand and Bayly (1971) observed clearly represents acclimation to altered salinities, as do the changes in tolerance of *Eurytemora affinis* reported by Bradley (1975b).

Acclimation by truly marine zooplankton has not been studied. Although some have fairly wide tolerances to salinity (Hopper, 1960), others have very narrow tolerances (Bernard *et al.,* 1967). Bernard and co-workers noted that eggs of the three species of copepods withstood greater variation in salinity than juveniles or adults. Kasahara *et al.* (1975b) found that the resting eggs of *Tortanus* forcipatus hatched normally over a range of salinities from 18 to 54‰. Freshwater and salinities of 72‰ were required to totally prevent hatching. The tolerance to extreme salinities of this coastal species is thus much higher than the Mediterranean species studied by Bernard *et al.* (1967).

VI. Conclusions

It is true that organisms are successful in a particular habitat because they are adapted to the conditions there. It is thus not surprising that the adaptations that have been described for marine zooplankton are those that allow species to cope successfully with the stresses that are most important in a particular habitat. Oceanic zooplankton are thus adapted to periodic and chronic scarcity of food, to low temperatures at high latitudes and great depths, and, in some cases, to seasonal variation in predation. Coastal and estuarine zooplankton, in contrast, are more highly adapted to changing physical conditions and high rates of predation, and are rather poorly adapted to scarcity of food. The recent discovery of resting eggs of a number of coastal species of copepods suggests that the adaptive strategies of coastal marine zooplankton (mostly copepods) are more similar to those of temperate freshwater forms (mostly rotifers and cladocerans) than to those of oceanic copepods.

Because of the small size and fragility of most species of marine zooplankton, the understanding of the physiological bases for adaptations will probably come from comparisons of responses of whole organisms with those of larger species that are better understood. Observations to date suggest that the physiology of marine zooplankton is no less complex than that of higher invertebrates.

Acknowledgement

I would like to thank Drs. J. D. Allan, R. J. Conover, and R. F. Lee for their thoughtful reviews of this paper. Contribution No. 987, Center for Environmental and Estuarine Studies, University of Maryland.

References

Ackman, R. G., Linke, B. A., and Hingley, J. (1974). Some details of fatty acids and alcohols in the lipids of North Atlantic copepods. *J. Fish. Res. Board Can.* **31,** 1812–1818.

Allan, J. D. (1976). Life history patterns in zooplankton. *Am. Nat.* **110,** 165–180.

Anraku, M. (1964a). Influence of the Cape Cod Canal on the hydrography and on the copepods in Buzzards Bay and Cape Cod Bay, Massachusetts. II. Respiration and feeding. *Limnol. Oceanogr.* **9,** 195–206.

Anraku, M. (1964b). Some technical problems encountered in quantitative studies of grazing and predation by marine planktonic copepods. *J. Oceanogr. Soc. Jpn.* **29,** 19–29.

Battaglia, B. (1958). Balanced polymorphism in *Tisbe reticulata,* a marine copepod. *Evolution* **12,** 358–364.

Battaglia, B. (1959). Il polimorfismo a dattativo e i fattori della selizione nel copepode *Tisbe reticulata* Bocquet. *Arch. Oceanogr. Limnol.* **11,** 305–355.

Battaglia, B. (1962). Controllo genetico della velocita di sviluppo in popolazioni geografiche del copepode marino *Tisbe furcata* (Baird). *Atti Ist. Veneto Sci., Lett. Arti, Cl. Sci. Mat. Nat.* **120,** 83–91.

Battaglia, B. (1967). Genetic aspects of benthic ecology in brackish waters. *In* "Estuaries" (G. H. Lauff, ed.), Publ. No. 83, pp. 574–577. Am. Assoc. Adv. Sci., Washington, D.C.

Battaglia, B., and Bryan, G. W. (1964). Some aspects of ionic and osmotic regulation in *Tisbe* [Copepoda Harpacticoida] in relation to polymorphism and geographical distribution. *J. Mar. Biol. Assoc. U.K.* **44,** 17–31.

Bayly, I. A. E. (1965). Ecological studies on the planktonic copepoda of the Brisbane River estuary with special reference to *Gladioferens pectinatus* (Brady) (Calanoida). *Aust. J. Mar. Freshwater Res.* **16,** 315–350.

Bayly, I. A. E. (1969). The body fluids of some centropagid copepods: Total concentration and amounts of sodium and magnesium. *Comp. Biochem. Physiol.* **28,** 1403–1409.

Beers, J. R. (1966). Studies on the chemical composition of the major zooplankton groups in the Sargasso Sea off Bermuda. *Limnol. Oceanogr.* **11,** 520–528.

Beklemishev, C. W. (1954). The feeding of some common plankton copepods in the Far Eastern seas. *Zool. Zh.* **33,** 1210–1230.

Beklemishev, C. W. (1962). Superfluous feeding of marine herbivorous zooplankton. *Rapp. P.-V. Reun., Cons. Int. Explor. Mer* **153,** 108–113.

Bělehrádek, J. (1935). "Temperature and Living Matter," Protoplasma Monogr. No. 8. Borntraeger, Berlin.

Bell, R. K., and Ward, F. J. (1970). Incorporation of organic carbon by *Daphnia pulex. Limnol. Oceanogr.* **15,** 713–726.

Benson, A. A., and Lee, R. F. (1975). The role of wax in oceanic food chains. *Sci. Am.* **232,** 77–86.

Bernard, M. (1970). Quelques aspectes de la biologie du copépode pélagique *Temora stylifera* en Mediterranee. *Pelagos* **11,** 1–196.

Bernard, M., Braci, M., Lalami, Y., and Moueza, M. (1967). Tolérances des oeufs de copépodes pélagiques aux variations de salinité. Note préliminaire. *Pelagos* **7,** 85–93.

Berner, A. (1962). Feeding and respiration in the copepod *Temora longicornis* (Muller). *J. Mar. Biol. Assoc. U.K.* **42,** 625–640.

Birch, L. C. (1948). The intrinsic rate of natural increase of an insect population. *J. Anim. Ecol.* **17,** 15–26.

Bogorov, V. G. (1939). Sex ratio in marine copepoda. *Dokl. Akad. Nauk SSSR* **23,** 705–708.

Bosch, H. F., and Taylor, R. W. (1973). Distribution of the cladoceran *Podon polyphemoides* in the Chesapeake Bay. *Mar. Biol.* **19,** 172–181.

Bottrell, H. H. (1975). The relationship between temperature and duration of egg development in some epiphytic cladocera and copepoda from the River Thames, Reading, with a discussion of temperature functions. *Oecologia* **18,** 63–84.

Bousfield, E. L. (1955). Ecological control of the occurrence of barnacles in the Miramichi estuary. *Natl. Mus. Can., Bull.* **137,** 1–69.

Boyd, C. M. (1973). Small scale spatial patterns of marine zooplankton examined by an electronic in situ zooplankton detecting device. *Neth. J. Sea Res.* **7,** 103–111.

Bradley, B. P. (1975a). Adaptation of copepod populations to thermal stress. *Tech. Rep., Water Resour. Res. Cent., Univ. Md.* **34.** Available from NTIS, Doc. No. PB244706/AS.

Bradley, B. P. (1975b). The anomalous influence of salinity on temperature tolerances of summer and winter populations of the copepod *Eurytemora affinis. Biol. Bull.* (*Woods Hole, Mass.*) **148,** 26–34.

Bradley, B. P. (1976). The measurement of temperature tolerance: Verification of an index. *Limnol. Oceanogr.* **21,** 596–599.

Bradley, B. P. (1978a). Increase in range of temperature tolerance by acclimation in the copepod *Eurytemora affinis. Biol. Bull.* (*Woods Hole, Mass.*) **154,** 177–187.

Bradley, B. P. (1978b). Genetic and physiological adaptation of the copepod *Eurytemora affinis* to seasonal temperatures. *Genetics* **90,** 193–205.

Brand, G. W., and Bayly, I. A. E. (1971). A comparative study of osmotic regulation in four species of calanoid copepods. *Comp. Biochem. Physiol. B.* **38,** 361–371.

Brodskii, K. A. (1950). Calanoida off the Far Eastern seas and polar basin of the USSR (in Russian.) *Opred. Fauni SSSR Zool. Inst. Akad. NAUK SSSR* **35,** 1–441. Translation program for Scientific Translations.

Butler, E. I., Corner, E. D. S., and Marshall, S. M. (1969). On the nutrition and metabolism of zooplankton. VI. Feeding efficiency of *Calanus* in terms of nitrogen and phosphorus. *J. Mar. Biol. Assoc. U.K.* **49,** 997–1003.

Butler, E. I., Corner, E. D. S., and Marshall, S. M. (1970). On the nutrition and metabolism of zooplankton. VII. Seasonal survey of nitrogen and phosphorus excretion by *Calanus* in the Clyde Sea-Area. *J. Mar. Biol. Assoc. U.K.* **50,** 525–560.

Cairns, A. A. (1967). The zooplankton of Tanquary Fjord, Ellsmere Island, with special reference to calanoid copepods. *J. Fish. Res. Board Can.* **24,** 555–568.

Campbell, M. H. (1934). The life history and post-embryonic development of the copepods *Calanus tonsus* (Brady) and *Euchaeta japonica* (Marukawa). *J. Biol. Board Can.* **1,** 1–65.

Carlisle, D. B. (1965). The effects of crustacean and locust ecdysons on moulting and proecdysis in juvenile shore crabs, *Carcinus maenas. Gen. Comp. Endocrinol.* **5,** 366–372.

Carter, J. C. H. (1965). The ecology of the calanoid copepod *Pseudocalanus minutus* Krøyer in Tessiarsuk, a coastal meromictic late of northern Labrador. *Limnol. Oceanogr.* **10,** 345–353.

Conover, R. J. (1956). Oceanography of Long Island Sound, 1952–1954. VI. Biology of *Acartia clausi* and *A. tonsa. Bull. Bingham Oceanogr. Collect.* **15,** 156–233.

Conover, R. J. (1959). Regional and seasonal variation in the respiratory rate of marine copepods. *Limnol. Oceanogr.* **4,** 259–268.

Conover, R. J. (1962). Metabolism and growth in *Calanus hyperboreus* in relation to its life cycle. *Rapp. P.-V. Reun., Cons. Int. Explor. Mer* **153,** 190–197.

Conover, R. J. (1964). Food and nutrition of zooplankton. *Univ. R. I., Mar. Publ. Series* **2,** 81–91.

Conover, R. J. (1966a). Assimilation of organic matter by zooplankton. *Limnol. Oceanogr.* **11,** 338–345.

Conover, R. J. (1966b). Factors affecting the assimilation of organic matter by zooplankton and the question of superfluous feeding. *Limnol. Oceanogr.* **11,** 346–354.

Conover, R. J. (1967). Reproduction cycle, early development, and fecundity in laboratory populations of the copepod *Calanus hyperboreus. Crustaceana (Leiden)* **13,** 61–72.

Conover, R. J. (1968). Zooplankton—Life in a nutritionally dilute environment. *Am. Zool.* **8,** 107–118.

Conover, R. J., and Corner, E. D. S. (1968). Respiration and nitrogen excretion by some marine zooplankton in relation to their life cycles. *J. Mar. Biol. Assoc. U.K.* **48,** 49–75.

Conover, R. J., and Francis, V. (1973). The use of radioactive isotopes to measure the transfer of materials in aquatic food chains. *Mar. Biol.* **18,** 272–283.

Corkett, C. J. (1970). Development rate of copepod eggs of the genus *Calanus. J. Exp. Mar. Biol. Ecol.* **10,** 171–175.

Corkett, C. J., and McLaren, I. A. (1969). Egg production and oil storage by the copepod *Pseudocalanus* in the laboratory. *J. Exp. Mar. Biol. Ecol.* **3,** 90–105.

Corkett, C. J., and McLaren, I. A. (1970). Relationships between development rate of eggs and older stages of copepods. *J. Mar. Biol. Assoc. U.K.* **50,** 161–168.

Corner, E. D. S. (1961). On the nutrition and metabolism of zooplankton. I. Preliminary observations on the feeding of the marine copepod, *Calanus helgolandicus* (Claus). *J. Mar. Biol. Assoc. U.K.* **41,** 147–167.

Corner, E. D. S. (1972). Laboratory studies related to zooplankton production in the sea. *Symp. Zool. Soc. London* **29,** 185–201.

Corner, E. D. S., and Cowey, C. B. (1968). Biochemical studies on the production of marine zooplankton. *Biol. Rev. Cambridge Philos. Soc.* **43,** 393–426.

Corner, E. D. S., and Davies, A. G. (1971). Plankton as a factor in the nitrogen and phosphorus cycles in the sea. *Adv. Mar. Biol.* **9,** 101-204.

Corner, E. D. S., and Newell, B. S. (1967). On the nutrition and metabolism of zooplankton. IV. The forms of nitrogen excreted by *Calanus. J. Mar. Biol. Assoc. U.K.* **47,** 113-120.

Corner, E. D. S., Cowey, C. B., and Marshall, S. M. (1965). On the nutrition and metabolism of zooplankton. III. Nitrogen excretion by *Calanus. J. Mar. Biol. Assoc. U.K.* **45,** 429-442.

Corner, E. D. S., Cowey, C. B., and Marshall, S. M. (1967). On the nutrition and metabolism of zooplankton. V. Feeding efficiency of *Calanus finmarchicus. J. Mar. Biol. Assoc. U.K.* **47,** 259-270.

Corner, E. D. S., Head, R. N., and Kilvington, C. C. (1972). On the nutrition and metabolism of zooplankton. VIII. The grazing of *Biddulphis* cells by *Calanus helgolandicus. J. Mar. Biol. Assoc. U.K.* **52,** 847-861.

Corner, E. D. S., Head, R. N., Kilvington, C. C., and Marshall, S. M. (1974). On the nutrition and metabolism of zooplankton. IX. Studies relating to the nutrition of overwintering *Calanus. J. Mar. Biol. Assoc. U.K.* **54,** 319-331.

Costlow, J. D., Jr., Bookhout, C. G., and Monroe, R. J. (1960). The effect of salinity and temperature on larval development of *Sesarma cinereum* (Bosc) reared in the laboratory. *Biol. Bull. (Woods Hole, Mass.)* **118,** 183-202.

Costlow, J. D., Jr., Bookhout, C. G., and Monroe, R. J. (1962). Salinity-temperature effects on the larval development of the crab, *Panopeus herbstii* Milne-Edwards, reared in the laboratory. *Physiol. Zool.* **35,** 79-93.

Costlow, J. D., Jr., Bookhout, C. G., and Monroe, R. J. (1966). Studies on the larval development of the carb, *Rithropanopeus harrisii* (Gould). I. The effect of salinity and temperature on larval development. *Physiol. Zool.* **39,** 81-100.

Cowey, C. B., and Corner, E. D. S. (1963). On the nutrition and metabolism of zooplankton. II. The relationship between the marine copepod *Calanus helgolandicus* and particulate matter in Plymouth seawater in terms of amino acid composition. *J. Mar. Biol. Assoc. U.K.* **43,** 495-511.

Cushing, D. H. (1951). The vertical migration of planktonic crustacea. *Biol. Rev. Cambridge Philos. Soc.* **26,** 158-192.

Davis, C. C. (1969). Seasonal distribution, constitution, and abundance of zooplankton in Lake Erie. *J. Fish. Res. Board Can.* **26,** 2459-2476.

Deevey, G. B. (1948). The zooplankton of Tisbury Great Pond. *Bull. Bingham Oceanogr. Collect.* **12,** 1-44.

Deevey, G. B. (1960). Relative effects of temperature and food on seasonal variations in length of marine copepods in some eastern American and western European waters. *Bull. Bingham Oceanogr. Collect.* **17,** 54-85.

Drost-Hansen, W. (1969). Allowable thermal pollution limits - A physicochemical approach. *Chesapeake Sci.* **10,** 281-288.

Edmondson, W. T., Comita, G. W., and Anderson, G. C. (1962). Reproductive rate of copepods in nature and its relation to phytoplankton population. *Ecology* **43,** 625-634.

English, T. S. (1961). Some biological oceanographic observations in the central North Polar Sea, Drift Station Alpha, 1957-58. *Arct. Inst. North Am., Res. Pap.* **13,** 1-80.

Enright, J. T. (1977a). Copepods in a hurry: Sustained high-speed upward migration. *Limnol. Oceanogr.* **22,** 118-125.

Enright, J. T. (1977b). Problems in estimating copepod velocity. *Limnol. Oceanogr.* **22,** 160-162.

Frank, J. R., Sulkin, S. D., and Morgan, R. P., II (1975). Biochemical changes during larval development of the xanthid crab *Rhithropanopeus harrasii.* I. Protein, total lipid, aklaline phosphatase, and glutamic oxaloacetic transaminase. *Mar. Biol.* **32,** 105-111.

Frolander, H. T. (1955). The biology of the zooplankton of the Narragansett Bay area. Ph.D. Dissertation, Brown University, Providence, Rhode Island.

Fulton, J. (1973). Some aspects of the life history of *Calanus plumchrus* in the Strait of Georgia. *J. Fish. Res. Board Can.* **30,** 811–815.

Gatten, R. R., and Sargent, J. R. (1973). Wax ester biosynthesis in calanoid copepods in relation to vertical migration. *Neth. J. Sea Res.* **7,** 150–158.

Geen, G. H., and Hargrave, B. T. (1965). Primary and secondary production of Bras d'Or Lake, Nova Scotia, Canada. *Verh., Int. Ver. Theor. Agnew. Limnol.* **16,** 333–340.

Gilfillan, E. (1972). Reactions of *Euphausia pacifica* Hansen (Crustacea) from oceanic, mixed oceanic-coastal and coastal waters of British Columbia to experimental changes in temperature and salinity. *J. Exp. Mar. Biol. Ecol.* **10,** 29–40.

Gilles, R. (1975a). Mechanisms of thermoregulation. *In* "Marine Ecology" (O. Kinne, ed.), Vol. 2, Part 1, pp. 251–258. Wiley, New York.

Gilles, R. (1975b). Mechanisms of ion and osmoregulation. *In* "Marine Ecology" (O. Kinne, ed.), Vol. 2, Part 1, pp. 259–348. Wiley, New York.

Gonzalez, J. C. (1974). Critical thermal maxima and upper lethal temperatures for the calanoid copepods *Acartia tonsa* and *A. clausi. Mar. Biol.* **27,** 219–223.

Grice, G. D., and Gibson, V. R. (1975). Occurrence, viability and significance of resting eggs of the calanoid copepod *Labidocera aestiva. Mar. Biol.* **31,** 335–337.

Haq, S. M. (1967). Nutritional physiology of *Metrida lucens* and *M. longa* from the Gulf of Maine. *Limnol. Oceanogr.* **12,** 40–51.

Harder, W. (1968). Reactions of plankton organisms to water stratification. *Limnol. Oceanogr.* **13,** 156–168.

Hargrave, B. T., and Geen, G. H. (1968). Phosphorus excretion by zooplankton. *Limnol. Oceanogr.* **13,** 332–342.

Harris, E. (1959).The nitrogen cycle in Long Island Sound. *Bull. Bingham Oceanogr. Collect.* **17,** 31–65.

Harris, J. E. (1953). Physical factors involved in the vertical migration of zooplankton. *Q. J. Microsc. Sci.* **94,** 537–550.

Harris, R. P. (1973). Feeding, growth, reproduction and nitrogen utilization of the harpacticoid copepod, *Tigriopus brevicornis. J. Mar. Biol. Assoc. U.K.* **53,** 785–800.

Hart, R. C., and McLaren, I. A. (1978). Temperature acclimation and other influences on embryonic duration in the copepod *Pseudocalanus* sp. *Mar. Biol.* **45,** 23–30.

Heinle, D. R. (1966). Production of a calanoid copepod, *Acartia tonsa,* in the Patuxent River estuary. *Chesapeake Sci.* **7,** 59–74.

Heinle, D. R. (1969a). Effects of temperature on the population dynamics of estuarine copepods. Ph.D. Dissertation, University of Maryland, College Park.

Heinle, D. R. (1969b). Temperature and zooplankton. *Chesapeake Sci.* **10,** 186–209.

Heinle, D. R. (1970). Population dynamics of exploited cultures of calanoic copepods. *Helgol. Wiss. Meeresunters.* **20,** 360–373.

Heinle, D. R. (1974). An alternate grazing hypothesis for the Patuxent estuary. *Chesapeake Sci.* **15,** 146–150.

Heinle, D. R., and Flemer, D. A. (1975). Carbon requirements of a population of the estuarine copepod, *Eurytemora affinis. Mar. Biol.* **31,** 235–247.

Heinle, D. R., Harris, R. P., Ustach, J. F., and Flemer, D. A. (1977). Detritus as food for estuarine copepods. *Mar. Biol.* **40,** 341–353.

Heinrich, A. K. (1962). The life histories of plankton animals and seasonal cycles of plankton communities in the oceans. *J. Cons., Cons. Int. Explor. Mer* **27,** 15–24.

Herman, S. S., Mihursky, J. A., and McErlean, A. J. (1968). Zooplankton and environmental characteristics of the Patuxent River estuary. *Chesapeake Sci.* **9,** 67–82.

Hirota, R. (1962). Species composition and seasonal changes of copepod fauna in the vicinity of Mukaishima. *J. Oceanogr. Soc. Jpn.* **18,** 35–40.

Hirota, R. (1964). Zooplankton investigations in Hiuchi-Nada in the Setonaiki (Inland Sea of Japan). I. The seasonal occurrence of copepods at three stations in Hiuchi-Nada. *J. Oceanogr. Soc. Jpn.* **20,** 24–31.

Hopper, A. F. (1960). The resistance of marine zooplankton of the Caribbean and South Atlantic to changes in salinity. *Limnol. Oceanogr.* **5,** 43–47.

Hutchinson, G. E. (1967). "A Treatise on Limnology," Vol. 2, Wiley, New York.

Ikeda, T. (1972). Chemical composition and nutrition of zooplankton in the Bering Sea. *In* "Biological Oceanography of the Northern North Pacific Ocean" (A. Y. Takenouchi, ed.), pp. 433–442. Idemitsu Shoten, Tokyo.

Ikeda, T. (1974). Nutritional ecology of marine zooplankton. *Mem. Fac. Fish., Hokkaido Univer.* **22,** 1–97.

Jacobs, J. (1961). Laboratory cultivation of the marine copepod *Pseudodiaptomus coronatus* Williams. *Limnol. Oceanogr.* **6,** 443–446.

Jawed, M. (1969). Body nitrogen and nitrogenous excretion in *Neomysis rayii* Murdoch and *Euphausia pacifica* Hansen. *Limnol. Oceanogr.* **14,** 748–754.

Jeffries, H. P. (1962). Succession of two *Acartia* species in estuaries. *Limnol. Oceanogr.* **7,** 354–364.

Johannes, R. E., and Webb, K. L. (1965). Release of dissolved amino acids by marine zooplankton. *Science* **150,** 7677.

Johnson, J. K. (1974). The dynamics of an isolated population of *Acartia tonsa* Dana (Copepoda) in Yaquina Bay, Oregon. M. S. Thesis, Oregon State University, Corvallis.

Kasahara, S., Uye, S., and Onbé, T. (1974). Calanoid copepod eggs from sea bottom muds. *Mar. Biol.* **26,** 167–171.

Kasahara, S., Uye, S., and Onbé, T. (1975a). Calanoid copepod eggs in sea-bottom muds. II. Seasonal cycles of abundance in the populations of several species of copepods and their eggs in the Inland Sea of Japan. *Mar. Biol.* **31,** 25–29.

Kasahara, S., Onbé, T., and Kamigaki, M. (1975b). Calanoid copepod eggs in sea-bottom muds. III. Effects of temperature, salinity and other factors on the hatching of resting eggs of *Tortanus forcipatus*. *Mar. Biol.* **31,** 31–35.

Katona, S. K. (1970). Growth characteristics of the copepod *Eurytemora affinis* and *E. herdmani* in laboratory cultures. *Helgol. Wiss. Meeresunters.* **20,** 373–384.

Katona, S. K. (1971). Ecological studies on some planktonic marine copepods. Ph.D. Dissertation, Harvard University, Cambridge.

Katona, S. K., and Moodie, C. F. (1969). Breeding of *Pseudocalanus elongatus* in the laboratory. *J. Mar. Biol. Assoc. U.K.* **49,** 743–747.

Kerfoot, W. B. (1970). Bioenergetics of vertical migration. *Am. Nat.* **104,** 529–546.

Kersting, K. (1973). The energy flow in a *Daphnia magna* population. Thesis, University of Amsterdam.

Kinne, O. (1963). The effects of temperature and salinity on marine and brackish water animals. I. Temperature. *Oceanogr. Mar. Biol.* **1,** 301–340.

Kinne, O. (1964a). Non-genetic adaptation to temperature and salinity. *Helgol. Wiss. Meeresunters.* **9,** 433–458.

Kinne, O. (1964b). The effects of temperature and salinity on marine and brackish water animals. II. Salinity and temperature salinity combinations. *Oceanogr. Mar. Biol.* **2,** 281–339.

Kinne, O. (1966). Physiological aspects of animal life in estuaries with special reference to salinity. *Neth. J. Sea Res.* **3,** 222–244.

Kinne, O. (1967). Physiology of estuarine copepods with special reference to salinity and temperature: General aspects. *In* "Estuaries" (G. H. Lauff, ed.), Publ. No. 83, pp. 525–540. Am. Assoc. Adv. Sci., Washington, D.C.

Kinne, O. (1970). Temperature, animals, invertebrates. *In* "Marine Ecology" (O. Kinne, ed.), Vol. 1, Part 1, pp. 407–514. Wiley, New York.

Krazhan, S. A. (1971). Species composition and growth of zooplankton in brackish ponds of the Sivash area. *Hydrobiol. J.* (*Engl. Transl.*) **7,** 33–39.

Kremer, P. M. (1975). The ecology of the ctenophore *Mnemiopsis leidyi* in Narragansett Bay. Ph.D. Dissertation, University of Rhode Island, Kingston.

Lalli, C. M. (1970). Structure and function of the buccal apparatus of *Clione limacina* (Phipps) with a review of feeding in gymnosomatous pteropods. *J. Exp. Mar. Biol. Ecol.* **4,** 101–118.

Lance, J. (1963). The salinity tolerance of some estuarine planktonic copepods. *Limnol. Oceanogr.* **8,** 440–449.

Lance, J. (1965). Respiration and osmotic behavior of the copepod *Acartia tonsa* in diluted sea water. *Comp. Biochem. Physiol.* **14,** 155–165.

Landry, M. R. (1975a). Dark inhibition of egg hatching of the marine copepod *Acartia clausi* Giesbr. *J. Exp. Mar. Biol. Ecol.* **20,** 43–47.

Landry, M. R. (1975b). Seasonal temperature effects and predicting development rates of marine copepod eggs. *Limnol. Oceanogr.* **20,** 434–440.

Lasker, R. (1960). Utilization of organic carbon by a marine crustacean: Analysis with ^{14}C. *Science* **131,** 1098–1100.

LeBrasseur, R. J., Barraclough, W. E., Kennedy, O. D., and Parsons, T. R. (1969). Production studies in the Strait of Georgia. Part III. Observations on the food of larval and juvenile fish in the Fraser River Plume, February to May 1967. *J. Exp. Mar. Biol. Ecol.* **3,** 51–61.

Lee, R. F. (1974a). Lipid composition of the copepod *Calanus hyperboreus* from the Arctic Ocean. Changes with depth and season. *Ma. Biol.* **26,** 313–318.

Lee, R. F. (1974b). Lipids of zooplankton from Bute Inlet, British Columbia. *J. Fish. Res. Board Can.* **31,** 1577–1582.

Lee, R. F., and Barnes, A. T. (1974). Lipids in the mesopelagic copepod, *Gaussia princeps.* Wax ester utilization during starvation. *Comp. Biochem. Physiol. B* **52,** 265–268.

Lee, R. F., Nevenzel, J. C., Paffenhöfer, G.-A., and Benson, A. A. (1970). The metabolism of wax esters and other lipids by the marine copepod. *Calanus helgolandicus. J. Lipid Res.* **11,** 237–240.

Lee, R. F., Hirota, J., and Barnett, A. M. (1971a). Distribution and importance of wax esters in marine copepods and other zooplankton. *Deep-Sea Res.* **18,** 1147–1165.

Lee, R. F., Nevenzel, J. C., and Paffenhöfer, G.-A. (1971b). Importance of wax esters and other lipids in the marine food chain: Phytoplankton and copepods. *Mar. Biol.* **9,** 99–108.

Lee, R. F., Nevenzel, J. C., and Paffenhöfer, G.-A. (1972). The presence of wax esters in marine planktonic copepods. *Naturwissenschaften* **59,** 406–411.

Lee, R. F., Nevenzel, J. C., and Lewis, A. G. (1974). Lipid changes during life cycle of marine copepod, *Euchaeta japonica* Marukawa. *Lipids* **9,** 891–898.

Lehman, J. T. (1977). On calculating drag characteristics for decelerating zooplankton. *Limnol. Oceanogr.* **22,** 170–172.

Leslie, P. H. (1945). On the use of matrices in certain population mathematics. *Biometrika* **33,** 183–212.

Lewis, A. G., and Ramnarine, A. (1969). Some chemical factors affecting the early developmental stages of *Euchaeta japonica* (Crustacea: Copepoda: Calanoida) in the laboratory. *J. Fish. Res. Board Can.* **26,** 1347–1362.

Lewis, A. G., Ramnarine, A., and Evans, M. S. (1971). Natural chelators - an indication of activity with the calanoid copepod *Euchaeta japonica. Mar. Biol.* **11,** 1–4.

Linford, E. (1965). Biochemical studies on marine zooplankton. II. Variations in the lipid content of some Mysidacea. *J. Cons., Cons. Int. Explor. Mer* **30,** 16–27.

McLaren, I. A. (1963). Effects of temperature on growth of zooplankton, and the adaptive value of vertical migration. *J. Fish. Res. Board Can* **20,** 685–727.

McLaren, I. A. (1966). Predicting development rate of copepod eggs. *Biol. Bull.* (*Woods Hole, Mass.*) **131,** 457–469.

McLaren, I. A. (1969). Population and production ecology of zooplankton in Ogac Lake, a land-locked fiord of Baffin Island. *J. Fish. Res. Board Can.* **26,** 1485–1559.

McLaren, I. A. (1974). Demographic strategy of vertical migration by a marine copepod. *Am. Nat.* **108,** 91–102.

McLaren, I. A., Walker, D. A., and Corkett, C. J. (1968). Effects of salinity on mortality and development rate of eggs of the copepod *Pseudocalanus minutus. Can. J. Zool.* **46,** 1267–1269.

McLaren, I. A., Corkett, C. J., and Zillioux, E. J. (1969). Temperature adaptations of copepod eggs from the Artic to the Tropics. *Biol. Bull.* (*Woods Hole, Mass.*) **137,** 486–493.

McLeese, D. W. (1956). Effects of temperature, salinity and oxygen on the survival of the American lobster. *J. Fish. Res. Board Can.* **13,** 247–272.

Maetz, J. (1974). Aspects of adaptation of hypo-osmotic and hyper-osmotic environments. *Biochem. Biophys. Perspect. Mar. Biol.* **1,** 1–167.

Marshall, S. M. (1973). Respiration and feeding in copepods. *In Adv. Mar. Biol.* **11,** 57–120.

Marshall, S. M., and Orr, A. P. (1955a). "The Biology of a Marine Copepod." Oliver & Boyd, London.

Marshall, S. M., and Orr, A. P. (1955b). On the biology of *Calanus finmarchicus.* VIII. Food uptake, assimilation and excretion in adult and stage V *Calanus. J. Mar. Biol. Assoc. U.K.* **34,** 495–529.

Marshall, S. M., and Orr, A. P. (1961). On the biology of *Calanus finmarchicus.* XII. The phosphorus cycle; excretion, egg production, autolysis. *J. Mar. Biol. Assoc. U.K.* **41,** 463–488.

Marshall, S. M., and Orr, A. P. (1966). Respiration and feeding in some small copepods. *J. Mar. Biol. Assoc. U.K.* **46,** 513–530.

Martin, J. H. (1965). Phytoplankton-zooplankton relationships in Narragansett Bay. *Limnol. Oceanogr.* **10,** 185–191.

Mattson, F. H., Volpenhein, R. A., and Benjamin, L. (1970). Inhibition of lipolysis by normal alcohols. *J. Biol. Chem.* 245, 5335–5340.

Mednikov, B. M. (1961). On the sex ratio in deep sea calanoida. *Crustaceanna* (*Leiden*) **3,** 105–109.

Mihursky, J. A., and Kennedy, V. S. (1967). Water temperature criteria to protect aquatic life. *Spec. Publ.—Am. Fish. Soc.* **4,** 20–32.

Moreira, G. S., and Vernberg, W. B. (1968). Comparative thermal metabolic patterns in *Euterprina acutifrons* dimorphic males. *Mar. Biol.* **1,** 282–284.

Mullin, M. M. (1963). Some factors affecting the feeding of marine copepods of the genus *Calanus. Limnol. Oceanogr.* **8,** 239–250.

Mullin, M. M., and Brooks, E. R. (1967). Laboratory culture, growth rate, and feeding behavior of a planktonic marine copepod. *Limnol. Oceanogr.* **12,** 657–666.

Mullin, M. M., and Brooks, E. R. (1970a). Growth and metabolism of two planktonic, marine copepods as influenced by temperature and type of food. *In* "Marine Food Chains" (J. H. Steele, ed.), pp. 74–95. Univ. of California Press, Berkeley.

Mullin, M. M., and Brooks, E. R. (1970b). The effect of concentration of food on body weight, cumulative ingestion, and rate of growth of the marine copepod *Calanus helgolandicus. Limnol. Oceanogr.* **15,** 748–755.

Nassogne, A. (1970). Influence of food organisms on the development and culture of pelagic copepods. *Helgol. Wiss. Meeresunters.* **20,** 333–345.

Nevenzel, J. C. (1970). Occurrence, function and biosynthesis of wax esters in marine organisms. *Lipids* **5,** 308–319.

O'Conner, J. D., and Gilbert, L. I. (1968). Aspects of lipid metabolism in crustaceans. *Am. Zool.* **8,** 529–540.

Omori, M. (1973). Cultivation of marine copepods. *Bull. Plankton Soc. Jpn.* **20,** 3–11.

Ong, J. E. (1969). The fine structure of the mandibular sensory receptors in the brackish water calanoid copepod *Gladioferens pectinatus* (Brady). *Z. Zellforsch. Mikrosk. Anat.* **97,** 178–195.

Paffenhöfer, G.-A. (1970). Cultivation of *Calanus helgolandicus* under controlled conditions. *Helgol. Wiss. Meeresunters.* **20,** 346–359.

Paffenhöfer, G.-A. (1976). Feeding, growth, and food conversion of the marine planktonic copepod *Calanus helgolandicus. Limnol. Oceanogr.* **21,** 39–50.

Parsons, T. R., and LeBrasseur, R. J. (1970). The availability of food to different trophic levels in the marine food chains. *In* "Marine Food Chains" (J. H. Steele, ed.), pp. 325–343. Univ. of California Press, Berkeley.

Parsons, T. R., and Takahaski, M. (1973). "Biological Oceanographic Processes." Pergamon, Oxford.

Petipa, T. S. (1966). Relationship between growth, energy metabolism, and ration in *Acartia clausi* Giesbrecht. *In* "Physiology of Marine Animals" (transl.), pp. 82–91. Akad. Nauk SSR, Moscow.

Petipa, T. S. (1967). Oxygen consumption and food requirements in the copepods *Acartia clausi* Giesbr. and *A. latisotosa* Kritcz. [Translation No. 901, *Fish. Res. Board Can.*] From *Zool. Zh.* **45,** 363–370 (1966).

Petipa, T. S., Pablova, E. V., and Midonov, G. N. (1970). The food web structure, utilization and transport of energy by trophic levels in the planktonic communities. *In* "Marine Food Chains" (J. H. Steele, ed.), pp. 142–167. Univ. of California Press, Berkeley.

Pinhey, K. F.(1923). Entomostraca of the Bell Isle Straight expedition, 1923, with notes on other planktonic species. *Contrib. Can. Biol. Fish.* [N.S.] **3,** 179–233.

Precht, H. (1958). Concepts of the temperature adaptation of unchanging reaction systems of cold-blooded animals. *In* "Physiological Adaptation" (C. L. Prosser, ed.), pp. 50–78. Ronald Press, New York.

Ranade, M. R. (1957). Observations on the resistance of *Tigriopus fulvus* (Fischer) to changes of temperature and salinity. *J. Mar. Biol. Assoc. U.K.* **36,** 115–119.

Raymont, J. E. G., and Conover, R. J. (1961). Further investigations on the carbohydrate content of marine zooplankton. *Limnol. Oceanogr.* **6,** 154–164.

Raymont, J. E. G., and Krishnaswamy, S. (1960). Carbohydrates in some marine planktonic animals. *J. Mar. Biol. Assoc. U. K.* **39,** 239–248.

Redfield, A. C., Ketchum, B. H., and Richards, F. A. (1963). The influence of organisms on the composition of sea water. *In* "The Sea: Ideas and Observations on Progress in the Study of the Seas" (M. N. Hill, ed.), Vol. II, pp. 26–77. Wiley (Interscience), New York.

Reeve, M. R., and Cosper, E. (1970). The acute effects of heated effluents on the copepod *Acartia tonsa* from a subtropical bay and some problems of assessment. *FAO Tech. Conf. Mar. Pollut. Effects Living Resour. Fishing, 1970* Ref. FIR:MP/70/E-59, pp. 1–5.

Reeve, M. R., Raymont, J. E. G., and Raymont, J. K. B. (1970). Seasonal biochemical composition and energy sources of *Saggita hispida. Mar. Biol.* **6,** 357–364.

Richman, S. (1958). The transformation of energy by *Daphnia pulex. Ecol. Monogr.* **28,** 273–291.

Roberts, M. H., Jr. (1971). Larval development of *Pagurus longicarpus* Say reared in the laboratory. II. Effects of reduced salinity on larval development. *Biol. Bull.* (*Woods Hole, Mass.*) **140,** 104–116.

Rodriguez, G. (1973). "El Sistema de Maracaibo." Inst. Venez. Invest. Cienti. Caracas.

Rose, M. (1933). Copepodes pélagiques. *Faune Fr.* **26,** 1–374.

Shushkina, E. A. (1968). Calculation of copepod production based on metabolic features and the coefficient of the utilization of assimilated food for growth. *Oceanology* (*Engl. Transl.*) **8,** 98–110.

Smyly, W. J. P. (1961). The life cycle of the freshwater copepod, *Cyclops leuckarti* Claus, in Estwaite Water. *J. Anim. Ecol.* **30,** 153–169.

Sorokin, Y. I. (1966). Carbon-14 method in the study of the nutrition of aquatic animals. *Int. Rev. Gesamten Hydriobiol.* **51,** 209–224.

Strickler, J. R. (1977). Observation of swimming performances of planktonic copepods. *Limnol. Oceanogr.* **22,** 165–170.

Sulkin, S. D. (1975). The significance of diet in the growth and development of larvae of the blue crab, *Callinectes sapidus* Rathbun, under laboratory conditions. *J. Exp. Mar. Biol. Ecol.* **20,** 119–135.

Sulkin, S. D., and Minasian, L. L. (1973). Synthetic sea water as a medium for raising crab larvae. *Helgol. Wiss. Meeresunters.* **25,** 126–134.

Sushchenya, L. M. (1970). Food rations, metabolism and growth of crustaceans. *In* "Marine Food Chains" (J. H. Steele, ed.), pp. 127–141. Univ. of California Press, Berkeley.

Vacquier, V. D., and Belser, W. L. (1965). Sex conversion induced by hydrostatic pressure in the marine copepod *Tigriopus californicus. Science* **150,** 1619–1621.

Vargo, S. L. (1974). Seasonal and vertical distribution of zooplankton in an estuarine anoxic basin and their tolerances to hydrogen sulfide and dissolved oxygen. Ph.D. Dissertation, University of Rhode Island, Kingston.

Vernberg, W. B., and Vernberg, F. J. (1972). "Environmental Physiology of Marine Animals." Springer-Verlag, Berlin and New York.

Vlymen, W. J. (1970). Energy expenditure of swimming copepods. *Limnol. Oceanogr.* **15,** 348–356.

Vlymen, W. J. (1977). Reply to comment by J. T. Enright. *Limnol. Oceanogr.* **22,** 162–165.

Walne, P. R. (1965). Observations on the influence of food supply and temperature on the feeding and growth of larvae of *Ostrea edulis* L. *Fish. Invest. Ser. 2 Mar. Fish. G. B., Minist. Agric., Fish Food,* **20**(9), 1–23.

Webb, K. L., and Johannes, R. E. (1967). Studies of the release of dissolved free amino acids by marine zooplankton. *Limnol. Oceanogr.* **12,** 376–382.

Webb, K. L., and Johannes, R. E. (1969). Do marine crustaceans release dissolved amino acids? *Comp. Biochem. Physiol.* **29,** 875–878.

Wilson, D. F., and Parrish, K. K. (1971). Remating in a planktonic marine calanoid copepod. *Mar. Biol.* **9,** 202–204.

Winberg, G. G. (1956). "Rate of Metabolism and Food Requirement of Fishes." Nauchn. Tr. Beloruss. Gos. Univ., Minsk. (*Fish. Res. Board Can., Transl. Ser.* No. 194).

Zillioux, E. J., and Gonzalez, J. G. (1972). Egg dormancy in a neritic calanoid copepod and its implications to overwintering in boreal waters. *Proc. Eur. Symp. Mar. Biol., 5th, 1970* pp. 217–330.

Zillioux, E. J., and Wilson, D. F. (1966). Culture of a planktonic calanoid copepod through multiple generators. *Science* **151,** 996–998.

Meiofauna 5

Winona B. Vernberg and Bruce C. Coull

I. Introduction

The term "meiobenthos" (meiofauna) was coined relatively recently (Mare, 1942), and at this time, meiofauna were defined as benthic metazoans of intermediate size. This fact, however, does not mean that these organisms were unknown before the 1940s, for scientists began conducting ecological research on meiofauna at the turn of the century (Remane, 1933).

The meiofauna may be characterized as those animals which are smaller than those traditionally called the "macrobenthos," but larger than the "microbenthos," i.e., algae, bacteria, and protozoa. It is generally accepted that "meiofauna" refers to those animals which pass through a 0.5 mm sieve and are retained on a sieve with mesh widths smaller than 0.1 mm. There is a wide diversity of habitats in which the meiofauna live. Some are interstitial, readily moving between sediment particles. Still others have an epibenthic or phytal existence. Some animals, usually larvae of the macrofauna, are a part of the meiobenthos for only a part of their life cycle, but many species are meiobenthic throughout their life cycle including Rotifera, Gastrotricha, Nematoda, Archiannelida, Tardigrada, Copepoda, Ostracoda, Mystacocarida, Turbellaria, Acarina, Gnathastomulida, and some specialized members of the Hydrozoa, Nemertina, Bryozoa, Gastropoda, Soelenogastres, Holothuroidea, Tunicata, Priapulida, Polychaeta, Oligochaeta, and Sipunculida.

FUNCTIONAL ADAPTATIONS OF MARINE ORGANISMS

ISBN 0-12-718280-2

From the high intertidal zone to the abyssal depths of the ocean, meiofauna are the most abundant animals in marine sediments. World wide average meiobenthic density is on the order of 10^6 m^{-2}, and the highest known densities are 25 × 10^6 m^{-2} of sea bottom (R. M. Warwick, personal communication). Thus it is not surprising that there has been an upsurge in interest in these organisms within recent years.

Energetically the meiofauna are the most important animals in the sediment; they are ubiquitous in nature, have relatively rapid generation time, are small in size, and have high metabolic rates. In fact, Gerlach (1971) has suggested that the meiofauna, although low in biomass at any one time, are responsible for five times as much metabolic activity as the more traditionally studied macrofauna. In those situations where macrofauna/meiofauna biomass ratios are close to one, as in the deep sea (Thiel, 1975) or in shallow water mud flats (B. C. Coull, unpublished), the meiofaunal role is even more significant.

One approach to determining the role of meiofauna in the benthic biocenose has been through the use of physiological measurements. In the following pages we will summarize the physiological ecology of the marine meiofauna. In actuality we understand little of the role of these omnipresent organisms; thus, physiological studies of these organisms must continue to be investigated if we are ever to understand their significance in marine ecosystems.

II. Physiological Tolerances

Biologists have recognized for years that environmental factors can be limiting. Obviously, organisms that persist in a specific geographical area are able to survive the environmental extremes found there, but if an environmental factor reaches a given intensity for a sufficient period of time it may limit species distribution. For example, Jansson (1966) suggested that low salinities were probably the most important factor in controlling the landward distribution of the mystacocarid, *Derocheilocaris remanei,* at Canet Plage, France, and that the limiting factor in the seaward distribution of this species was turbulence.

One of the problems in trying to determine which factor is the significant limiting one in a particular environment is that the tolerance of some organisms to one particular environmental factor is much greater than the animal will normally experience in its habitat. However, if various combinations of environmental factors are studied simultaneously, the resultant tolerance limits more nearly approximate the environmental fluctuations experienced by the animal in the field. Generally, the interaction of two or more factors will kill an organism at intensity levels, which, if taken separately would not be lethal (Vernberg and Vernberg, 1972).

A. TEMPERATURE TOLERANCE

Of all the factors in the environment, temperature is one of the most obviously changable, since it fluctuates both temporally and geographically. There may also be gradients within a particular environment, and there have been numerous studies documenting the relationship between the distribution (both seasonally and geographically) of marine meiofauna and their thermal tolerances.

Tolerance to thermal stress can often be correlated with the vertical distribution of intertidal meiofauna (Wieser *et al.*, 1974; Wieser, 1975). As might be expected, surface dwelling species tend to have higher thermal lethal limits than do species living 1-5 cm in the sediment, where temperatures are not only lower, but also tend to be more constant (Fig. 1). Similarly, those species living high in

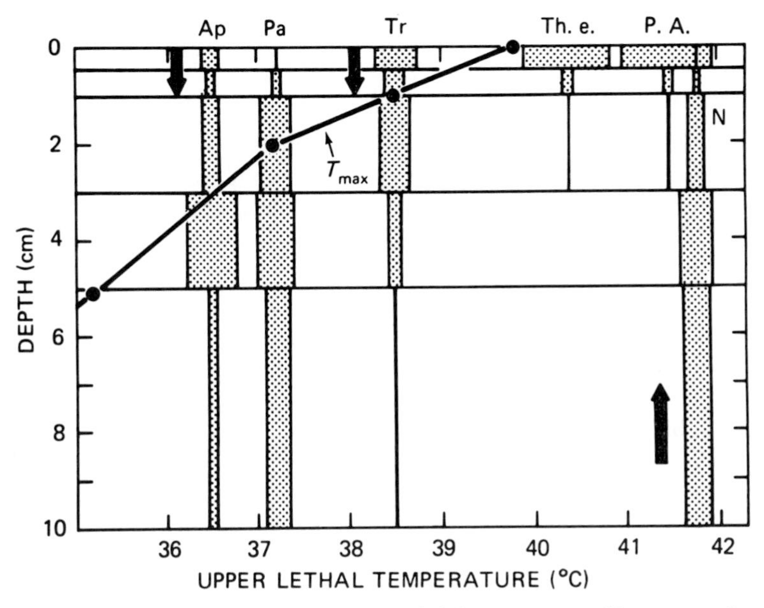

Fig. 1. Vertical distribution plotted against upper lethal temperatures at 5 hr exposure of several meiofauna species from Tuckers Town Beach, Bermuda. The maximum temperatures recorded during the investigation period (August, 1973) at various depths are also indicated (T_{max}), using the abscissa as a normal temperature scale. In several species, a discrepancy was observed between vertical distribution and LT_{50}. Two species occurred in greater numbers in zones in which the temperature may rise appreciably above their experimentally determined LT_{50}. One species, according to its LT_{50}, could have occurred much higher up than it was actually found. In these species then there exists a potential or a requirement for vertical migration, the direction of which is indicated by arrows. Ap, *Apodopsyllus bermudensis* (Copepod); Pa, *Paramonhystera* sp. (Nematode); Tr, *Tripyloides* sp. (Nematode); Th.e., *Theristus erectus;* N, *Nannolaimoides decoratus;* P. A., *Psyllocamptus, Amphiascoides subdebelis* (copepods, not separated). (From Wieser, 1975.)

the intertidal zone tolerate high temperatures better than those living near the water's edge (Boaden and Erwin, 1971). Little is known of low thermal tolerances, although Wieser (1975) has noted that the relationship between low temperature tolerances and habitat temperature is not as evident as it is with high thermal tolerance.

B. OXYGEN

There are three generalizations which can be made about the distribution of meiofauna and their response to oxygen deprivation (Wieser, 1975).

(1) An animal that cannot tolerate oxygen lack is not apt to migrate into anoxic sediments.
(2) Those species which migrate between surface and deeper layers are most likely unaffected by oxygen deprivation.
(3) Species which are harmed by the presence of oxygen are limited to the anoxic layers (obligate anaerobes).

Evidence suggests that those species which are highly mobile are less tolerant to oxygen lack than are less agile species which may be trapped from time to time in low oxygen environments. This is well illustrated by Wieser and Kanwisher (1959). In their study, one group of animals with low mobility and no special respiratory organs (the nematode, *Enopulus communis,* and the sea mites, *Rhombognathides seahami* and *Halacaris basteri*) were compared with three amphipods which were highly agile and equipped with gills (*Calliopius laeviusculus, Gammarus oceanicus,* and *Hyale prevosti*). In laboratory studies they found that the nematode and the mites became quiescent when oxygen was removed, but even after a 12-hr exposure, these animals recovered when oxygen was made available. This response correlates well with their mode of existence, since these animals are often trapped under large seaweeds at low tide, and since oxygen tensions drop to very low levels during the time span between high and low tide. In contrast, the amphipods quickly became inactive under anaerobic stress and died. Obviously, to ensure survival these animals must escape the seaweed habitat when the tide recedes and the oxygen levels drop.

Results of tolerance experiments on several species of intertidal zone nematodes from both the oxidized surface layer and deeper anoxic sediments showed striking differences between organisms living in different layers in their ability to tolerate oxygen lack (Ott and Schiemer, 1973). Whereas those which lived near the surface could survive only 6–12 hr, those species which normally live in the deeper sediments survived 4–13 days.

Subtidally, the lack of oxygen and/or low oxygen concentrations limits the vertical distribution of meiofauna. Numerous investigators have studied the vertical distribution of subtidal meiofaunal assemblages and the general conclusion has been that about 70% of all the fauna is located in the upper 2 cm and 95% in the upper 5 cm. Of course, the depth of oxygen penetration depends on the sedimentary characteristics, and thus oxygen will occur deeper in sands than in muds. Intertidally there is some evidence that oxygen availability is at least one factor limiting the vertical distribution of meiofauna. For example, Jansson (1967) reported that the vertical distribution of the harpacticoid copepod, *Parastenocaris vicesima*, was related to organic material, oxygen availability, and grain size, and Giere (1973) stresses an oxygen-dependent vertical distribution of intertidal oligochaetes.

Some meiofaunal species which are found in the reduced layers of the sediment appear to be sensitive to ambient oxygen levels (Wieser *et al.*, 1974), and, in fact, some "sulfide biome" meiofauna are obligate anaerobes (Fenchel and Riedl, 1970). Maguire and Boaden (1975), working with the gastrotrich, *Thiodasys sterreri*, suggest that this thiobiotic animal fixes CO_2, which is then incorporated into a reversed Krebs cycle sequence. The actual biochemical mechanisms are not entirely known, but oxygen is not necessarily required for all meiofauna.

Coull (1969), de Bovée (1975), and Elmgren (1975) have reported on meiofaunal assemblages controlled by naturally occurring anoxia. In Bermuda, Coull (1969) found that areas which regularly became anoxic in the summers were devoid of meiofauna during this period and during the following 3–4 month period (Fig. 2). In the Antarctic, de Bovée (1975) reported that low nematode numbers and diversity were correlated with increasing sulfide ion and low dissolved oxygen values. He attributed the high sulfide content to extremely high

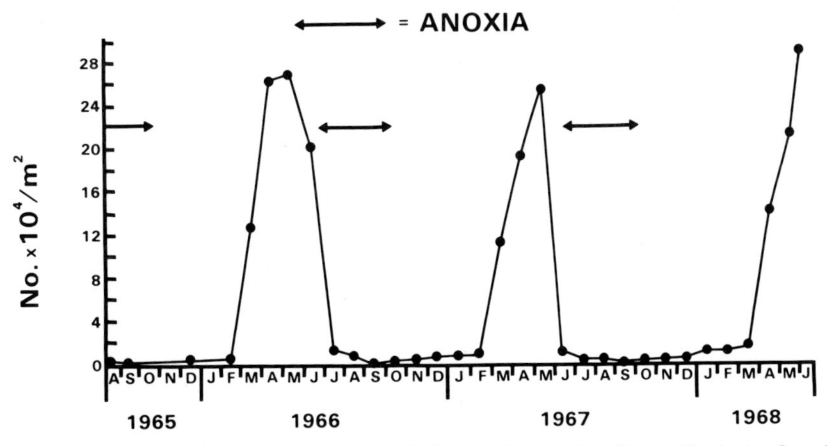

Fig. 2. Meiofauna density in relation to regularly recurring anoxia at 26 m in Harrington Sound, Bermuda. Note that after anoxic periods the meiofauna disappear and do not repopulate until 3–4 months following the reoxygenation. (Adopted from Coull 1969.)

deposition of *Yoldia isonota* (bivalve) feces and pseudofeces, which required all the available oxygen for degradation. Elmgren (1975) was able to show that the meiofaunal communities in the Baltic were distinctly zoned by depth and oxygen content; the deep oxygen-poor zone (80 m) was extremely impoverished with only a few nematode species present.

C. SALINITY TOLERANCE

As is true for other marine organisms living in estuarine and coastal environments, many species of estuarine meiofauna exhibit some degree of euryhalinity, although the range of tolerance among them is quite large. Gray (1966a) speculated that most interstitial species are indeed euryhaline, but Wieser (1975) noted that this is probably true only for species that are subjected to wide salinity fluctuations, such as those meiofauna found on high energy beaches living in undersaturated sediments. In this type of habitat, salinity fluctuations tend to be largely due to a combination of factors, including evaporation, rainfall, and freshwater run off (Jansson, 1967). Hummon (1972) found that two marine gastrotrichs (*Chaetonotus testiculophorus* and *Turbanella cornuta*), which commonly dwell on a sandy beach subjected to wide salinity fluctuations, could withstand 6 and 8‰, respectively, for 24 hr, although the "normal" salinity of the area was 34‰. Species that live in high salinity estuaries where salinities seldom fall below 28‰, often cannot tolerate salinities below 20–25‰ (Vernberg and Coull, 1975). Generally, low intertidal species are less tolerant of low salinities than are high intertidal ones (Jansson, 1966; Boaden and Irwin, 1971), although there are some exceptions such as the mystacocarid, *Derocheilocaris remanei,* which can tolerate salinities ranging from 5.4 to 40.8‰ without apparent damage (Delamare Deboutteville, 1960).

As with macrofauna, the number of meiofauna species in an estuary decreases with a decrease in salinity. Bilio (1967) lists 60, 59, 2, and 4 meiofaunal species from high salinity to freshwater, respectively. Brickman (1972) also noted an increase in the number of species approaching the sea; however, he felt a more significant factor was the change from one or two dominant species in the low salinity areas to a more equal representation of 5–6 species in the higher salinity areas. Reid (1970) also found the number of benthic copepod species increased from one to seven as higher salinities were approached.

From the small amount of data available, there also appears to be a decrease in the number of meiofaunal animals per unit area proceeding from estuarine conditions to freshwater. However, other parameters may distort this idealized picture. Warwick (1971), for example, recorded his lowest population densities at his most saline station, but attributed this to sediment granulometry and not the salinity conditions.

D. THE INTERACTION OF MULTIPLE FACTORS

In nature, meiofauna are not exposed to only one environmental factor at a time, and exposure to the interaction of one or more factors nearing tolerance limits usually produces a more drastic effect on organisms than does exposure to a single factor. Since one of the chief characteristics of the marine environment is fluctuation in salinity and, in the temperate zones, temperature, it would seem logical to try to determine the effects of these two factors acting in concert on the tolerance limits of meiofauna. In a study on the euryhaline and eurythermal mystacocarid, *Derocheilocaris typica,* Kraus and Found (1975) reported that it survived least well under conditions of high temperature and low salinity. At any single salinity, greatest survival was observed at 7°C, while at any single temperature highest tolerance levels were noted when the animals were maintained at 31 ‰. Distribution of this species has been reported from intersitial water which ranges in temperature from −1.5°C in Massachusetts to 28°C in Florida, in salinities ranging from 11 to 36‰. These observations, plus the salinity–temperature studies, led Kraus and Found to suggest that the depth at which the mystacocarid population is found (20–45 cm at Woods Hole) may reflect an avoidance to salinity–thermal extremes which would more likely be encountered in surface waters. Since both seasonal and tidal fluctuations alter distributional patterns this offers further evidence that selection does occur and that it can be linked to tolerance limits.

Hummon (1975), working with salinity–temperature interactions with several intertidal gastrotichs, found that "salinity did not appear to be limiting," and that temperature accounted for most of the observed mortality. Earlier, Hummon (1972) found that temperature–salinity tolerances increased with increasing temperature, again suggesting that salinity was not the limiting factor. In almost all instances he was able to correlate the multiple factor effect with environmental conditions and the distribution of the species.

Thermal–salinity tolerances of meiofaunal species have also been investigated in populations of meiofauna on the South Carolina (USA) coast (Vernberg and Coull, 1975). For three species of copepods (*Thompsonula hyaenae, Pseudobradya pulchera,* and *Scottolana canadensis*), temperature was the only statistically significant cause of death when the animals were maintained under aerobic conditions. Neither salinity alone nor temperature–salinity interactions were significant factors as a cause of death. There did seem to be a relationship between salinity and high temperature tolerance in all three species of copepods, however, since increased survival of these species was noted in salinities of 30‰. In one species of copepod, *P. pulchera,* and in the ciliate, *Tracheloraphis* sp., salinity as well as temperature was a significant factor as a cause of mortality when these animals were maintained under anaerobic conditions.

Seasonal distributional patterns of *Thompsonula hyaenae* probably can be correlated with observed responses to a combination of temperature, salinity, and lack of oxygen. This species disappears during the summer months, but then reappears during the early fall before the thermal regime of the waters has changed. The clue to their disappearance and reappearance probably lies in their inability to tolerate oxygen lack at high temperatures. During the summer months, the prevailing winds are from the southwest, pushing in silt and clays from a nearby estuary, lowering oxygen levels, and moving the redox layer closer to the surface. In early fall the prevailing winds are from the northeast open oceanside, making the substrata sandy and well-oxygenated with a deeper redox layer. Since *T. hyaenea* has a relatively high thermal tolerance under well-oxygenated conditions, it seems likely that the low oxygen content in the substrata in combination with high temperature during summer months is responsible for the seasonal disappearance of this species.

In contrast to the copepods, the ciliate, *Tracheloraphis* sp., survived anaerobically almost as well as it did aerobically at 23°C and 30‰ S. However, both temperature and salinity fluctuations did contribute significantly to observed mortalities of this species. Thus, their seasonal occurrence was probably related not so much to oxygen lack, but rather to the seasonal thermal–salinity fluctuations.

E. POLLUTANTS

Over the past few years, there has been increasing recognition that man-induced environmental changes are now a part of what classically has been called a "normal environment." Meiofauna are no less affected by such environmental perturbations than are macrofauna. A good example of how meiofaunal populations are affected can be found in the study by Bleakley and Boaden (1974), who assessed the effects of an oil spill remover on the meiofauna of a sandy beach. They found that the meiofauna were more resistant than were the planktonic species; but following treatment with a surfactant there was a marked decline in numbers of meiofauna. Harpacticoid copepods were particularly affected, and their numbers dropped immediately and drastically. After the initial mortalities, the number of animals increased slightly, but this was followed by a rise in morbidity and further population decline. At the end of 28 days, the populations still had not approached their original numbers. The authors speculated that the observed population decline was due to the action of the oil spill remover on the meiofauna themselves and not the result of the action of this agent on their bacterial food source, since the number of bacteria present remained constant.

On a beach adjacent to the River Tees, England, Gray and Ventilla (1971) found significantly lower meiofauna densities at the station where there were visible signs of sewage pollution than at nonpolluted adjacent sites, and they concluded that "the meiofauna was significantly affected by the pollution here." However, in a subsequent paper dealing with the meiofauna of the Tees Estuary and surrounding beaches, Gray (1976) could find no correlation between total meiofauna density and pollution, although there were changes in taxa dominance related to the level of pollution. In Swedish estuaries, Olsson *et al.* (1973) determined the type of meiofaunal communities in four diverse estuarine systems, all of which were polluted to some degree with domestic and industrial sewage. The communities in each of the estuaries seemed to be a function of the combined effects of salinity and pollutants, although the authors concluded that the meiofaunal species composition reflected differences in pollution rather than differences in salinity. Marcotte and Coull (1974), working on an organic (domestic) pollution gradient in the North Adriatic, reported that, in winter, harpacticoid copepods dominated in the heavily polluted areas, whereas in the summer, nematodes were the most abundant taxon; this suggested a temperature–pollution interaction. Furthermore, they found that the polluted copepod assemblages were dominated by one species (low diversity), but that the nonpolluted areas had a more even distribution of several species (higher diversity). The authors concluded that pollution had a marked effect on the community structure, a conclusion also supported by McIntyre (1977), who found that Scottish "polluted" beaches were dominated by nematodes and the "clean" beaches by harpacticoid copepods. Further, he reports that the copepod assemblage on the polluted beaches were dominated by one to two species, whereas the nonpolluted beaches had several species of rather equal abundance. However, Tietjen (1977) working in Long Island Sound, found no differences in nematode species diversity, population densities, and species composition when the fauna of sediments with and without heavy metal contamination were compared. Perhaps nematodes are more resistant to "impaction" than the harpacticoid copepods of Marcotte and Coull (1974) and McIntyre (1977).

III. Behavioral Responses

Relatively little information is available on behavioral responses of meiofauna, and most of the studies which have been done to date relate to environmental selection. Jansson (1967) demonstrated a relationship between temperature and salinity preferences and field distribution of the harpacticoid copepod, *Parastenocaris vicesima*. In the salinity experiments, the copepods were given a choice between two connecting sand layers, each of which was saturated with water of a distinct salinity. Clearly the preference zone of *P. vicesima* species lies between 0.1 and 2.5‰ S (Fig. 3). He also observed that the locomotor activity of animals was greatest in 0.5‰. These results fit well with field observations, for in nature these animals live in low salinity waters, usually below 5‰ S.

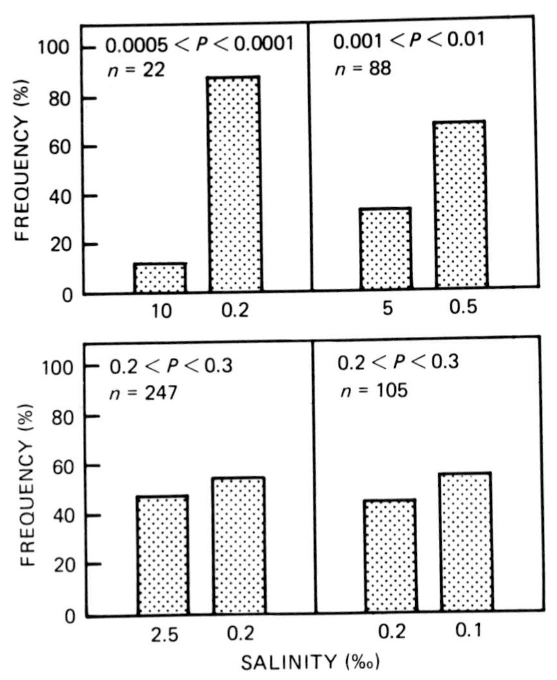

Fig. 3. Salinity preference of harpacticoid copepod, *Parastenocaris vicesima*. *n* Represents the total number of living specimens at the end of the experiment. Note that in all cases the species preferred the lower salinity water. (From Jansson, 1967.)

Among sand-dwelling meiofauna, many species are able to distinguish between sand grain sizes (Boaden, 1962). For example, the interstitial archiannelid, *Protodriloides symbioticus* (*sensu* Boaden and Irwin, 1971) demonstrated a preference for sand of 200–300 μm diameter particles (Gray, 1966a). However, not all sand grains within this size range were equally attractive. In further studies on *P. symbioticus,* Gray (1966b) noted that the type and number of bacteria coating the sand grain surface were also important. If the bacteria were removed from the sand by acid washing or heating to a high temperature, the attractiveness of the sand grains was destroyed. The attractiveness of the bacteria-laden sand grains, however, lies not so much in their numbers per unit area but rather in the bacterial species present (Gray and Johnson, 1970). They found that specific coccoid bacteria are responsible for the attractiveness of certain sediments. Their experimental organism, the gastrotrich, *Turbanella hyalina,* was most attracted to bacteria that were coccoid in shape, and sensitive to EDTA and lysozyme treatment; these bacteria were still attractive after the treatments, evidence that the attractive character resides in the cell wall and not in the internal structure.

There has been some question as to whether the response of meiofaunal organisms is a chemotactic one or simply a tactile response. Gray (1966a, 1967a,b) postulated that since some organisms show avoidance of a treated sand grain only after direct contact, the response is tactile; for if it were a chemotactic response, unattractive sand grains could be detected from a distance.

Behavioral factors other than sand grain size and preferred bacterial coatings can also play a role in environment selection. In a study on the interstitial harpacticoid copepod, *Leptastacus constrictus,* Gray (1968) found that geotactic and photoresponses were important in determining vertical distribution. When these copepods were placed in a light–dark choice apparatus, *L. constrictus* proved to be highly photonegative (Table I). They were also shown to be negatively geotactic. Thus the vertical distribution of this organism in sand is controlled by a balance between a negative geotaxis, which brings the animals to the sand surface, and a photonegative response, which keeps the animals just below the sand surface.

It also appears that chemoreception can play a role in environment selection in some species of meiofauna as illustrated by the work of Boaden and Erwin (1971) with the archiannelid, *Protodriloides symbioticus,* and the gastrotrich, *Turbanella hyalina*. These two species occur on the same beach, but the populations do not overlap despite a similarity in their zones of lethality as shown in laboratory studies. In an attempt to explain the distributional patterns of these two species, choice experiments were designed in which each species was presented heat-treated sand and sand containing the other species. These experiments offered clear-cut evidence that *T. hyalina* will avoid sand inhabited by *P. symbioticus,* thus indicating environment selection by negative chemotaxis.

TABLE I

Light Response of the Interstitial Harpacticoid Copepod, *Leptastacus constrictus*[a,b]

Experiment	Number of animals	
	In light	In dark
I	23	124
II	6	77
II	9	118
Total	38	319
Control (all light)	49	38

[a] From Gray (1968).

[b] After 2 hr in sand exposed to light and sand without light.

IV. Feeding

The bioenergetics of an animal depends on an input of energy and chemical necessities in the form of food. Early workers speculated that the meiofauna were primarily detrital feeders or indiscriminate feeders on benthic diatoms and bacteria (Mare, 1942; Perkins, 1958), but more recent studies indicate that there is great diversity in feeding patterns and food preferences among these organisms (Coull, 1973; Ivester, 1975; Marcotte, 1977; Tietjen and Lee, 1977a). There is also some evidence that certain species of meiobenthic nematodes can utilize dissolved organic matter (Chia and Warwick, 1969). In the laboratory, meiofaunal species have been cultured on bacteria, yeasts, algae, and ciliates (Sellner, 1976; Tietjen and Lee, 1977a).

How does a meiofaunal species select a particular type of food? There have been very few studies that experimentally demonstrate what stimuli initiate feeding activity. In two species of intertidal amphipods, *Bathyporeia pilosa* and *B. sarsi,* the feeding stimulus appeared to be mechanical (Nicholaisen and Kunneworff, 1969). These amphipods lived on food material or matter adhering to sand grains. The authors observed that the animals thoroughly worked over some of the sand grains with their mouth parts, while other sand grains were ignored. To determine if feeding was initiated chemically or mechanically, a number of animals were allowed to burrow in sand mixed with particles of an inert fluorescent pigment which adhered to sand grains. The amphipods were permitted to feed for a short while, then removed and examined. In all of the animals, the foregut was completely filled with particles of the compound. Thus it appeared that the attractiveness of the sand grain was determined by the amount of material adhering to the grain rather than through any chemical cue.

Predator–prey relationships between some meiofaunal species appears to be primarily a function of predator density. In a study of the predator–prey system between the polyp, *Protohydra leuckarti* (predator species) and the harpacticoid copepod, *Tachidius discipes* (prey species), Heip and Smol (1976a) observed no active search for the prey nor active avoidance of the predator. Whether or not a prey organism was captured depended on the probability of contact with the predator, which in turn was primarily related to density (Fig. 4). They did note, however, that it was the movement of the individual prey that evoked the feeding response of *P. leuckarti*. One of us has observed and evoked predatory responses from proseriata turbellarians by killing (squeezing with tweezers) harpacticoid copepods cohabiting with the flatworms. Upon killing the copepod, the flatworms swarmed from all areas of the 8.9 cm diameter experimental petri dish to the dead copepod, everted their pharynxes, and rapidly sucked out the internal juices, leaving only the copepod carapace. The turbellarians then randomly scattered over the dish. Live copepods were never attacked, but as soon as one was killed, the flatworms swarmed to the carcass and immediately devoured the dead

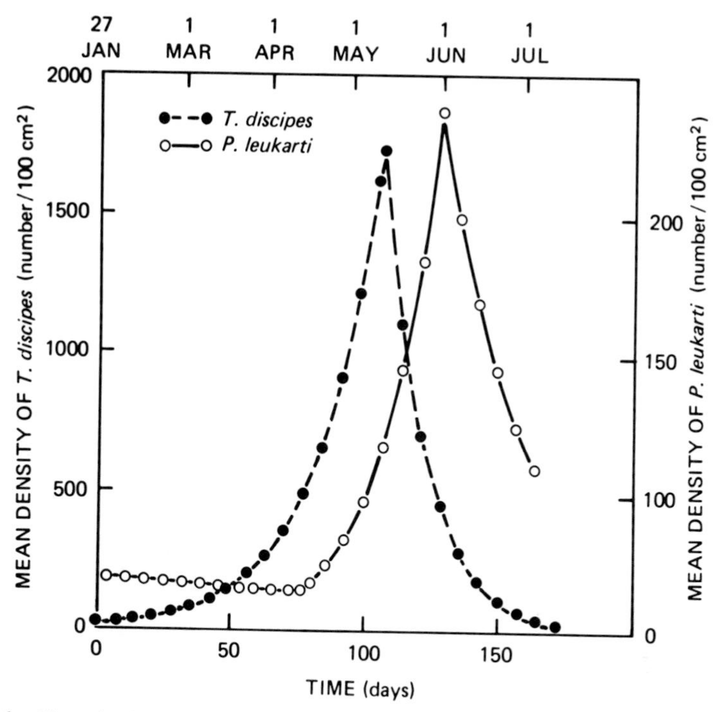

Fig. 4. Mean density of simulated model of the polyp, *Protohydra leuckati* (predator species), and the harpacticoid copepod, *Tachidius discipes* (prey species). Note that the predator population peak immediately follows the prey peak. (From Heip and Smol, 1976a.)

copepod (B. C. Coull, personal observation). This type of behavioral response suggests chemoreception, since copepods and turbellarians regularly "bumped into" one another while alive, but a feeding response was only elicited when internal body fluids pervaded the water.

Much of the work on feeding in meiofauna has been done on nematodes. Wieser (1953) proposed a classification of the types of feeding among the nematodes based on size of buccal cavity and whether armed or unarmed with teeth, and he suggested that there was a correlation between the type of buccal cavity and feeding pattern. Thus, species with large buccal cavities and no teeth may be nonselective deposit feeders; those with small buccal cavities and no teeth, selective deposit feeders; those with teeth and small buccal cavities, epigrowth feeders; and those with large buccal cavities and teeth, omnivores and/or predators.

In a comparative study on feeding habits of nematodes, Tietjen and Lee (1977a) worked with four species. Two of the species, *Chromadora macrolaimoides* and *Chromadora germanica,* are epigrowth algae feeders; i.e., they have three small teeth in the buccal cavity and they either scrape food materials off larger particles or use a piercing and sucking method for obtaining cell liquid. In contrast, the other two species, *Rhabditis marina* and *Monhystera denticulata* [deposit feeders based on Wieser's (1953) classification], feed by means of the sucking power of the esophagus. Under controlled laboratory conditions, Tietjen and Lee found that *C. macrolaimoides* and *C. germanica* ingested significantly more bacteria and algae than did the other two species. A comparison of the percentage of algae and bacteria grazed by *C. germanica, M. denticulata,* and *R. marina* relative to the total number of algae and bacteria available for consumption showed striking species differences. *Chromadora germanica* grazed only one bacterium to any extent, but the daily ingestion rate of algae was at or nearly equal to the body weight of the animal. On the other hand, *M. denticulata* and *R. marina* ingested far more bacterial cells than algal ones. Although they did consume some algae, these nematodes do not have any buccal armature, and only algae cells that could be handled by the stoma were processed. Thus, cell size and shape are of prime importance to these animals.

In nematodes, digestive secretions are produced by a few esophageal gland cells and a single layer of gut cells, and the low number of secretory cells may limit the kinds and amount of digestive enzymes that are secreted. In addition, the secretory gut cells also function in the absorption process (Lee, 1965). With the limited number of digestive enzymes secreted, it would seem logical to expect a close correlation between mode of life, type of diet, and specific digestive enzymes. Jennings and Deutsch (1975) illustrated such a correlation with β-glucuronidase activity in two marine nematodes, *Chromodorina germanica* and *Monhystera denticulata*. As mentioned above, *C. germanica* feeds primarily on algae and *M. denticulata* on bacteria. β-Glucuronidase is found only in the intestine of adult specimens of *M. denticulata*. Jennings and Deutsch speculated that *M. denticulata* utilized β-glucuronidase in their feeding since this enzyme hydrolyzes components of mucopolysaccharides and since the bacteria consumed by this species had a polysaccharide coat. Ecologically, of course, the ability of the nematodes to selectively ingest and digest food material may be one way to reduce interspecific competition.

Marcotte (1977) using slow motion video taping and SEM has recently devised a scheme defining four feeding groups in harpacticoid copepods.

(1) Point feeders: animals that locate a food site on a large mineral or organic particle and feed only from that point.
(2) Line feeders: those harpacticoids that clutch rectilinear solids and cylinders with grappling hook-like first legs and move up and down the line gleaning particles with tiny hooks and claws of the oral appendages.
(3) Plane sweepers and sand filers: animals that sweep food into their mouth from large organic detritus particles with the second antennae or scrape their food from pits and faults on sand grains.

(4) Solid feeders of which there are three types: (a) prey crushers who grasp and devour prey (e.g., nematodes); (b) sphere cleaners who twirl spheres and organic floccules in their mouths cleaning food from the entire surface; and (c) rubble sorters which pass organic debris over and through their mouth parts, cleaning food off the particles and passing the residue through a vaulted space running through swimming legs.

If Marcotte's designations are correct, then it is obvious that a wide variety of food is used by harpacticoid copepods.

The amount of interaction between the meiofauna and higher trophic levels has been the subject of considerable discussion. Both McIntyre and Murison (1973) and Heip and Smol (1976a) suggest that meiobenthic prey species are consumed primarily by meiobenthic predators and thus are not available to higher trophic levels. Lasserre *et al.* (1976) speculated that there may be competition for food between species of detrital feeding meiofauna and the grey mullet, *Chelon labrosus,* and McIntyre (1964) and Marshall (1970) concluded that there was competition for food between macrobenthic filter-feeders and meiofauna, and that meiofauna serve primarily as rapid metazoan nutrient regenerators. However, Feller and Kaczynski (1975) and Sibert *et al.* (1977) showed quite conclusively that juvenile salmon fed almost exclusively on meiobenthic copepods, and Odum and Heald (1972) reported that meiobenthic copepods made up 45% of grey mullet gut contents. Sikora (1977) has recently reported that nematodes provide a significant portion of the *in situ* food of the grazing grass shrimp, *Paleamonetes pugio*.

It is probably true, however, that meiofauna serve as food for higher trophic levels more in muds than in sands. Table II lists those papers that discuss the functional role of meiofauna *in situ;* in almost all cases where meiofauna are known to be food for higher trophic levels, the study has been conducted in a muddy or detrital substrate. Conversely those papers which suggest that the meiofauna are not food for higher trophic levels are based primarily on work done in sandy environments. In muddy/detrital substrates most of the meiofauna are restricted to the top-most layers of sediment where an indiscriminant browser/ingestor would inevitably collect them. With all the available interstitial space in sands, however, meiofauna generally go deeper into the sediment and would thus not be as susceptible to browsing predation. Figure 5 schematically illustrates mud and sand cores and associated redox layers with the mean values of numbers of animals and biomass from 5 years of monthly subtidal data in North Inlet, South Carolina. The mud core has approximately twice the meiofaunal biomass as the sand core, and in the mud all the fauna is concentrated in the upper cm, whereas in the sand the fauna is distributed to a depth of 10–15 cm. Obviously, if higher trophic levels feed on meiofauna it seems most reasonable to expect that predation pressures are most pronounced in muddy/detrital substrates and not of great significance in sands. This, of course, does not preclude higher

TABLE II

Summary of Papers Which Mention Functional Role of Meiofauna *in Situ*

Food for higher trophic levels		Nutrient remineralization/competition with macrofauna	
Author	Substrate (predator)	Author	Substrate
Smidt (1951)	mud/sand (plaice)	McIntyre (1964)[a]	sand/mud
Bregnballe (1961)	mud/sand (plaice)	Tietjen (1967)	muddy sand
Teal (1962)	mud (*Uca*)	McIntyre (1969)	sand
Odum and Heald (1972)	detritus (several fish)	Marshall (1970)	sand
Braber and De Groot (1973)	? (larval plaice)	McIntyre (1971)[a]	sand
Walter (1973)	mud (sipunculan)	McIntyre and Murison (1973)	sand
Feller and Kaczynski (1975)	epibenthic (larval salmon)	Lasserre *et al.* (1976)[a]	detritus–mud
Sibert *et al.* (1977)	detritus (larval salmon)	Giere (1975)	sand
		Feller (1977)	sand
Sikora (1977)	mud (grass shrimp)		
		McLachlan (1977)	sand
Bell and Coull (1978)	mud (grass shrimp)		
		Tenore *et al.* (1977)	sand (lab)

[a] These authors suggest competition with macrofauna.

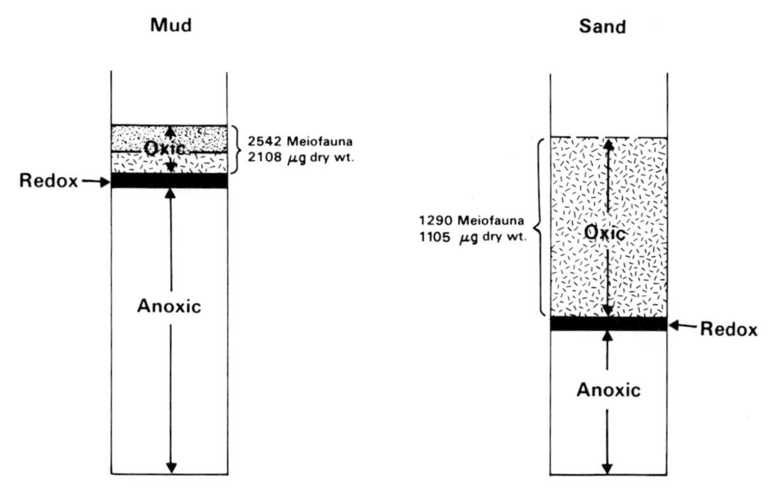

Fig. 5. A schematic diagram of the sediment at a muddy and sandy site in North Inlet, South Carolina, with numbers and biomass at the two sites. The oxidized layer in the mud is limited to 1 cm; in the sand it varies between 10 and 15 cm.

trophic level predation in sands or meiofauna nutrient regeneration/ mineralization/bioturbation, etc. in muds. However, it does emphasize the primary role in each biotope. As a hypothesis then, we suggest that the mud/detrital meiofauna do serve as a significant source of food for higher trophic levels (primarily natant browsers), whereas in sandy substrates the meiofauna primarily serve as rapid metazoan nutrient regenerators.

There is also some evidence that meiofauna in the marine food web may play a role in making detritus available to the macrobenthos (Tenore *et al.*, 1977). Net incorporation rates of 5 month aged eel-grass detritus by *Nephtys*, cultured with and without meiobenthos, were nearly doubled in cultures containing meiofaunal organisms. The authors suggest that the observed increase in net incorporation of the aged detritus could be due to a combination of factors, such as "meiofaunal enhancement of microbial activity and subsequent polychaete utilization and/or ingestion of the meiofauna themselves."

V. Respiratory Adaptations

As noted previously, some meiofaunal species can live anaerobically for relatively long periods of time, but most require molecular oxygen at some time in their life cycle to generate the energy needed to survive. The data on respiratory adaptations in meiofauna are not as extensive as they are for the macrofauna, but it is clear from the available studies that the meiofauna have evolved widely varying respiratory adaptations to their various environments. Because of the great abundance of meiofauna, these adaptations, as well as the actual respiratory rates, take on even greater importance as we attempt to quantify energy flow in marine systems.

Both intrinsic and extrinsic factors can influence respiratory rates. One of the most important of the intrinsic factors is body size, and generally meiofauna follow the dictum that smaller-sized organisms have a higher rate of oxygen consumption on a weight-specific basis (Lasserre, 1970, 1971, 1976; Atkinson, 1973b; Ott and Schiemer, 1973; Vernberg and Coull, 1974). Data from these studies are commonly expressed as a power function of body size. For weight specific data, the formula is:

$$O_2 = aW\ (b-1) \text{ or } \log O_2/W = \log a + (b-1) \log W$$

where O_2 is the oxygen consumed per unit time; W is the Weight of animal, which may be expressed as wet weight, dry weight, nitrogen, etc.; and a and b are coefficients in which a represents the intercept of the y-axis and b represents the slope of the function in the logarithmic plot.

The relationship of oxygen consumption to body size can be determined from the slope of the line (the constant b). Using data from early studies, Zeuthen (1953) postulated that the metabolism of organisms larger than 40 mg is proportional to 0.75 (b value) of the body weight, but in organisms ranging in size from 1 to 40 mg, the b value is 0.95. However, Hemmingson (1960) was able to demonstrate convincingly that the proportionality of metabolism to body size is the same among all organisms and that this value is approximately 0.75. Using respiratory data on meiofauna from a number of studies (Fig. 6), Vernberg and Coull (1974) calculated a value of 0.74, which supports Hemmingson's thesis and the data of Banse *et al.* (1971) and Klekowski *et al.* (1972).

While the evidence is clear that the meiofauna maintain the same metabolic rate–body size proportionality as do the macrofauna, factors other than size can influence the metabolic rate of these organisms, including temperature, season of the year, activity (Coull and Vernberg, 1970), and mode of collection/extraction (Vernberg *et al.*, 1977).

As in chemical reactions, an increase in temperature is typically accompanied by an increase in metabolic rate in poikilotherms up to a certain temperature. At this temperature, which usually is within the thermal boundaries to which the organisms are exposed in nature, the respiratory rate tends to level off in a plateau. Such responses in meiofauna have been reported by Hagerman (1969), Lasserre (1971), Lasserre and Renaud-Mornant (1971), Hummon (1975), and Wieser and Schiemer (1977).

Seasonal changes in respiratory rates have also been observed in meiofauna. When oxygen uptake rates of the copepod, *Thompsonula hyaenae,* were measured at 10° and 25°C at different seasons of the year, the rate at 10°C was more

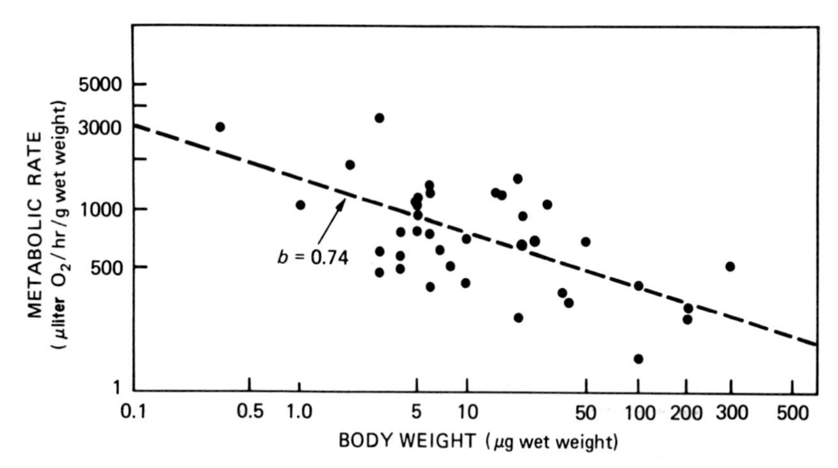

Fig. 6. The relation of metabolism to body weight of some meiofauna and the ciliate *Tracheloraphis* sp. Oxygen uptake rates of the meiofaunal data are from Gerlach (1971). All data were corrected to 20°C using the correction factor of Winberg (1971). (From Vernberg and Coull, 1975.)

than twice as high during the spring as it was during the summer (Fig. 7). Conversely at 25°C, the rate was lowest during the spring (W. B. Vernberg and B. C. Coull, unpublished data). Such acclimation is typical of many temperate zone organisms. Wieser and Schiemer (1977) demonstrated a seasonal shift in metabolic rate in the nematode, *Trefusia schiemeri*. No seasonal acclimation was observed in the nematode, *Theristus floridanus,* although populations of both species have been reported on Bermudian beaches. However, *T. floridanus* is found only seasonally and over a narrower thermal range than *T. schiemeri,* demonstrating the close correlation between metabolic adaptation and the thermal microenvironment of meiofaunal species.

It is difficult to generalize about the respiratory response of meiofauna to different salinities. Most of the studies on metabolic response to different salinities have been done on estuarine species, and since changing salinities are characteristic of estuaries, it is not surprising that there is a wide diversity of metabolic responses to variable salinities. For example, the respiratory rate of the gastrotrich, *Turbanella ocellata,* was maximum at summer ambient salinities; at greater or lesser salinities, oxygen consumption rates tended to be lower (Hummon, 1975). In contrast, both the oligochaeta, *Marionina achaeta,* and the mystacocarid, *D. remanei* had peak respiratory rates at both high and low salinities (Lasserre, 1971, 1975; Lasserre and Renaud-Mornant, 1971). Interpretation of these results on intact meiofauna is speculative at best. Lasserre (1969, 1975, 1976) suggested an energetic relationship between ionic regulation and metabolic rate, while Hummon (1975) speculated that the observed metabolic

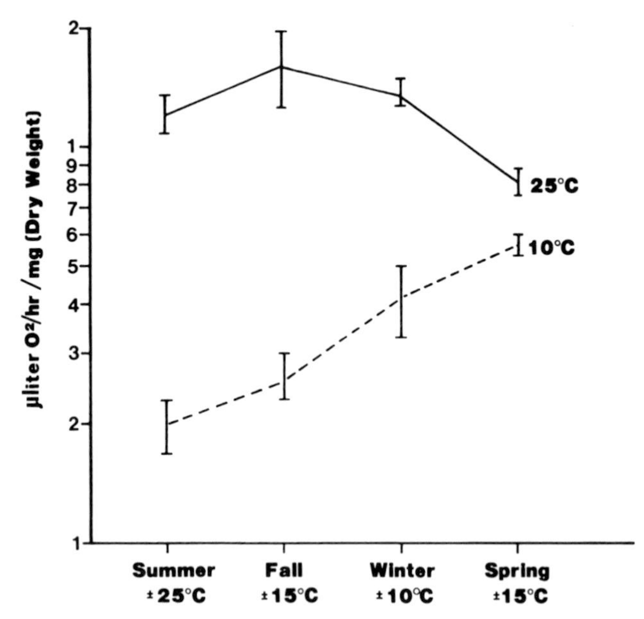

Fig. 7. Seasonal respiratory rates of the harpacticoid copepod *Thompsonula hyaenae* measured at 10° and 25° C. Ambient seasonal temperatures are noted on the abcissa.

response was dependent on the salinity at which the organism was acclimated. It is also possible that a change in salinity may alter the locomotor ability of the animal so that observed changes in metabolic rate are the result of behavioral changes rather than the effect of salinity on a basic metabolic process.

While the metabolic response of intact meiofauna to changing salinities cannot be clearly linked to habitat preference, results of studies on metabolic response to different oxygen tensions show a very good correlation with the oxygen levels encountered by the meiofauna in their habitat. Lasserre (1976) proposed a classification of metabolic responses to different oxygen levels.

(1) Euroxybiontic regulators: species able to regulate their metabolic rate down to very low oxygen tension levels.
(2) Stenoxybionts: the metabolic rate of these species characteristically conforms to ambient oxygen levels.
(3) Complex metabolic types: species in this category cannot clearly be defined as either conformers or regulators.

Generally, euroxybiontic regulators are defined as species that live in oxygen-deficient habitats; stenoxybiontic species are those that require well-oxygenated environments; and the more complex metabolic types tend to be species that can survive poorly-oxygenated habitats, but appear to be better adapted to fluctuating or high oxygen level habitats.

Data for species living in habitats where oxygen levels are very low indicate that the oxygen consumption rates of such species are generally lower than the rates of species living in habitats where oxygen levels are high (Ott and Schiemer, 1973; Schiemer, 1973; Lasserre and Renaud-Mornant, 1973). The physiological adaptations that allow meiofaunal organisms to live at very low oxygen levels are not fully understood, but several adaptations have been suggested. Hemoglobin that seems to function at low P_{O_2} has been reported in the nematode, *Enoplus communis* (Ellenby and Smith, 1966; Atkinson, 1973a), and Wieser *et al.* (1974) observed coloration changes in *Paramonhystera* at low oxygen tensions. Another mechanism, a reverse Krebs cycle sequence, has been suggested by Maguire and Boaden (1975) for the gastrotrich, *Thiodasys sterreri*.

One of the questions that can be raised about any attempt to measure a rate function in the laboratory is whether or not the responses of the animals are similar to what they would be in the natural environment, and thus whether or not the measured values are quantitatively valid. Such questions are particularly relevant to meiofauna since collection, extraction, and sorting often involves techniques requiring conditions which the organisms do not usually encounter. Growing enough animals for experimental purposes and efficiently extracting them from the substrate is very real problem, particularly if the substrate is mud. Further, since meiofauna live in close contact with the substrate, does removal of the animal from the substrate have metabolic consequences? Data indicate that both the method of extraction and the presence or absence of substrate can affect the outcome of respiratory determinations (Vernberg *et al.*, 1977). When the mud-dwelling copepod, *Nannopus palustris,* was extracted using the highly efficient sieving, centrifugation, and sucrose flotation method, measured respiration rates were significantly lower than when the copepods were laboriously hand-sorted from the sediment. Similarly when metabolic determinations were made on the interstitial copepod, *Hastigerella leptoderma,* in the absence of sand grains the rates were significantly higher than when sand grains were provided as substrate. The higher metabolic rate observed in those copepods lacking a substrate can undoubtedly be linked to increased activity, for in the presence of a substrate these animals curl around a particle and are relatively quiescent, whereas in the absence of a substrate they flail about vigorously.

VI. Osmoregulatory Adaptations

Much of the available data on osmoregulatory capability of marine meiofauna has been obtained by observing volume changes in intact organisms following exposure to different osmotic concentrations of various ions. Hummon (1975) working with the gastrotrich, *Turbanella ocellata,* can serve to illustrate such a study. Specimens of *T. ocellata,* an estuarine lower intertidal zone species, were placed in salinities ranging from 2 to 80%, and the changes in their body width were monitored. The immediate response of *T. ocellata* to low salinity water was volume increase and to high salinity water, volume decrease; thus, this species proved to be a volume osmotic regulator. With time the body volume returned to normal, probably by means of an isosmotic solute–solvent flux. Hummon speculated that there was active transport since osmoregulatory capability was lessened in the presence of cyanide.

Croll and Viglierchio (1969) also used changes in body volume to measure the osmoregulatory capability of the marine nematode, *Deontostoma californicum*. This species was able to osmoregulate in hypertonic solutions but not in hypotonic ones. Such observations suggest that this species is limited to the intertidal zone in high salinity areas. An interesting aspect of this study was the evidence from ligaturing experiments which indicated that the uptake of ions from hypertonic solutions occurred through the integument immediately under the cuticle. Neither the gut nor localized receptors appeared to be involved in the osmoregulatory process.

The oligochaete, *Marionina achaeta,* also has the ability to osmoregulate under hypo- and hyperosmotic conditions. Lasserre (1975) reported that this species shows good volume regulation over a wide range of salinities. He also presented evidence of active transport mechanisms and suggested that the nephridia and epidermal cells were involved in this process.

Since so very little is known of the osmoregulatory ability of marine meiofauna, it is obviously a fertile source of study to pursue and worthy of much more attention by meiofaunal ecophysiologists.

VII. Reproduction and Growth

It is not our purpose to review all the literature on meiofaunal reproduction and growth but rather to discuss what is known about the physiology of meiofaunal reproduction/growth and the environmental cues that trigger such activities. Henley (1974), Higgins (1974), Hummon (1974a), Hope (1974), and Pollock (1974) have recently reviewed the reproductive mechanisms of several meiobenthic taxa, i.e., Turbellaria, Kinorhyncha, Gastrotricha, Nematoda, and Tardigrada, respectively, and we see no need to reiterate them here.

Reproduction and growth are just two components intimately tied into the whole life history pattern, a pattern which has evolved to allow survival in the meiofaunal habitat. In general most meiobenthic taxa have life history patterns that appear to minimize the energetic cost of reproduction and to optimize the number of offspring in succeeding generations. Several adaptations appear to be particularly well suited for success in the meiofaunal habitat. Most meiofauna have a low number of eggs and many of them (particularly in the interstitial habitat) are "sticky." Thus, when they are released into the environment they remain in close proximity to the population center. Copulation, oviposition, or spermatophore transfer are the rule rather than the exception in meiobenthic taxa, thus ensuring that fertilization occurs. Except for the crustacean elements of the permanent meiofauna, direct development predominates and few produce larval stages. Crustaceans (e.g., copepods, ostracods) do have true larval stages, but in the majority of the species (>90%) all larval development takes place in the sediment and not in the water column, again insuring relatively little loss from the system. In fact, almost all meiofaunal development takes place within the sediment, whether direct or indirect. An exception can be found among the archiannelids; only one of four interstitial species listed by Schroeder and Hermans (1975) had direct development, the other three were pelagic.

In addition to minimizing loss of reproductive output, reproductive potential of many meiofaunal taxa is increased through hermaphroditism (see Hummon, 1974a) and parthenogenesis (particularly in freshwater meiobenthic forms). Generation times many also be relatively short and some species are able to reproduce throughout the year (Gerlach, 1971; Coull and Vernberg 1975). Furthermore, the ability of some meiofauna to delay development with resting eggs (primarily in freshwater habitats) or larval stage delay (Coull and Dudley , 1976) is another unique attribute that promotes maximum utilization of the biotope with minimized reproductive loss. Even within taxa that are not predominantly meiobenthic but which have some meiobenthic representatives (e.g., Mollusca, Hydrozoa, Echinodermata), ''shrinking'' has also required mechanisms to optimize reproductive output. Interstitial molluscs are known to have cuticular fertilization, attached cocoons, and epithelial spermatophores (Swedmark, 1968, 1971), whereas interstitial hydrozoa are known to be hermaphroditic and exhibit polyp fragmentation, incubation, brooding, direct development, and viviparity (Clausen, 1971).

These attributes, while not entirely restricted to the meiofauna, are life history adaptions of particular importance to meiofaunal organisms. In many, small size and cell constancy certainly limit the number of cells available for reproduction. Furthermore, the relatively high cost of maintenance in small animals (i.e., the typical regression where respiratory cost per unit of body weight decreases with increasing body size) leaves proportionately less energy available for reproduction. The high population densities of the meiofauna can indeed be attributed to their very effective reproductive adaptations.

Of all the environmental cues that may trigger reproductive and/or developmental activity in meiofauna, temperature is by far the most studied, which explains why it is most frequently cited as the most responsible in triggering the event. In almost all species studied it has been found that increased temperature (not exceeding or approaching the upper lethal limit) decreases egg development time (Heip and Smol, 1976b), length of postembryonic development (Apelt, 1969; Gerlach, 1971; Gerlach and Schrage, 1971; Hopper *et al.*, 1973; Tietjen and Lee, 1972, 1977b) and generation time (Gerlach and Schrage, 1971; Heip and Smol, 1976b) and increases reproductive potential (Heip, 1977; Heip and Smol, 1976b; Tietjen and Lee, 1977b). There are limits, of course, beyond which increasing temperature is counter productive. Hopper *et al.* (1973) conclusively showed that even with tropical nematodes, copepods, foraminifera, and the capitellid polychaete, *Capitellides giardi,* high temperatures >33°C caused elongation of the life cycle in some and stopped reproduction entirely in others. Numerous studies on meiofauna taxa, as with most other organisms, come to the same conclusions when dealing with temperature, i.e., life processes are slowest at low temperatures, reach maximum rates at some optimum temperature range, and decrease beyond the optimum.

What other factors play a significant role in growth and reproduction of meiofauna besides temperature? Although every potential environmental factor is a possibility, data are only available on three; salinity, food quality, and crowding. Tietjen and Lee (1972, 1977b) report that life history parameters of two marine nematodes, *Monhystera denticulata* and *Chromadorina germanica,* appeared to be as sensitive to salinity as they were to temperature. With *M. denticulata,* as conditions varied from the optimum of 25°C and 26‰ S, an increase or decrease in salinity of 13‰ resulted in a doubling of the generation time. With *C. germanica,* however, a drop of 13‰ from the optimum salinity resulted in a doubling of the generation time, but an increase in salinity did not change the growth characteristics. Tietjen and Lee (1977b) suggested that since marine nematodes osmoregulate better in hypertonic solution compared to hypotonic solution (Croll and Viglierchio, 1969), much energy is expended in maintenance at the low salinities, and thus the lower reproductive rate. They offered no explanation for the *M. denticulata* generation time increases at high salinity.

Several authors have hinted at the possible change of food quality and its role in controlling life histories (Apelt, 1969; Gerlach and Schrage, 1971). Sellner (1976) showed that hatching success of the harpacticoid copepod, *Thompsonula hyaenae,* was highest when the gravid female was fed the diatom, *Navicula pelliculosa* (Fig. 8). In contrast, naupliar survival was best when fed *Navicula* sp. Sellner concluded that a mixed diet would probably be best for the overall well-being of a species, and her results clearly indicate the need for more detailed analysis of the relationship between food quality/quantity and meiofaunal life history patterns.

Heitkamp (1972) reported on egg development and formation in an interstitial turbellarian that may have a resting egg or the normal subcutaneous, live-bearing egg. In his German populations he found that the eggs were thermally influenced; at low temperatures only subcutaneous eggs were produced, and at high temperatures only resting eggs were produced. Most intriguing, however, was the finding that in Finnish and French populations, resting eggs occurred only in over-

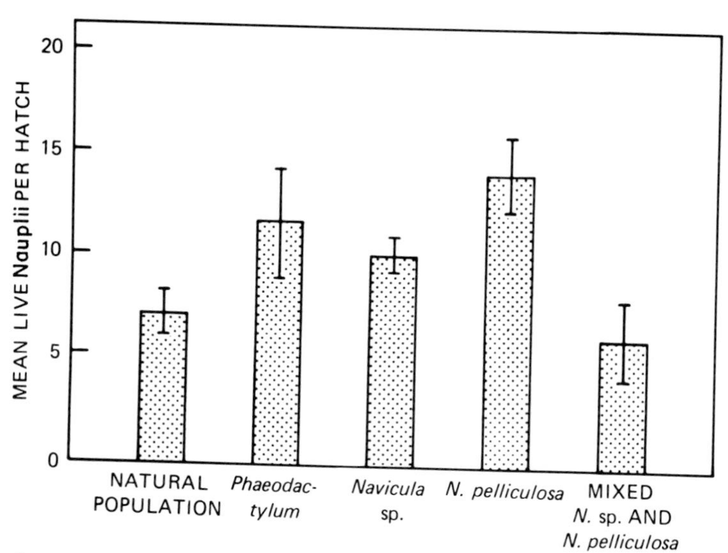

Fig. 8. Mean number of live nauplii per hatch produced by the harpacticoid copepod, *Thompsonula hyaenae*, grown of five food combinations.

crowded conditions, Heitkamp (1972) attributed formation of the resting egg in the overcrowded conditions to a "substance" secreted by the animals, which then triggered the resting egg formation. Since the few meiofauna developmental studies have usually been done in noncrowded, ideal laboratory conditions, it is impossible to tell how widespread such effects may be. The development of a resting egg, however, is obviously a tactic to delay development. Coull and Dudley (1976) reported on a unique delaying response to benthic copepods, where some of the nauplii would "grow up" immediately while others (from the same hatch) would become supine, attach to the substrate, and be inactive. On some "cue," up to 50 days after the hatch, these "resting" nauplii would begin normal larval development. The authors did not know the reason for the prolongation, but suggested that by not overcrowding the habitat with a single size class, a more equitable utilization of the available resources would be allowed.

Environmental contaminants are often thought to alter the life histories and/or reproductive capabilities of marine animals, and although there have been several studies dealing with the effects of contaminants on population or community structure of meiofauna, there are very few that deal with contaminant effects on reproduction and growth. Dalla Venezia and Fossato (1977) showed that there was no significant difference in eggs produced, number of nauplii, and hatching success of the harpacticoid copepod, *Tisbe bulbisetosa,* whether exposed to oil or not. The only data on the effects of contaminants on life history phenomena in meiofauna of which we are aware are those of Faucon and Hummon (1976),

Hummon (1974) and Hummon, and Hummon (1975), who studied the effects of DDT on the freshwater gastrotrich, *Lepidodermella squammata*. Although *L. squammata* is not a marine species, the data available are the most complete. In these studies, three life table parameters, mean life expectency (e_x), net reproductive rate (R_0) and intrinisic rate of increase (R_{max}), all decreased with increasing concentrations of DDT, illustrating the profound effects of at least one contaminant on the reproductive ability of this species. Obviously, such work in the marine environment is required if we are ever to estimate potential contaminant effects on the meiofauna.

References

Apelt, G. (1969). Fortpflanzungbiologie, Entwicklungszklen und vergleichende Fruhentwicklung acoeler Turbellarien. *Mar. Biol.* **4,** 267–325.

Atkinson, H. J. (1973a). The respiratory physiology of the marine nematodes *Enoplus brevis* (Bastian) and *E. communis* (Bastian). I. The influence of oxygen tension and body size. *J. Exp. Biol.* **59,** 255–266.

Atkinson, H. J. (1973b). The respiratory physiology of the marine nematodes *Enoplus brevis* (Bastian) and *E. communis* (Bastian). II. The effects of changes in the imposed oxygen regime. *J. Exp. Biol.* **59,** 267–264.

Banse, K., Nichols, F. M., and May, D. R. (1971). Oxygen consumption of the sea bed. III. On the role of the macrofauna at three stations. *Vie Milieu, Suppl.* **22,** 31–52.

Bell, S. S., and Coull, B. C. (1978). Field evidence that shrimp predation regulates meiofauna. *Oecologia* **35,** 141–148.

Bilio, M. (1967). Die aquatische Bodenfauna von Saxzwiesen der Nord-und Ostee. III. Die biotopeinflusse auf die Faunenverteilung. *Int. Rev. Gesamten Hydrobiol.* **5,** 487–533.

Bleakley, R. J., and Boaden, P. J. S. (1974). Effects of an oil spill remover on beach meiofauna. *Ann. Inst. Oceanogr. (Paris).* **50,** 51–58.

Boaden, P. J. S. (1962). Colonization of graded sand by an interstitial fauna. *Cah. Biol. Mar.* **3,** 245–248.

Boaden, P. J. S. (1971). *Turbanella hyalina* versus *Protodriloides symbioticus:* A study in interstitial ecology. *Vie Milieu, Suppl.* **22,** 479–492.

Braber, L., and De Groot, S. J. (1973). The food of five flatfish species (Pleuronecticormes) in the Southern North Sea. *Neth. J. Sea Res.* **6,** 163–172.

Bregnballe, F. (1961). Plaice and flounder as consumers of the microscopic bottom fauna. *Medd. Komiss. Dan, Fisk. Havunders.* [N.S.] **3,** 133–182.

Brickman, L. M. (1972). Base food chain relationships in coastal salt marsh ecosystems, Ph.D. Thesis, Lehigh University, Bethlehem, Pennsylvania.

Chia, F. S., and Warwick, R. M. (1969). Assimilation of labelled glucose from seawater by marine nematodes. *Nature (London).* **224,** 720–721.

Clausen, C. (1971). Interstitial Cnidaria: Present status of their systematics and ecology. *Smithson. Contrib. Zool.* **76,** 41–67.

Coull, B. C. (1969). Hydrographic control of meiobenthos in Bermuda. *Limnol. Oceanogr.* **14,** 953–957.

Coull, B. C. (1973). Estuarine meiofauna: A review, tropic relationships and microbial interactions. *In* "Estuarine Microbiology and Ecology" (L. H. Stevenson and R. R. Colwell, eds.), Univ. South Carolina Press, Columbia, South Carolina, pp. 499–511.

Coull, B. C., and Dudley, B. W. (1976). Delayed nauplier development of Meiobenthic copepods. *Biol. Bull. (Woods Hole, Mass.)* **150,** 38–46.

Coull, B. C., and Vernberg, W. B. (1970). Harpacticoid copepod respiration: *Enhydrosoma propinquum* and *Longipedia helgolandica*. *Mar. Biol.* **5**, 341–344.

Coull, B. C., and Vernberg, W. B. (1975). Reproductive periodicity of meiobenthic copepods: Seasonal or continuous? *Mar. Biol.* **32**, 289–293.

Croll, N. A., and Viglierchio, D. R. (1969). Osmoregulation and the uptake of ions in a marine nematode. *Proc. Helminthol. Soc. Wash.* **36**, 1–9.

Dallas Venzia, L., and Fossato, V. U. (1977). Characteristics of suspensions of Kuwait oil and Corexit 7664 and their short and long-term effects of *Tisbe bulbisetosa* (Copepoda: Harpacticoida). *Mar. Biol.* **42**, 233–327.

Deboutteville, C. D. (1960). "Biologie des eaux souterraines littorales et continentales." Hermann, Paris.

de Bovée, F. (1975). La nematofaune des vases autopolluées des îles Kerguelen (Terres Australes et Antarctiques françaises). *Cah. Biol. Mar.* **16**, 711–720.

Ellenby, C., and Smith, L. (1966). Haemoglobin in *Mermis subnigrescens* (Cobb) *Enoplus brevis* (Bastian), and *E. communis* (Bastian). *Comp. Biochem. Physiol.* **19**, 871–877.

Elmgren, R. (1975). Benthic meiofauna as indicator of oxygen conditions in the northern Baltic proper. *Merentutkimuslaitksen Julk.* **239**, 265–271.

Faucon, A. S., and Hummon, W. D. (1976). Effects of mine acid on the longevity and reproductive rate of the Gastrotricha *Lepidodermella squammata* (Dujardin). *Hydrobiologia* **50**, 265–269.

Feller, R. J. (1977). Life history and production of meiobenthic harpacticoid copepods in Puget Sound. Ph.D. Thesis, University of Washington, Seattle.

Feller, R. J., and Kaczynski, V. W. (1975). Size selective predation by juvenile chum salmon (*Oncorhynchus keta*) on epibenthic prey in Puget Sound. *J. Fish. Res. Board Can.* **32**, 1419–1429.

Fenchel, T., and Riedl, R. J. (1970). The sulfide system: A new biotic community underneath the oxidized layer of marine sand bottoms. *Mar. Biol.* **7**, 255–268.

Gerlach, S. A. (1971). On the importance of marine meiofauna for benthos communities. *Oecologia* **6**, 176–190.

Gerlach, S. A., and Schrage, M. (1971). Life cycles in marine meiobenthos. Experiments at various temperatures with *Monohystera disjuncta* and *Theristus pertenuis* (Nematoda). *Mar. Biol.* **9**, 274–280.

Giere, O. (1973). Oxygen in the marine hygropsammal and the vertical distribution of oligochaetes. *Mar. Biol.* **21**, 180–189.

Giere, O. (1975). Population structure, food relations and ecological role of marine oligochaetes, with special reference to meiobenthic species. *Mar. Biol.* **31**, 139–156.

Gray, J. S. (1966a). The response of *Protodrilus symbioticus* (Giard) (Archiannelida) to light. *J. Anim. Ecol.* **35**, 55–64.

Gray, J. S. (1966b). The attractive factor of intertidal sands to *Protodrilus symbioticus*. *J. Mar. Biol. Assoc. U.K.* **46**, 627–645.

Gray, J. S. (1966c). Selection of sand by *Protodrilus symbioticus* (Giard). *Veroeff. Inst. Meeresforch. Bremerhaven* **2**, 105–116.

Gray, J. S. (1967a). Substrate selection by the archiannelid *Protodrilus rubropharyngeus*. *Helgol. Wiss. Meeresunters.* **15**, 253–269.

Gray, J. S. (1967b). Substrate selection by the Arachiannelid *Protodrilus hypoleucus armenante*. *J. Exp. Mar. Biol. Ecol.* **1**, 47–54.

Gray, J. S. (1968). An experimental approach to the ecology of the harpacticid *Leptastacus constrictus* Lang. *J. Exp. Mar. Biol. Ecol.* **2**, 278–292.

Gray, J. S. (1976). The fauna of the polluted River Tees Estuary. *Estuarine Coastal Mar. Sci.* **4**, 653–676.

Gray, J. S., and Johnson, R. M. (1970). The bacteria of a sandy beach as an ecological factor

affecting the interstitial gastrotrich *Turbanella hyalina* Schultz. *J. Exp. Mar. Biol. Ecol.* **4,** 119–133.
Gray, J. S., and Ventilla, R. (1971). Pollution effects on micro- and meiofauna of sand. *Mar. Pollut. Bull.* **2,** 39–43.
Hagerman, L. (1969). Respiration, anaerobic survival and diel locomotory periodicity in *Hirschmannia viridis* Muller (Ostracoda). *Oikos* **30,** 384–391.
Heip, C. (1977). On the evolution of reproductive potentials in a brackish water meiobenthic community. *Microfauna Meeresboden* **61,** 105–112.
Heip, C., and Smol, N. (1976a) On the importance of *Protohydra leuckarti* as a predator of meiobenthic populations. *Proc. Eur. Symp. Mar. Biol., 10th, 1975* Vol. 2, pp. 285–296.
Heip, C., and Smol, N. (1976b). Influence of temperature on the reproductive potential of two brackish-water harpacticoids (Crustacea, Copepoda). *Mar. Biol.* **35,** 327–334.
Heitkamp, U. (1972). Die Mechanismen der Subitan-und Dauereibildung bei *Mesostoma linqua* (Abildgaard, 1789) (Turbellaria, Neorhabdocoela). *Z. Morphol. Tiere* **71,** 203–289.
Hemmingson, A. M. (1960). Energy metabolism as related to body size and respiratory surface and its evolution. *Rep. Steno Mem. Hosp. Nord. Insulinlab.* **9,** 7–11.
Henley, C. (1974). Turbellaria. *In* "Reproduction of Marine Invertebrates" (A. C. Giese and J. S. Pearse, eds.), Vol. 1, pp. 267–343. Academic Press, New York.
Higgins, R. P. (1974). Kinorhyncha. *In* "Reproduction of Marine Invertebrates." (A. C. Giese and J. S. Pearse, eds.), Vol. 1, pp. 507–518. Academic Press, New York.
Hope, W. D. (1974). Nematoda. *In* "Reproduction of Marine Invertebrates" (A. C. Giese and J. S. Pearse, eds.), Vol. 1, pp. 391–469. Academic Press, New York.
Hopper, B. E., Fell, J. W., and Cefalu, R. C. (1973). Effect of temperature on life cycles of nematodes associated with the mangrove (*Rhizophora mangle*) detrital system. *Mar. Biol.* **23,** 293–296.
Hummon, W. D. (1972). Dispersion of Gastrotricha in a marine beach of the San Juan Archipelago, Washington, *Mar. Biol.* **16,** 349–355.
Hummon, W. D. (1974a). Gastrotricha. *In* "Reproduction of Marine Invertebrates" (A. C. Giese and J. S. Pearse, eds.), Vol. 1, pp. 485–506. Academic Press, New York.
Hummon, W. D. (1974b). Effects of DDT on longevity and reproductive rate in *Lepidodumella squammata* (Gastrotricha, Chaetonotida). *Am. Midl. Nat.* **92,** 329–339.
Hummon, W. D. (1975). Respiratory and osmoregulatory physiology of a meiobenthic gastrotrich, *Turbanella ocellata*. *Cah. Biol. Mar.* **16,** 255–268.
Hummon, W. D., and Hummon, M. R. (1975). Use of life table data in tolerance experiments. *Cah. Biol. Mar.* **16,** 743–749.
Ivester, M. S. (1975). Ecological Diversification within benthic harpacticoid copepods. Ph.D. Thesis, University of South Carolina, Columbia.
Jansson, B. O. (1966). Microdistribution of factors and fauna in marine sandy beaches. *Veroeff. Inst. Meeresforsch. Bremerhaven* **2,** 77–86.
Jansson, B. O. (1967). The importance of tolerance and preference experiments for interpretation of mesopsmmon field experiments. *Helgol. Wiss. Meeresunters.* **15,**41–58.
Jennings, J. B., and Deutsch, A. (1975). Occurrence and possible adaptive significance of B-glucuronidase and arylamidase ("leucine amino peptidose") activity in two species of marine nematodes. *Comp. Biochem. Physiol. A* **52,** 611–614.
Klekowski, R. Z., Wasilewska, L., and Paplinska, E. (1972). Oxygen consumption by soil-inhibiting nematodes. *Nematologica* **18,** 391–403.
Kraus, M. G., and Found, B. W. (1975). Preliminary observations on the salinity and temperature tolerances and salinity preferences of *Derocheilocaris typica* Pennak and Zinn 1943. *Cah. Biol. Mar.* **16,** 751–762.
Lasserre, P. (1969). Relations énergétiques entre le métabolisme respiratoire et la régulation ionique

chez une annelide oligochète euryhaline *Maionina achaeta* (Hagen). *C. R. Hebd. Seances Acad. Sci., Ser. D* **268,** 1541–1544.

Lasserre, P. (1970). Action des variations de salinité sur le métabolisme respiratoire d'oligochète euryhalins du genre *Marionina* Michaelsen. *J. Exp. Mar. Biol. Ecol.* **4,** 150–155.

Lasserre, P. (1971). Données écoloquques sur la répartition des oligochètes marins Méiobenthiques. Incidence des parametres salinité - température sur le métabolisme respiratoire de deux espēces euryhalines du genre *Marionina* Michaelsen (1889) (Enchytraidea, Oligochaeta). *Vie Milieu, Suppl.* **22,** 523–540.

Lasserre, P. (1975). Métabolisme et osmorégulation chez une annelide oligochète de la méiofaune: *Marionina achaeta* Lasserre. *Cah. Biol. Mar.* **16,** 765–799.

Lasserre, P. (1976). Metabolic activities of benthic macrofauna and meiofauna: Recent advances and review of suitable methods of analysis. *In* "The Benthic Boundary Layer" (A. C. McCave, ed.), Chapter 6, pp. 95–142. Plenum, New York.

Lasserre, P., and Renaud-Mornant, J. (1971). Interpretation écophysiologique des effets de température et de salinité sur l'intensité respiratoire de *Derocheilocaris remanei biscayensis* Delamare, 1953. (Crustacea, Mystacocarida). *C. R. Hebd. Seances Acad. Sci., Ser. D* **272,** 1159–1162.

Lasserre, P., and Renaud-Mornant, J. (1973). Resistance and respiratory physiology of intertidal meiofauna to oxygen-deficiency. *Neth. J. Sea Res.* **7,** 290–302.

Lasserre, P., Renaud-Mornant J., and Castel, J. (1976). Metabolic activities of meiofaunal communities in a semi-enclosed lagoon. Possibilities of trophic competition between meiofauna and mugilid fish. *Proc. Eur. Symp. Mar. Biol., 10th, 1975* pp. 393–414.

Lee, D. L. (1965). "The Physiology of Nematodes." Freeman, San Francisco, California.

McIntyre, A. D. (1964). Meiobenthos of sublittoral muds. *J. Mar. Biol. Assoc. U.K.* **44,** 665–674.

McIntyre, A. D. (1969). Ecology of marine meiobenthos. *Biol. Rev.* Cambridge *Philos. Soc.* **44,** 245–290.

McIntyre, A. D. (1971). Control factors on meiofauna populations. *Thalassia Jugosl.* **7,** 209–215.

McIntyre, A. D. (1977). Effects of pollution on inshore benthos. *In* "Ecology of Marine Benthos" (B. C. Coull, ed.), Univ. South Carolina Press, Columbia, South Carolina, pp. 301–318.

McIntyre, A. D., and Murison, D. J. (1973). The meiofauna of a flatfish nursery ground. *J. Mar. Biol. Assoc. U.K.* **53,** 93–118.

McLachlan, A. (1977). Studies on the psammolittoral meiofauna of Algoa Bay, South Africa. II. The distribution, composition and biomass of the meiofauna and macrofauna. *Zool. Afr.* **12,** 33–60.

Maguire, C., and Boaden, P. J. S. (1975). Energy and evolution in the thiobios: An extrapolation from the marine gastrotrich *Thiodasys sterreri*. *Cah. Biol. Mar.* **16,** 635–646.

Marcotte, B. M. (1977). The ecology of meiobenthic harpacticoids (Crustacea: Copepoda) in West Lawrencetown, Nova Scotia. Ph.D. Thesis, Dalhousie University, Nova Scotia.

Marcotte, B. M., and Coull, B. C. (1974). Pollution, diversity and meiobenthic communities in the North Adriatic (Bay of Piran, Yugoslavia). Vie Milieu **24,** 281–300.

Mare, M. F. (1942). A study of a marine benthic community with special reference to the micro-organisms. *J. Mar. Biol. Assoc. U.K.* **25,** 517–574.

Marshall, N. (1970). Food transfer through the lower trophic levels of the benthic environment. *In* "Marine Food Chains" (J. H. Steele, ed.), pp. 52–66. Oliver & Boyd, Edinburgh.

Nicolaisen, W., and Kunneworff, E. (1969). On the burrowing and feeding habits of the amphipods *Bathyporeia pilosa* Lindstrom and *Bathyporeia sarsi* Watkin. *Ophelia* **6,** 231–250.

Odum, W. E., and Heald, E. T. (1972). Trophic analyses of an estuarine mangrove community. *Bull. Mar. Sci.* **22,** 671–738.

Olsson, I., Rosenberg, R., and Olundh, E. (1973). Benthic fauna and zooplankton in some polluted Swedish estuaries. *Ambio* **2,** 158–163.

Ott, J., and Schiemer, F. (1973). Respiration and anaerobiosis of free-living menatodes from marine and limnic sediments. *Neth. J. Sea Res.* **7,** 233–243.

Perkins, E. J. (1958). The food relationships of the microbenthos, with particular reference to that found at Whitstable, Kent. *Ann. Mag. Nat. Hist.* [13], **1,** 64–77.

Pollock, L. (1974). Tardigrada. *In* "Reproduction of Marine Invertebrates" (A. C. Giese and J. S. Pearse, eds.), Vol. 2, pp. 43–53. Academic Press, New York.

Reid, J. W. (1970). The summer meiobenthos of the Pamlico River Estuary, North Carolina, with particular reference to the harpacticoid copepods. M.S. Thesis, North Carolina State University, Raleigh.

Remane, A. (1933). Verteilung und Organisation der benthonischen Mikrofauna der Kieler Bucht. *Wiss. Meeresunters.* (*Abt. Kiel*) **21,** 161–221.

Schiemer, F. (1973). Respiration of two species of Gnathostomulids. *Oecologia* **13,** 403–406.

Schroeder, P. C., and Hermans, C. O. (1975). Annelida: Polychaeta. *In* "Reproduction of Marine Invertebrates" (A. C. Giese, J. S. Pearse, eds.), Vol. 3, pp. 1–214. Academic Press, New York.

Sellner, B. W. (1976). Survival and metabolism of the Harpacticoid Copepod *Thompsonula hyaenae* (Thompson) fed on different diatoms. *Hydrobiologia* **50,** 233–238.

Sibert, J., Brown, T. J., Healy, M. C., Kask, B. A., and Naiman, R. J. (1977). Detritus-based food webs: Exploitation by juvenile chum salmon (*Oncorhynchus keta*). *Science* **196,** 649–650.

Sikora, W. B. (1977). The ecology of *Palaemonetes pugio* in a southeastern salt marsh ecosystem with particular emphasis on production and trophic relationships. Ph.D. Thesis, University of South Carolina, Columbia.

Smidt, E. L. B. (1951). Animal production in the Danish Waddensea. *Medd. Kommiss. Dan. Fisk. Havunders., Ser. Fisk.* **11,** 1–151.

Swedmark, B. (1968). The biology of interstitial mollusca. *Symp. Zool. Soc. London* **22,** 135–149.

Swedmark, B. (1971). A review of Gastropoda, Brachiopoda and Echinodermata in marine meiobenthos. *Smithson. Contrib. Zool.* **76,** 41–47.

Teal, J. M. (1962). Energy flow in one salt marsh ecosystem of Georgia. *Ecology* **43,** 614–624.

Tenore, K. R., Tietjen, J. H., and Lee, J. J. (1977). Effect of meiofauna on incorporation of aged eelgrass, *Zostera marine,* detritus by the polychaete *Nephthys incisa*. *J. Fish. Res. Board Can.* **34,** 563–567.

Thiel, H. (1975). The size structure of the deep sea benthos. *Int. Rev. Gesamten Hydrobiol.* **60,** 575–606.

Tietjen, J. H. (1967). Observations on the ecology of the marine nematode *Monhystera filicaudata Allgen* 1929. *Trans. Am. Microsc. Soc.* **86,** 304–306.

Tietjen, J. H. (1977). Population distribution and structure of the free-living nematodes of Long Island Sound. *Mar. Biol.* **43,** 123–136.

Tietjen, J. H., and Lee, J. J. (1972). Life cycles of marine nematodes. Influence of temperature and salinity on the development of *Monhystera denticulata* Timm. *Oecologia* **10,** 167–176.

Tietjen, J. H., and Lee, J. J. (1977a). Feeding behavior of marine meiofauna. *In* "Ecology of Marine Benthos" (B. C. Coull, ed.), Univ. South Carolina Press, Columbia, South Carolina, pp. 21–35.

Tietjen, J. H., and Lee, J. J. (1977b). Life histories of marine nematodes. Influence of temperature and salinity on the reproductive potential of *Chromodorina germanica* Butschli. *Mikrofauna Meeresboden* **61,** 263–270.

Vernberg, W. B., and Coull, B. C. (1974). Respiration of an interstitial ciliate and benthic energy relationships. *Oecologia* **16,** 259–264.

Vernberg, W. B., and Coull, B. C. (1975). Multiple factor effects of environmental parameters on the physiology, ecology, and distribution of some marine meiofauna. *Cah. Biol. Mar.* **16,** 721–732.

Vernberg, W. B., and Vernberg, J. (1972). The synergistic effects of temperature, salinity, and mercury on the survival and metabolism of the adult fiddler crab, *Uca pugilator*. *Fish. Bull* **70,** 415–420.

Vernberg, W. B., Coull, B. C., and Jorgensen, D. D. (1977). Reliability of laboratory metabolic measurements of meiofauna. *J. Fish. Res. Board Can.* **34,** 164–167.

Walter, M. D. (1973). Fressverhalten und Darminhaltsuntersuchungen bei Sipunculiden. *Helgol. Wiss. Meeresunters.* **25,** 486–494.

Warwick, R. M. (1971). Nematode associations in the Exe Estuary. *J. Mar. Biol. Assoc. U.K.* **51,** 439–454.

Wieser, W. (1953). Die Beziehung zwischen Mundhohlengestalt, Ernahrungsweise und Vorkommen bei freilebenden Marinen Nematoden. Eine okologischmorphologische Studie. *Ark. Zool.* [2] **4,** 439–484.

Wieser, W. (1975). The meiofauna as a tool in the study of habitat heterogeneity: Ecophysiological aspects, A review. *Cah. Biol. Mar.* **16,** 647–670.

Wieser, W., and Kanwisher, J. (1959). Respiration and anaerobic survival in some sea weed inhabiting invertebrates. *Biol. Bull.* (*Woods Hole, Mass.*) **117,** 594–600.

Wieser, W., and Schiemer, F. (1977). The ecophysiology of some marine nematodes from Bermuda: Seasonal aspects. *J. Exp. Mar. Biol. Ecol.* **26,** 97–106.

Wieser, W., Ott, J., Schiemer, J., and Gnaiger, E. (1974). An ecophysiological study of some meiofauna inhabiting a sandy beach at Bermuda. *Mar. Biol.* **26,** 235–249.

Zeuthen, E. (1953). Oxygen uptake as related to body size in organisms. *Q. Rev. Biol.* **28,** 1–12.

Benthic Macrofauna

6

F. John Vernberg

I. Introduction

The benthic region of the sea is extensive both in terms of area and varied habitat types. Thus, it is not surprising that there is an abundant, diversified benthic fauna that exhibits a wide spectrum of physiological responses to differing sets of environmental factors. Based on the distribution of marine-associated organisms and the influence of the abiotic component of the marine environment, the landward limits of the marine benthic region are arbitrarily set as follows: (1)

FUNCTIONAL ADAPTATIONS OF MARINE ORGANISMS

ISBN 0-12-718280-2

the terrestrial boundary is that area immediately landward of the high tide mark, which is strongly influenced by seawater in the form of salt spray and/or irregular flooding; and (2) the boundary between freshwater and marine benthic communities, which typically experiences a salinity above 5–8‰. This range appears to represent an ecophysiological barrier (Khlebovich, 1969; Remane and Schlieper, 1971).

Extending from the land seaward, the benthic region is typically comprised of the following regions as proposed by Hedgpeth (1957): (1) supralittoral zone, the area immediately above the high tide mark which is influenced by the sea; (2) littoral or intertidal zone, the area between the high and low tide mark; (3) sublittoral zone, all benthic regions from the low tide mark to the edge of the continental shelf at about 200 m; and (4) bathyal zone, beginning at the edge of the shelf and extending to a depth of about 400 m. The abyssal and hadal zones are at greater depths. The precise boundaries between the various zones are still being actively discussed by marine ecologists (see Chapter 8).

Macrobenthic animals may be divided into two principal groups, i.e., the infauna and the epifauna. The infauna includes all animals living in the substratum of the ocean's floor. In contrast, the epifauna includes animals that are found on the surface of the various types of substrata which make up the floor of the sea, including rocks, pilings, shells, sand, mud, and coral reef. Remane (1933) proposed a third group, the phytal, to include organisms associated with large growths of plants (e.g., kelps) and sessile animals (e.g., hydroids, bryozoans, and corals). Although some animals spend their entire life in the benthic region, others live part of their life cycle in the water column before becoming benthos. If they are planktonic, they are called meroplanktonic; if nektonic, demersal.

Macrobenthic animals occupying these different regions of the sea are confronted with a different combination of environmental factors to which they must adapt. In the deeper waters, the macrobenthos live in relatively stable environments; in contrast, littoral animals generally experience wide fluctuations in such ecological factors as temperature, light, and salinity. In South Carolina, a temperate zone environment, daily temperature variations of 20°C are not uncommon in the intertidal zone, but in contrast, temperatures at a depth of 100 m varies slightly on a daily basis and may vary only 5°C seasonally (Stefansson and Atkinson, 1967). Salinity in estuaries may vary with periodic tidal changes and as a result of aperiodic heavy rains. In the deeper water, fixed vegetation is absent, whereas in the intertidal zone, plants, such as cordgrass, eel grass, and macroalgae, may dominate the landscape and influence the expression of other environmental factors. Light, hydrostatic pressure, gases, nutrients, and other important abiotic factors in the various benthic zones may differ not only quantitatively but also in the rate and magnitude of change. The ecological characteristics of specific habitats are described in various publications: Swedmark (1964), sandy beaches; Remane and Schlieper (1971), brackish waters; Bliss

(1968) and Newell (1970), intertidal zone; Stephenson and Stephenson (1972), rocky shores; Thorson (1957) level bottom communities; and Hedgpeth (1957), Friedrich (1969), Vernberg and Vernberg (1972), Coull (1977), and Keegan *et al.* (1977), various benthic regions.

In this chapter the physiological ecology of macrobenthic invertebrates found at depths of 4000 m or less is discussed. Adaptations of benthic organisms found at greater depths are discussed in Chapter 8, which deals with the deep sea, and the meiobenthos is treated in Chapter 5.

Since the benthic region is very extensive and the fauna diversified, the physiological ecology of the macrobenthos will be discussed under the following categories: (1) resistance adaptations, the lethal effects of various environmental factors acting singly and in concert; and (2) capacity adaptations, the sublethal response of the following physiological systems: feeding, digestion, and assimilation; respiration; energy budgets; ionic and osmoregulation; and reproduction and growth. Because of the large number of published papers dealing with these topics, this chapter will highlight basic contributions and some current studies which exemplify recent research trends. Thus, rather than being a comprehensive literature review, the goal of this presentation is to present a general background and an update on adaptative mechanisms of macrobenthic animals.

II. Resistance Adaptations

Any environmental factor may fluctuate in its expressions over a gradient, and at either end of this gradient an organism may not survive. The lethal effect of a specific factor acting independently or in concert with one or more other factors has marked implications to understanding the ecology of benthic organisms. The concept of limiting factors and the law of the minimum were elucidated early in the history of physiological ecology (see review of Allee *et al.*, 1949). Many earlier studies emphasized the lethality of a single factor, while a recent research trend has been to study multiple factor interactions. Both types of study are valuable, although multiple factor studies may more nearly approximate the environmental complex in which most organisms live. Thus, the data from carefully designed multiple factor studies should permit a more realistic appraisal of the functional capabilities of organisms to respond to the stresses of their habitat. Examples of the influence of single factors will be followed by a discussion of multiple factor interaction.

A. TEMPERATURE

Generally, benthic organisms from cold-water environments survive lower temperatures better than animals living in warmer climates, whereas warm-water animals do better at elevated thermal points (Vernberg, 1962). Animals exposed to a widely ranging seasonal thermal regime typically are better able to survive a wider range of temperatures than organisms living in a region of either relatively constant high or low temperatures. In addition, animals living in habitats which experience widely ranging temperatures are more labile in that their lethal limits may be shifted dramatically by thermal acclimation up to a genetically imposed limit. An excellent example of this principle can be seen in Fig. 1 where the lethal limits of tropical and temperate zone fiddler crabs are compared (Vernberg and Vernberg, 1972).

At one latitudinal site, animals living higher in the intertidal zone may survive high temperatures better than closely related species living subtidally or lower in

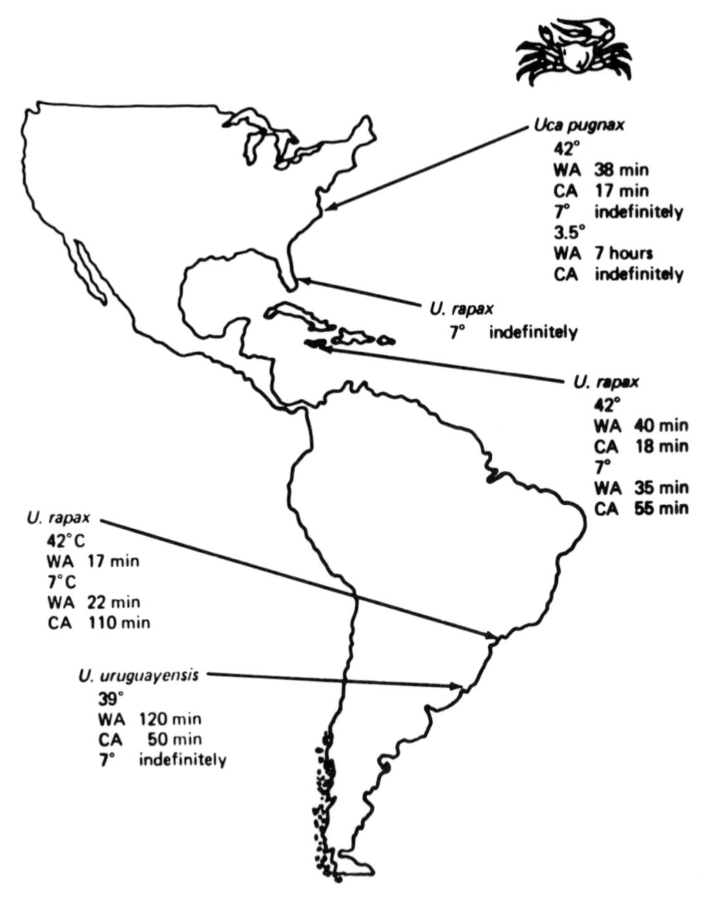

Fig. 1. Temperature tolerances of temperate and tropical zone species of fiddler crabs (genus *Uca*). Each LD_{50} given in minutes. WA, warm-acclimated; CA, cold-acclimated. (From Vernberg and Vernberg, 1972.)

the intertidal zone (bivalves, Vernberg *et al.*, 1963; snails, Fraenkel, 1966; amphipods, Sameoto, 1969; various classes, Kennedy and Mihursky, 1971). Recently, Wilson (1978) demonstrated this phenomenon with two molluscan species, *Tellina tenuis* and *T. fabula*. The higher intertidal species, *T. tenuis*, had a heat tolerance level approximately 5°C higher than that of the more subtidal species. Both species were more tolerant of high temperature after thermal acclimation. Although the lethal temperature of the intertidal snail, *Littorina littorea*, could only be shifted by 1°–2°C as a result of thermal acclimation, the threshold for entering a reversible state of heat coma could be shifted as much as 8.5°C (Hamby, 1975). Further, Hamby demonstrated a good correlation between cessation of spontaneous activity of the central nervous system and temperature of heat coma, and Grainger (1975) suggested that ionic disturbances at high temperature are the prime causes of heat death.

Cold tolerance of the temperate zone barnacle, *Balanus balanoides*, varied seasonally from −18.6°C in winter to −6°C in summer. At this winter lethal temperature, 80% of the body water was frozen, while only 40–45% was frozen at the summer lethal temperature. This difference suggests that different adaptive mechanisms are operative seasonally (Crisp *et al.*, 1977). When comparing freezing resistance of antarctic invertebrates, Rukusa-Suszczewski and McWhinnie (1976) found pelagic species to be more tolerant to supercooling than benthic species.

In another study of antarctic fauna, the limpet, *Patinigera polaris*, was found to be frozen in intertidal scale ice or anchor ice. Those animals which secrete a surrounding envelope of mucus survive low temperatures of −10°C better than limpets without mucus. Potentially, mucus could inhibit the growing surface of any ice crystal with which it comes into contact (Hargens and Shabica, 1973). In British Columbia, Canada, Roland and Ring (1977) found that 50% of a population sample of the limpet, *Acmaea digitalis*, survived exposure to −10° and −12°C for at least 24 hr. Approximately 60–80% of the body water froze, and the solutes were concentrated between 350–500%. These authors suggested that cold death could result from increased solute concentration, which would upset osmotic gradients across membranes, causing cellular damage. As demonstrated in other intertidal molluscs (Kanwisher, 1966), no glycerol was detected which could function as a cryoprotectant. However, Cook and Gabbott (1972) found an increase in glycerol content of barnacles; but this increase, which was far below that reported for cold-hardy insects, was assumed to be insufficient to promote super-cooling of the barnacles' body water.

Acclimation to low temperatures or high salinity increased the tolerance of the intertidal bivalve, *Modiolus demissus*, to subfreezing temperatures (Murphy and Pierce, 1975). Salinity acclimation acted by reducing the amount of tissue water frozen, thereby preventing a lethal dehydration level. However, the mechanisms by which thermal acclimation increases resistance to low temperature are unknown, although it may be linked to increasing resistance to high concentrations of toxic salts or greater levels of dehydration. DeVries (1974) reviewed survival at freezing temperatures in various organisms from a number of environments, including some benthic species.

When potentially lethal thermal extremes are being experienced by a mobile animal, it may seek a more preferred temperature or, if it is sessile, it may use some functional ploy to survive. Boddeke (1975), based on 14 years of laboratory research and field work, reported that the autumn migration of the shrimp, *Crangon crangon,* from the coastal waters of the Netherlands to the North Sea is induced by the fluctuations of water temperatures, especially those in the tidal zone. This migration parallels a period of increasing sexual activity. The sexually mature shrimp, which are more sensitive to temperature fluctuation, migrate first followed by waves of shrimp in decreasing stages of sexual maturity. This behavior leaves the more immature animals in the regions with the richest food supplies for the longest period of time. With the approach of stressful high temperature conditions, less mobile and sessile animals living in the intertidal zone may use either physiological mechanisms such as evaporative cooling to reduce their body temperatures, or behavioral responses such as clustering, where the temperature of the animals in the center may be 5°–10°C lower than ambient level (Rhode and Sandland, 1975; see Vernberg and Vernberg, 1972, for review).

Not only have the lethal effects of temperature been determined on adults, but the developmental stages have also been studied. In a number of marine invertebrates, Andronikov (1975) found that the heat resistance of gametes could be correlated with the degree of thermophily of the adult. Species from similar thermal environments exhibited similar resistance levels. During ontogenesis, eggs, zygotes, and early embryonic development take place at temperatures conducive for successful completion of the life cycle, and the distribution of a species is limited by this relationship. Therefore, temperatures that exceed the upper or lower thermal limits will inhibit population growth. Earlier, Patel and Crisp (1960) demonstrated this correlation with barnacles (Table I) more recently, Bayne *et al.* (1975) observed that stress in the adults led to diminished viability in the larvae.

Larvae of many benthic species survive over a more restricted temperature range than adults (Vernberg and Vernberg, 1972); however, there are exceptions to this generalization. For example, larvae of tropical zone fiddler crabs survive longer at low temperature than do adults (F. J. Vernberg and Vernberg, 1975), and juvenile amphipods (*Gammarus palustris*) live closer to their upper lethal temperatures than do the other life stages. The over-wintering immature stages showed the best tolerance to low temperatures and salinities typical of late winter and early spring (Gable and Croker, 1978).

Temperature may influence the resistance capability of larvae to other en-

TABLE I

Upper Limits of Temperature Range During Breeding Season and the Range within Which Embryonic Development Proceeded at Maximum Rate *in Vitro* in Different Species of Barnacles[a]

Species and geographical range in Europe	Breeding season	Range of monthly mean sea temperatures at southern limit of range during breeding season (°C)	Upper temperature limit (°C)
B. balanoides			
Arctic–N. Spain	November–February	11–13	14–16
B. balanus			
Arctic–English Channel	February–April	9–10	13–14
V. stroemia			
Iceland–Mediterranean	January–April	13–15	21–23
B. crenatus			
Arctic–W. France	January–June	11–16	22–24
B. perforatus			
S. Wales–W. Africa	April–August	21–27	25–27
E. modestus			
S. W. Scotland–S. Portugal	Throughout year	14–19	23–25
C. stellatus			
N. Scotland–W. Africa	Probably throughout year	19–27	29–31
B. amphitrite			
W. France–Equator	June–August	27–29	>32

[a] Based on data of Patel and Crisp, 1960, in Vernberg and Vernberg, 1972.

vironmental factors. Larvae of the mudflat snail, *Nassarius obsoletus,* can tolerate low salinity and low dissolved oxygen best in combination with low temperature. This response may have ecological significance in that during the time of larval release in the spring, low salinity and low temperature often prevail (W. Vernberg, and Vernberg, 1975).

B. SALINITY

Numerous studies have demonstrated a correlation between the salinity tolerance limit and distribution of benthic and pelagic species (see reviews of Gunter, 1945; Wells, 1961; Kinne, 1964, 1971; Remane and Schlieper, 1971; Vernberg and Vernberg, 1972; Theede, 1975; Dorgelo, 1976). Based on a review of much of the literature, Dorgelo (1976) proposed that the five main types of salt tolerance are represented by the following groups.

I. Polystenohaline, true oceanic species.
II. More or less euryhaline species from hypersaline, marine, intertidal, or estuarine environments.
III. Extremely euryhaline species that tolerate the entire range from marine to limnetic conditions equally well.
IV. More or less euryhaline, genuine brackish-water species.
V. Oligostenohaline, true freshwater species.

A recent example of the correlation of salinity and species distribution is that of Norse and Estevez (1977). These authors demonstrated that differences in salinity tolerance may account for the marked differences in distribution of 10 species of portunid crabs along a salinity gradient extending from low salinity reaches of an estuary to the open ocean. However, other factors may also be involved, such as competition, predation by fishes, and food availability.

Even within a species, latitudinally separated populations may exhibit differences in salinity limits. For example, Stancyk and Shaffer (1977) reported that a population of the echinoderm, *Ophiothrix angulata,* from Florida was more tolerant to low salinity than those from a population from North Inlet, South Carolina. This difference might be the result of different selection pressures for salinity tolerance in the two study estuaries; the Florida estuary has a lower average salinity. In another study, salinity tolerance of populations of a gastropod, *Theodoxus fluviatilis,* from freshwater and different salinity regimes in the Baltic Sea were found to differ significantly (Kangas and Skoog, 1978). Populations from higher salinity habitats were most tolerant of higher salinity while freshwater populations were more tolerant of reduced salinity waters. Genetic differences in salinity tolerance of larvae from two populations of oysters from low and intermediate salinity habitats were reported by Newkirk *et al.* (1977). These differences could be correlated with the salinity in which the parental stock lived.

C. OTHER FACTORS

Organisms inhabiting the upper portions of the intertidal zone are generally more resistant to desiccation than organisms distributed lower in the intertidal zone or the subtidal region. These differences are the result of structural and/or physiological adaptations (Broekhuysen, 1940; Barnes *et al.,* 1963; Herreid, 1969; Foster, 1971; Vernberg and Vernberg, 1972; Wolcott, 1973; Newell, 1976; Skoog, 1976). The upper level of distribution of an organism may be determined by the interplay between rate of water loss and the time required to regain the water loss when the organism is recovered by the incoming tide

(Davies, 1969). Achituv and Borut (1975) found a relationship between the landward distribution of a snail and the size of the shell aperture; the smaller the aperture the more terrestrial the habitat. Ahsanullah and Newell (1977) substantiated what had been reported by others; relative humidity, temperature, and body size influence the rate of water loss in crabs. The subtidal crab, *Portunus marmoreus,* has a higher evaporation rate than the shore crab, *Carcinus maenas.* The most terrestrial of three species of hermit crabs found in North Inlet, South Carolina is most resistant to desiccation (Young, 1978), and can also lose more body water before death occurs than can the more subtidal species.

Oxygen levels also greatly influence species distribution. Organisms inhabiting a soft substratum are typically exposed to low oxygen levels and high hydrogen sulfide concentrations. Theede *et al.* (1969) found a positive correlation between species distribution and the ability of the species to withstand elevated amounts of H_2S and reduced O_2 concentrations. Similar results have recently been reported by Hunter and Arthur (1978), who found that the common estuarine oligochaete, *Peloscolex benedini,* was resistant to low O_2 levels. In some species, the larvae of adults that exhibit a high degree of anoxic resistance are also relatively resistant. However, the larvae of *Nassarius* are not as resistant as the adult, which suggests that different metabolic patterns develop during ontogeny. Vernberg (1972) and Vernberg and Vernberg (1972) reviewed much of the earlier literature on this subject. Additional discussion is found in the section on anaerobiosis (Section IV, F).

Special emphasis has been placed on the influence of various pollutants on the physiology and ecology of benthic organisms, especially in the past 10 years. Acute and chronic toxicity studies have defined the lethal limits for many types of substances acting singly or in concert with other "normal" environmental parameters. Some of this extensive literature is summarized in a series of symposia proceedings and the reader is referred to them for specific details (Vernberg and Vernberg, 1974; F. J. Vernberg *et al.*, 1977; W. Vernberg *et al.*, 1979).

D. MULTIPLE FACTOR INTERACTION

Although benthic organisms are exposed to numerous factors in their field habitat, more data exists on the lethal effect on one environmental factor than on multiple factor interaction. Therefore, to gain a more encompassing insight into how organisms survive in "natural" conditions, greater emphasis has recently been placed on the multiple factor interactions that define the lethal zone of various species.

A classic paper demonstrating the effects of three factors (salinity, temperature, and dissolved oxygen concentration) on the survival of an adult aquatic animal, the American lobster, is that of McLeese (1956). A sublethal but stressful exposure to one factor may become lethal when an animal is exposed simultaneously to a second sublethal but stressful factor. The net result is a reduction in the size of the zone of compatibility. For example, when salinity is optimal, the lobster can survive at a higher temperature than it can when exposed to low salinity and elevated temperature.

Simpson (1976) studied the effects of multiple physical and biotic factor interactions on the distribution and abundance of six species of littoral molluscs on the subantarctic Macquarie Island. He concluded that it is dangerous to place too much emphasis on data from studies of a single factor. Interaction of physical (temperature, desiccation, and salinity) and biotic factors (food, predation, and reproduction) are limiting the upper distribution of these species, but the effective combinations differ for each species. Any one factor, such as temperature, desiccation, or salinity, may not reach a critical lethal level in the field, but the debilitating sublethal effects in combination with other adverse factors can be lethal, especially a combination of high temperature and predation. In general, a correlation between distribution and tolerances was noted, with animals in the high intertidal zone being more resistant than organisms found lower in the distribution gradient; however, the heat resistance of chitons is greater than their limited distribution would suggest.

Larval stages are also sensitive to multiple factor exposure, as seen in the work of Vernberg *et al.* (1974). In this study the multiple effects of temperature, salinity, and cadmium reduced the size of the compatibility zone for larval fiddler crabs (Fig. 2).

Vargo and Sastry (1977) analyzed the combined effects of two other environmental factors, dissolved oxygen and temperature. They determined the tolerance limits of five zoeal stages and megalops of the crab, *Cancer irroratus*. Interstage variation was observed in that the first, second, and fourth zoeal stages showed similar responses, but the responses were different from those of the third and fifth stages, which were similar to each other. Although the larval stages did not show a progressive increase in tolerance to temperature or low dissolved oxygen with development, the megalops is relatively insensitive to changes in oxygen concentration with temperature. In general, these larval stages appear to have the capacity to tolerate a wider range of temperature and oxygen conditions than they encounter in the natural environment.

The levels of various environmental factors may fluctuate with time, and Thorp and Hoss (1975) studied the effects of salinity and cyclic temperatures on the survival of two sympatric species of grass shrimp (*Palaemonetes pugio* and *P. vulgaris*). Cyclic temperatures in combination with reduced salinity appeared to be more detrimental than reduced salinity and constant low temperature. Initially, the physiological tolerance to winter temperature and salinity conditions was determined so as to provide a basis for examining the ecological interaction between these two species. However, neither salinity nor temperature tolerances appear to be of primary importance in habitat partitioning of these sympatric species.

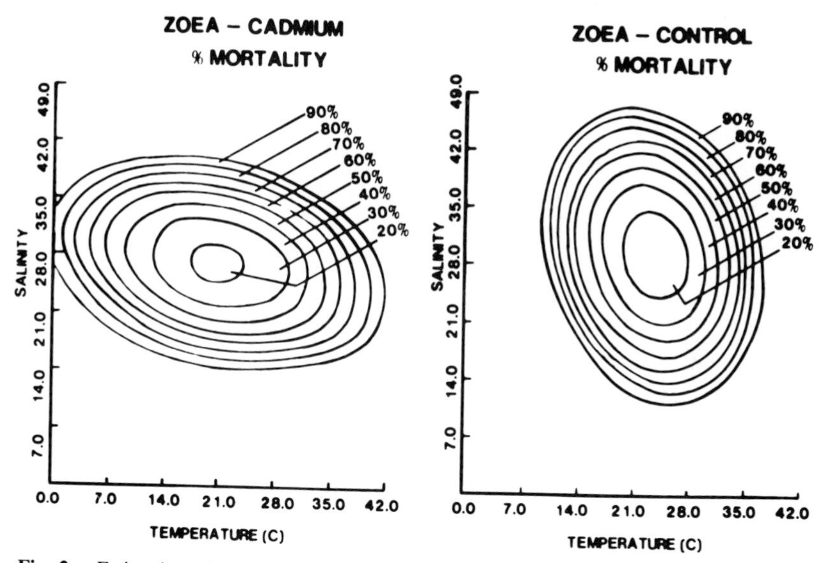

Fig. 2. Estimation of percentage mortality of first-stage *Uca pugilator* zoea based on response surface fitted to observed mortality under 13 combinations of salinity and temperature (a) with and (b) without the addition of 1 ppb Cd. (From Vernberg *et al.*, 1974.)

Wallis (1975), using probit analysis based on the dosage–mortality response curve, reported that thermal resistance of the blue mussel, *Mytilus edulis,* was influenced statistically by size, photoperiod, and salinity.

Temperature and salinity interact to influence the distribution of animals. Dorgelo (1976), reviewing this subject, stated that acclimation to salinity can affect salt as well as temperature tolerance, and the same holds for thermal acclimation. Although several authors have suggested that high temperature regimes of the tropics facilitated colonization of low salinity waters, Dorgelo found the evidence rather poor and concluded that in warm and temperate regions euryhaline animals from high salinity regimes are equally well able to invade low saline waters, at least in terms of salt tolerance.

The combined effects of salinity and temperature on a tropical mussel that might be exposed to thermal additions were described by response surface analysis (Wallis, 1976). Higher salinities increased thermal resistance, but thermal acclimation had little effect on the response surface in the upper thermal region.

Three factors (temperature, light, and salinity) influence the viability of reef coral, *Montipora verrucosa*. Low salinity decreased the coral's ability to survive short-term exposure to high temperature. High light intensity coupled with either high or low sublethal temperatures had deleterious effects on growth and increased the mortality rate (Coles and Jokiel, 1978).

Response to multiple factor exposure may be observed at the tissue level. This fact has been well demonstrated by a comparison of effect of temperature acclimation and salinity on isolated gill tissue of two estuarine species of bivalves. The gill of the oyster, *Crassostrea virginica,* is more resistant to low salinity and high and low temperature than is the gill from scallops (Vernberg *et al.*, 1963). Oysters are sessile, intertidal animals which are unable to leave their habitat, whereas scallops are motile, subtidal animals, which are able to move to other regions when conditions become stressful.

Some of the earlier literature on multiple factor effects are found in the reviews of Alderdice, 1972; W. B. Vernberg, 1975; F. J. Vernberg, 1979. Various mathematical techniques have been developed to design multiple factor experiments and to analyze and interpret the data (see review of Alderdice, 1972). Wallis (1976) published a simple multifactorial model using response–surface analysis for studies relating to power station cooling systems. He incorporated salinity, temperature, temperature shock, exposure time, and mortality responses as well as some sublethal effects.

III. Feeding, Digestion, and Assimilation

Basic to maintaining the energetic machinery of benthic organisms is the necessity for them to obtain energy and vital chemical compounds in the form of food or dissolved substances in seawater. The diversity of habitats and the multiplicity of animal species occupying these habitats have resulted in the evolution of a wide spectrum of adaptive feeding mechanisms in benthic animals.

The process of feeding involves a complex series of interrelated responses which can be broadly grouped under perception and capture of food and ingestion. Some attractant in the environment, either a chemical substance or a physical factor, will cause the organism to be oriented to the source of food. This behavior may be complex, as exhibited by rapidly moving predators, or apparently less complex, as in the case of sessile species. The mechanisms of intake or ingestion of food vary with food size preference and the animal's mode of existence. Following the introduction of food into the digestive system, it is broken into smaller chemical units both by the action of physical means and by digestive enzymes (lipases, proteases, and carbohydrases). These smaller chemical units are absorbed from the digestive tract, transported to various sites in the body, and assimilated. Although these various processes are common to most macrobenthic organisms, it is of particular interest to the physiological ecologist that many variations on this scheme have evolved which can be more closely correlated with specific niche requirements than with taxonomic affinities. A clear relationship exists between the morphology, physiology, and ecology of an organism when considering the essential process of obtaining energy. Because of the large number of published papers on certain phases of these energy gathering processes, an all-encompassing review is not possible in this chapter. Instead, principles will be emphasized with certain papers being cited as illustrations.

A. FEEDING STIMULI

Benthic animals may use a number of sensory modalities to detect and initiate feeding behavior, including sight, chemical receptors, electromagnetic field receptors, light receptors, sound receptors, and detection of various physical factors such as temperature, salinity, and wave action.

Mobile intertidal zone snails are known to use chemoreception as the primary method of detecting food *Ilyanassa obsoleta* (=*Nassarius obsoletus*), the mudflat snail, can detect chemical substances in the water from distances of 2–3 ft (Carr, 1967a,b). While moving toward the potential food resource, its siphon moves horizontally until the food is approached, and a previously unexposed proboscis is extended in a searching reaction until feeding begins. Carr determined that this feeding response was stimulated by chemical substances similar to amino acids and certain other nonvolatile, nitrogenous compounds of low molecular weight. Similar types of compounds have been suggested as a stimulus for other benthic organisms (crabs, Laverack, 1963; polychaetes, Mangum and Cox, 1971; sea anemones, Reimer, 1973).

Recently, Trott and Dimock (1978) reported that the ability of *Ilyanassa obsoleta* to detect and follow mucous trails of conspecifics and of *Nassarius vibex* may have significance in feeding. Chemical cues may account for this behavior. Since *I. obsoleta* is primarily a deposit feeder, following the trail of *N. vibex,* a scavenger, could lead to potential food. In contrast, *N. vibex* does not follow mucous trails of *I. obsoleta* because of differences in feeding strategies. However, not all chemical cues stimulate feeding; a number of invertebrates are noxious or toxic to certain predatory fishes and emit chemicals to chase them away (Bakus, 1969). Ecological differences in threshold response levels are noted in that animals from relatively nutrient-rich shallow water environments are much less sensitive to amino acids than are animals from nutrient-poor environments.

Based on various studies, Lindstedt (1971) proposed the following types of feeding stimuli: (1) attractant, a long distance stimulus that orients the animal toward food; (2) arrestant, a stimulus that inhibits locomotor movement when the animal is in contact with a potential food substance; (3) repellent, a stimulus that forces the animal to retreat from the potential food source; (4) incitant, a stimulus that initiates ingestion of food; (5) suppressant, a stimulus that inhibits ingestion of food; (6) stimulant, a stimulus that results in continued feeding; and (7) deterrent, a stimulus that hastens termination of feeding. Little is known of the physical and/or chemical nature of these various stimuli. Besides the chemical characterization studies mentioned above, the role of physical factors needs clarification although some data are available. For example, feeding activities of various species of barnacles is differentially stimulated by water currents. Those species living in sheltered regions with higher silt loads begin feeding at lower rates of water flow than those species living in the surf zone exposed to strong currents (Crisp, 1964). Increased silt loads depress the feeding rate of oysters (Loosanoff and Tommers, 1948). The adaptive feeding response of benthic animals from different climatic regions to reduced temperatures was shown in polychaetes by Mangum (1969). A boreal species from Alaskan waters could feed at a lower temperature than a closely related temperate zone species. For example, 80% of the boreal species fed at 5°C, whereas the species from lower latitudes could not feed at 10°C.

Other abiotic factors influence the onset and duration of feeding. Reduced salinity inhibits feeding in many benthic animals, which could limit the distribution of oceanic species into estuaries. Interstitial amphipods apparently fed selectively on sand grains to which food material or matter has adhered (Nicolaisen and Kanneworff, 1969).

The feeding response of the intertidal gastropod, *Hydrobia ulvae*, was correlated with a variety of sediment types (Barnes and Greenwood, 1978). These animals showed a marked preference for fine sediments both in regard to different natural sediments and to different particle size fractions isolated from a single natural sediment. However, the laboratory-determined behavioral preference did not account for their field abundance in that *Hydrobia* occurred in greatest abundance in some of the local sandy sediments. This suggests that there are other environmental factors that play a role in the distribution of various species.

Light may inhibit or elicit a feeding response, as demonstrated by the echinoid, *Diadema setosum*, which feeds only at reduced light conditions (Lawrence and Hughes-Games, 1972). A decrease in oxygen content will inhibit pumping rates and feeding in tubiculous crustaceans, polychaetes, molluscs, and lower chordates (Vernberg, 1972).

Two sensory modalities are used by the hermit crab, *Clibanarius vittatus*. A chemical stimulus attracts the animal to the food source and then visual cues orient it to begin ingestion (Hazlett, 1968). Visual stimuli are important to feeding in benthic organisms, especially in animals who actively move about in search of prey. This is well demonstrated in the intertidal fiddler crabs, genus *Uca*, and the subtidal blue crab, *Callinectes sapidus*. In the aphotic zone found in deeper waters, the dependence on visual cues is thought to be greatly reduced, whereas chemoreception is more important.

In various predator–prey models, a crucial step is the ability of the predator to choose their diets carefully. The shore crab, *Carcinus maenas*, will manipulate a mussel in its chelae for 1–2 sec before accepting or rejecting it (Elner and Hughes, 1978). These authors proposed that for a given sized crab there is an optimal mussel size at which energy content per handling time is at a maximum. This optimal size increases with crab size. However, this crab can rapidly adjust to changes in the availability of different sized prey.

B. CONVERSION EFFICIENCIES

The mere ingestion of food does not indicate that sufficient energy has been made available to sustain the metabolic machinery of the organism. For food to be beneficial, benthic organisms must convert ingested nutrients into protoplasm. Recently Calow (1977) reviewed the various conversion efficiencies which have been used to express this essential metabolic process. The ratio of G/I (growth of protoplasm = G; nutrient input = I) was used to express conversion efficiency. To distinguish between apparent input (food consumed) and the actual input (food absorbed across the gut wall), Calow used the ratios of G/C and G/A. The former is referred to as gross efficiency (K_1) and the latter as net efficiency (K_2). Based on data representing various benthic species, as well as animals from other habitats, certain generalizations were made. Gross efficiencies may achieve max-

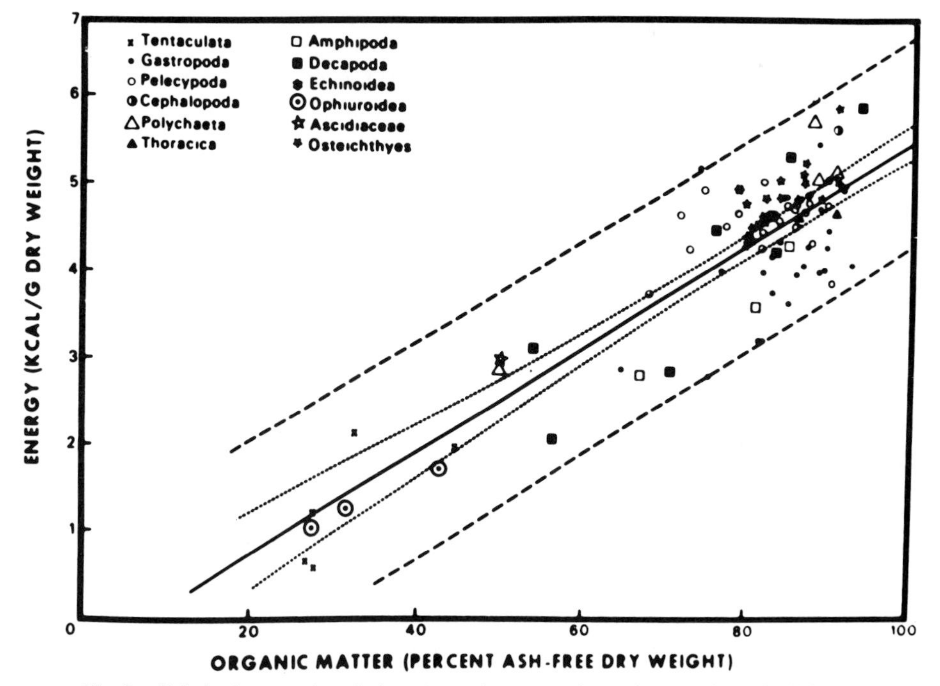

Fig. 3. Relation between the caloric and organic content of estuarine organisms. Solid line in the linear regression line; dashed line represents the 95% confidence limits for the prediction equation; dotted line represents the 95% confidence limits for the regression line. (From Thayer *et al.*. 1973.)

ima between 35 and 50% and net efficiencies between 50 and 80%. During development, Calow (1977) suggested that there is an initial period in which efficiency rises as the system becomes more metabolically coordinated. The net efficiency of metazoans, which are "sit and wait" predators, is relatively high compared with mobile carnivores.

Bayne and Scullard (1978) determined that the assimilation efficiency of the predacious snail, *Thais,* when feeding on the flesh of the intertidal bivalve, *Mytilus,* was 66%.

One problem in determining feeding conversion efficiency has been the lack of data on the caloric content of benthic organisms. To fill this void and to provide data needed for energy flow analyses, Thayer *et al.* (1973) determined the caloric content of 51 species from a shallow water estuary in North Carolina. Significant differences in caloric values of species grouped by phyla were noted. Also there appeared to be an evolutionary trend toward increasing energy content per gram live weight in more highly evolved species (Fig. 3).

C. FEEDING MECHANISMS

An early and continuing interest has been shown in the feeding mechanisms of benthic invertebrates. In 1928, Young proposed a classification of feeding mechanisms based on ecologic rather than systematic considerations. These mechanisms range from those in animals capable of absorbing nutrients directly from seawater through their body covering to organisms feeding on small particles (microphages) by means of pseudopods, cilia, mucus, setae, and/or eating substrates, to species which devour large-sized food particles (macrophages). The similarity in feeding methods of various species from one habitat is an example of convergent evolution and illustrates the close interrelationship between ecology and physiology. Jennings (1965), Jorgensen (1966), and Pandian (1975) have updated but did not dramatically alter Young's classification. In recent years, a number of papers have described new aspects of feeding mechanisms.

Early in this century, Putter proposed that many aquatic animals feed mainly on dissolved organic matter (DOM) in the surrounding water, an idea which provoked much controversy (see review of Jorgensen, 1976). The recent development of more sensitive analytical tools, especially radioisotopic, histochemical, and electron microscopic techniques, has permitted a more critical analysis of the role of DOM in the nutrition and energetics of benthic organisms. For example, the early work of Stephens and co-workers cited below rekindled interest in this problem. Stephens and Schinske (1957) reported that glycine and other amino acids were removed from seawater by three benthic species, a limpet, a mussel, and a coral. In 1961, they found the same response in 35 other species, many of which were not restricted to ciliary–mucoid feeding, which suggested that uptake was not restricted to organic molecules being absorbed to mucus. Since these studies, numerous papers have been published on different types of animals from various environmental conditions (see reviews of Stephens, 1972; Jorgensen, 1976). Although uptake can be demonstrated, the question can be raised as to whether it is meaningful to the physiological ecology of benthic organisms. Table II presents some results on infaunal, epifaunal, and pelagic species. Infaunal polychaetes can supply a major portion of their energy requirements from the rich supply of dissolved amino acids present in their environment. In contrast, when epifaunal species were exposed to concentrations of amino acids typically found in their natural environment, only a small percentage of their energy requirements was provided by amino acid uptake. Thus the role of dissolved DOM in nutrition appears to be correlated with the chemical nature of an organism's environment. It has been suggested that uptake of DOM might be important to functional processes other than nutrition, such as growth-controlling factors, vitamins, and pheromones (Jørgensen, 1976). Detailed studies on the role of various environmental parameters, such as temperature and ionic composition of seawater, on the dynamics of DOM flux are needed in order to assess the significance of DOM to benthic organisms.

Not only is the role of dissolved organic material of vital concern to students of benthic organisms, but attention has also been directed toward understanding the importance of suspended and particulate particles. Typically sessile animals or those with a limited locomotory capability feed on small particles, but there are exceptions, such as the actively burrowing bivalve, *Donax,* which is a filter-

TABLE II

Role of Dissolved Amino Acids in Energy Budgets of Some Aquatic Organisms[a]

Species	Body weight (g[b])	Type	Experimental medium	Influx (μM/ghr[b])	Oxygen requirements (%)
Capitella capitata	0.01	Infaunal	50 μM glycine	0.8–0.9	60–90
Nereis diversicolor	0.1–0.2	Infaunal	equivalents	0.75	80
	0.3	Infaunal	25 μM alanine	0.21	20
Mytilus edulis	1	Epifaunal	0.5–1 μM alanine equivalent	0.94[c]	9
			1 μM alanine	3.0[c]	30
Aurelia aurita polyps	40–80 μg dry wt	Epifaunal	1 μM glycine	1.2[c]	10
Strongylocentrotus purpuratus embryo	—	Pelagic	1 μM amino acid	4×10^{-13} mol/embryo/h	40–15

[a] From Jørgensen, 1976.
[b] Wet weight when nothing else is stated.
[c] Per g dry weight.

feeder (Ansell and Trevallion, 1969). Kirby-Smith (1976) suggested one way of assessing the relative importance of various potential foods is to measure the growth rate of animals on a variety of diets. Using the bay scallop, *Argopecten irradians,* he found that they ingest a mixture of particulate organic carbon which is 80% detritus and microheterotrophs and 20% phytoplankton. However, the growth rate of scallops is greatest on a diet of mixed phytoplankton; it is slower when they are fed fresh detritus, and they lose weight on a diet of aged detritus or protein-rich fish food.

Certain xanthid crabs possess large claws which are used to crush the shells of hermit crabs and snails. Typically these crabs sever the spire of their prey or make a gash in the body whorl. In contrast to these xanthid crabs, which are found in the tropical waters of Guam, some temperate zone species of the genus *Cancer* commonly do not crush their prey with the larger of their two claws, but work both claws together in breaking open their victims. Zipser and Vermeij (1978) reported that the gastropods also show adaptations to confer resistance to the crushing behavior of crabs. These adaptations include a thick shell, narrow or otherwise small aperture, thickened outer lip, strong sculpture, and a low spire. They reported an evolutionary trend in that an equatorward increase in the expression of the characteristics of crushing resistance parallels an increase in the crushing power of the crabs.

Until recently most of the accounts of feeding by benthic holothurians have largely been limited to descriptions of the capturing of particles in the tentacles of suspension feeders and tentacular raking or shoveling by deposit feeders. Roberts (1979) reported on the deposit feeding mechanisms of a number of holothurians living in the tropics, an area of high species diversity. He found that aspidochirote holothurians living on tropical reef flats feed on particle deposits which form a variety of substrata. The synaptid holothurian, *Opheodesoma grisea,* feeds in a similar manner by scraping deposits from the surface of sea grasses. Using the scanning electron microscopy technique, he found that the tentacles of *Aspido chirotes* have a nodular surface, whereas those of *O. grisea* have a tessellated surface structure. Of the 12 species examined, he reported a difference in tentacular surface textures which bore an apparent relationship with the mean particle size selected by the different species. In addition, by examining the distributional gut content analysis, he felt that species spatial partitioning is on the basis of substratum and particle size preference.

Although some of the polychaetes are known to be deposit feeders, relatively little was known of their ecology until Kudenov (1978) studied one species (*Axiothella rubrocincta*) living in California. This worm inhabits a U-shaped tube of agglutinated sand grains and mucus. Morphological adaptations prevent the tube from becoming clogged. Foreign debris entering the tube is either consumed or incorporated into the tube wall. This species combined feeding and burrowing activities to form the funnel and complete construction of the tube.

Apparently it does not ingest sediment while burrowing. *Axiothella* consumes food from the upper 2 cm of the substrata and is 4.6% efficient as a bottom feeder. It also has the capability of feeding within the funnel and can ingest large quantities of food. During the feeding it irrigates its tube at a rate of 5.1 ml seawater/g/hr and briefly reverses this current to a rate of 0.1 ml/g/hr when defecating.

Interesting interspecific feeding relationships have been established between the predatory snail, *Thais lapillus* (L.) and its prey, such as species of barnacles and mussels. The snail penetrates its prey by drilling a circular hole through the shell using both mechanical and chemical means. Bayne and Scullard (1978) determined that the time spent in drilling by *Thais* and ingesting its prey did not vary for snails of different sizes, however, the time was shorter at higher temperatures. Although winter feeding was less intense than during the summer months, the authors felt that this was not entirely related to the higher temperatures in the summer, but may have reflected some basic seasonal physiological change.

D. DIGESTION

Typically in macrobenthonic animals, ingested materials are introduced into a cavity in the body where digestion occurs. This cavity may vary from an incomplete digestive tract with only one opening to a highly complex complete digestive tract with a mouth and an anus. In these animals digestion is internal and may be extracellular or intracellular with the more highly evolved animals having well-developed extracellular mechanisms. However, external digestion has been reported in organisms having internal digestive systems. Sponges, echinoderms, and molluscs have the ability to incorporate food particles through the outer body wall. In sponges, dermal cells phagocytize food particles (Rasmont, 1968). Ferguson (1971) reported that the epidermis of echinoderms is somewhat isolated from food stores located internally and therefore must depend primarily on procuring organic materials from the surrounding seawater. Using various histochemical and radioisotopic techniques, Pequignat (1972) not only clearly demonstrated digestion and absorption by the skin of echinoderms, but he showed that some absorbed materials reached the muscle tissue and were not restricted to the epidermis. Similarly the epitheium of various bivalves plays a significant role in the uptake of materials dissolved in seawater (Pequignat, 1973). Unfortunately, the influence of various environmental factors on external digestion is poorly known.

In recent years a number of studies have been published on the anatomical and functional features of digestion in various species. A recent example is that of Barker and Gibson (1978). Working with the mud crab, *Scyllaserrata forskal,* they reported that a complete digestive cycle took place in approximately 12 hr. As food is ingested, it is lubricated by a mucoid secretion. Digestion is principally extracellular with digestive enzymes being synthesized by the hepatopancreatic F cells. Enzymes are released in waves of activity at 0.5–1 hr, 3 hr, and 8 hr after a meal. Initially there is a discharge of IA- and NA-esterases and arylamidases. After 3 hr only NA-esterases are demonstrable within the B-cells, but by 8 hr further IA-esterases and arylamidases are secreted. Both acid and alkaline phosphatases occur in the hepatopancreatic cells and are apparently concerned with several stages in the digestive cycle, including the synthesis and secretion of digestive enzymes, the subsequent absorption of the products of extracellular digestion, and the active transport of metabolites across the cell membrane. Recently, Palmer (1979) described the histology and enzyme histochemistry of the stomach and digestive gland of a bivalve (*Arctica islandica*). Studies such as these are needed to provide the basic understanding of digestion so that it can be studied under various environmental conditions, thus allowing the physiological ecologists to better understand this aspect of benthic organisms in their natural environment.

IV. Respiration

Basic to life is the need for energy, and most of the energy-liberating systems depend on oxidation. In most animals, but not all, oxidation eventually involves oxygen. Unlike many molecules, oxygen cannot be stored by animals in great quantities; therefore, the demand is relatively constant, and each species must have a mechanism to meet their cellular demands.

Benthic organisms represent a wide spectrum of respiration adaptations because of the widely varying habitat types, ranging from oxygen-poor mud bottoms to oxygen-rich rocky intertidal shores.

The dynamic process of obtaining oxygen and releasing CO_2, which is characteristic of aerobic respiring organisms, is influenced by extra- and intraorganismic factors and by the interactive effects of both types of factors.

The principal internal factors are body size, locomotor activity, sex, starvation, diet, level of cellular metabolism activity, growth, and the internal physical–chemical milieu. Typically, smaller-sized organisms of the same species have a higher rate of oxygen consumption than larger-sized organisms when results are expressed per unit weight (weight-specific respiration). In general this relationship may be broadly applied on an interspecific basis, but numerous exceptions exist which can be attributed to the differential effort of temperature and/or salinity on large and small sized organisms and also to the behavioral patterns of the species being compared. For example, active, rapidly moving species tend to have higher weight-specific rates of oxygen consumption than do sluggish species, even though the sluggish species may be smaller in size and expected to have the higher rate (Vernberg and Vernberg, 1970). Within one species, different sized organisms may expend different amounts of energy on obtaining and digesting food. Within one organism, the energetic cost of obtaining food will vary with intensity of feeding. For example, Bayne *et al.* (1976) reported that oxygen consumption increased in proportion to the amount of water filtered by the bivalve, *Mytilus californianus*. Males have been found to have higher, lower, or the same metabolic rate as females. Recently fed animals tend to have higher metabolic rates than starved animals and the type of diet may influence the metabolic rate. Rhythmic cycles of oxygen consumption have been reported in various benthic species: a few examples are gastropods (Zann, 1973; Shirley and Findley, 1978), crabs (Aldrich, 1975), and polychaetes (Mangum and Miyamoto, 1970; Sander, 1973). In some cases the variation in metabolic rate is reported to be associated with diurnal locomotor activity rhythms, whereas in others it appears to be activity independent.

Differences exist in oxygen demands by tissues from various species; tissues from active species tend to have higher metabolic rates than that of those of sluggish species (Vernberg and Vernberg, 1970). The chemical–physical environment within the organism, such as cellular pH, oxygen levels, cellular temperatures, and the presence of hormones, also influences oxygen uptake.

The external environmental factors influencing respiration that have been most studied are temperature, oxygen levels, salinity, photoperiod, and chemical composition of the external milieu.

A. OXYGEN CONCENTRATION

An obvious factor which could influence respiration is the ambient oxygen concentration [see Vernberg (1972) review of the voluminous literature on the response of various groups of marine animals from various habitats to decreasing oxygen levels]. In general, two types of respiratory response can occur when the oxygen tension drops: (1) oxyconformation, i.e., the rate of oxygen uptake decreases in relation to a drop in oxygen content; and (2) oxyregulation, i.e., the rate of oxygen uptake is relatively constant over a wide range of ambient oxygen tensions until some critical value (P_c) is reached, after which the rate declines. The P_c value is variable depending upon such factors as temperature, food, molting, locomotor activity, body size, and acclimation to various levels of ambient oxygen. Dimock (1977) reported on still another variable influencing the P_c values in that eviscerated *Stichopus parvimensis* (a holothurian) had higher P_c values than intact animals.

The comparative responses of free-swimming and tubicolous crustaceans by Gamble (1970) will illustrate the physiological–ecological implication of this type of study. At low tide very low oxygen tensions are found in the tubes of intertidal burrowing animals. After exposure to low oxygen levels, the ventilation rate increased for the free-swimming species, remaining unchanged for the tubicolous amphipods. Moreover, a marked change in the ventilation rhythm was found; at air-saturated oxygen tensions, only the tubicolous amphipods exhibited an intermittent rhythm of ventilation, but when the oxygen levels dropped, a continuous rhythm resulted. Coupled with Gamble's work on anaerobic resistance, these findings suggest that tubicolous amphipods are more resistant to anoxia than other crustaceans of similar size, and that they do not attempt to hyperventilate at reduced oxygen tension. Because of the viscous drag resistance of the tube, water movement costs the organism energy, and hyperventilation would be energetically expensive. However, a minimal ventilation rate apparently is maintained, since the organisms show a continuous ventilation rhythm.

Two species of burrowing crabs, *Callianassa californiensis* and *Upogebia pugettensis,* regulate their metabolic rates over a wide range of oxygen concentrations. However, Thompson and Pritchard (1969) found species differences that correlated with habitat differences; *Callianassa* lives under more hypoxic conditions than *Upogebia* and has the lower metabolic rate, the lower P_c value, and is more resistant to anoxia.

Unlike these tube-dwelling crustaceans, fiddler crabs, genus *Uca,* are active on the surface of the intertidal zone when the tide is out, but retreat to burrows as the tide rises. Since these crabs do not pump water through their burrows, low oxygen tension may be experienced. They are adapted to this low oxygen environment by being resistant to anoxia and having a low critical oxygen tension; i.e., 1–3% of an atmosphere when animals are inactive and 3–6% when active. Further, *Uca* continued to consume oxygen down to a level of 0.4% of an atmosphere, whereas the nonburrowing wharf crab, *Sesarma cinereum,* stopped respiring at a higher value (Teal and Carey, 1967).

B. TEMPERATURE

Since benthic invertebrates do not have a well developed capability to regulate their body temperature, respiration rate may be expected to be drastically altered by thermal changes. In general, the rate decreases with decreasing temperature and increases with increasing temperature. However, the metabolic rate of some benthic organisms is thermally insensitive (low Q_{10} values) over a wide temperature range (Vernberg and Vernberg, 1972; Newell, 1976). In *Uca* (fiddler crabs), this range of insensitivity (20°–30°C) coincided with the thermal range in which the animal is typically active in the intertidal zone (Vernberg, 1969). This response could result in the conservation of energy. However, a simple correlation between metabolism and a range of insensitivity to temperature is not clear. Some variations shown by various species are as follows. (1) Q_{10} values are low (<2) over the normal thermal range in which animals are active, e.g., *Uca* cited above. These animals experience wide changes in ambient temperature. (2) Q_{10} values are >2, e.g., anthozoan, *Anemonia natalensis* (Griffiths, 1977a). These animals live in relatively thermal-stable environments. However, variations on both of these responses have been reported; animals from relatively thermally stable environments may have low Q_{10} values (polychaete, Mangum, 1969), and animals from unstable environments may have high Q_{10} values (limpet, Brance and Newell, 1978).

The laboratory precedure under which animals are maintained can affect the metabolic response to temperature. It has been known for a long time that thermal acclimation will influence metabolic-temperature (M–T) interaction, and alter the slope and/or position of the M–T curve. Prosser and Precht, in 1958, independently proposed two methods of classifying metabolic–temperature interactions. Some intertidal species have the ability to shift their metabolic response after thermal acclimation. In general, animals from thermally stable environments show little or no ability to alter their M–T curves after thermal acclimation, but those species which experience rapid temperature changes are more metabolically labile and exhibit changes in the shape and/or position of the M–T curve (Vernberg and Vernberg, 1972).

In many of the earlier acclimation studies, animals were subjected to a new but stable thermal level, and their metabolic rate was measured over time. This experimental approach gives valuable insight into metabolic responses to temperature, but might not result in a precise ecological interpretation of the metabolic capability of the animal. For example, the metabolic response of fiddler crabs exposed to a fluctuating thermal regime corresponding to a typical daily range was lower than that of animals maintained at different but constant temperatures (Dame and Vernberg, 1978). The difference in response was ecologically significant, especially when determining the energy budget of a population of fiddler crabs for the entire year.

An example of applying metabolic data to an ecological setting is the work of Bayne *et al.* (1976). They demonstrated in the bivalve, *Mytilus californianus,* that the high temperatures experienced when the animal was exposed at low tide resulted in a metabolic deficit which had to be recovered at each subsequent high tide.

Earlier studies of Fox, Sparck, and Thorson (see review of Vernberg, 1962) demonstrated the correlation of metabolism and the biogeographical distribution of a species. Benthic polar animals exhibited little or no metabolic control at elevated but sublethal temperatures, with the result that they could not eat enough to supply their energy demands. In contrast, animals from warmer environments could not maintain high metabolic rates to be active and feed at lower temperatures. Therefore, the equatorial spread of polar animals is limited by increasing metabolic demands at higher temperatures, and the polar migration of warm water animals is restricted by metabolic insufficency at low temperatures. In general, tropical animals consume oxygen at similar rates at their typical ambient temperatures as do related cold-water species at their ambient temperatures.

Animals from both polar and tropical regions are not greatly influenced by thermal acclimation, whereas animals from intermediate latitudes experience wider temperature changes, and many species have greater metabolic acclimatory ability to temperature.

C. SALINITY

No one pattern of metabolic response to altered salinity is exhibited by benthic animals. In most reported studies, exposure to low salinity increases the metabolic rate; in contrast, a few investigations found the rate to be unaffected, while in some cases, the oxygen consumption rate is decreased. Some authors suggest that any elevated oxygen consumption rate reflects the energetic cost of osmoregulation [as Engel and Eggert (1974) reported for the gill respiration of the blue crab, *Callinectes sapidus*], although the metabolic rate of intact organisms is not changed (Laird and Haefner, 1976). However, other workers have suggested that an altered metabolic rate in response to a changing salinity is the result of changes in locomotor activity, molting, temperature, or other interactive parameters.

D. PHOTOPERIOD

Earlier, Dehnel (1958) reported that a change in photoperiod influenced the seasonal metabolic rate of an intertidal crab, *Hemigrapsus oregonensis*. On a daily basis, Mangum and Miyamota (1970) found that the polychaete, *Glycera dibranchiata,* consumed more oxygen in light than in the dark. They felt that this change did not reflect a change in locomotor activity, although Sander (1973) suggested that in two other polychaetes the increased rates reflected changes in behavior. Recently, Shirley and Findley (1978) reported that heightened nocturnal metabolic rates of an intertidal snail appeared to be independent of locomotor activity.

E. ADAPTATIONS TO LAND

Animals originated in the sea and have successfully invaded land following different routes. One avenue includes movement from the sea to estuaries to freshwater and finally to land. A second route to land is directly from the sea across the intertidal zone. Following either route, the invading organism faces a new environmental complex. The oxygen content of an aquatic environment may vary markedly, from supersaturation to anoxic, whereas oxygen levels in air are relatively constant. In contrast, temperature fluctuations are more rapid and more extreme on land than in water. Furthermore, as animals leave the water and migrate to land, they may experience greater desiccation which, in turn, tends to dry out their respiratory membranes.

Numerous morphological differences exist between animals occupying a series of habitats ranging from sea to land. Some of the best examples of these morphophysiological adaptations to habitat are exhibited by crabs. The number, total volume, and gill surface of gills of crabs living on land are less than in aquatic species. Also the structure of the gills has undergone modification to compensate for leaving the sea. In the ocean, the gill leaves, which are subunits of the gills, can be held apart by the influence of water currents, whereas on land they tend to adhere together. To overcome this problem the gills of some intertidal animals have become highy sclerotized with the subunits being rigid and supported. The gills also may be at right angles to the gill bar, and thus the gill leaves are not easily closed. Also, the behavior of crabs affords a degree of protection to their respiratory system against environmental stress, such as desiccation and thermal extremes. If the aerial environment becomes stressful, animals may retreat to the sea or to cooler, moist burrows.

Not only were there behavioral and morphological adaptations as animals invaded land, but metabolic changes occured. The more terrestrial species of crabs tended to have higher weight-specific metabolic rates than subtidal species or those lower in the intertidal zone. This correlation was noted for the intact organism and for certain tissues (Vernberg, 1956). At the enzyme level, cytochrome *c* oxidase activity of gill tissue demonstrated a similar trend (Vernberg and Vernberg, 1968).

A shift in the relative importance of oxygen and carbon dioxide as a respiratory stimulant in terrestrial and aquatic species can be observed. In seawater, the level of oxygen may vary markedly; marine amphipods and isopods are sensitive to reduced oxygen levels. In contrast, oxygen levels in air change slightly, so that terrestrial species are more sensitive to increased CO_2 levels and relatively insensitive to reduced oxygen levels (Walshe-Maetz, 1956).

F. ANAEROBIOSIS

Both epi- and infaunal species may rely on anaerobic respiration for varying periods of time. This adaptive response results when animals are confronted with hypoxic or anaerobic conditions or when an intertidal species exposed to air may "withdraw" from its ambient environment and become anaerobic, such as demonstrated by barnacles and certain intertidal bivalves.

As the external oxygen tension decreases, there is a critical concentration below which rapid oxygen uptake is reduced or inhibited and the organism may become anaerobic (see review of Vernberg, 1972). Mangum and Van Winkle (1973) reported on the influence of declining oxygen conditions on 31 species of marine invertebrates. Using several analytical models, they concluded that the quadratic polynominal was the best. Further discussion of analysis of data is found in Van Winkle and Mangum (1975). In the 1973 paper, they also report the oxygen level at which oxygen uptake ceases for 16 invertebrate species. Those species without mechanisms to store oxygen, such as gas bubbles or pools of high oxygen affinity respiratory pigments, switch to anaerobic pathways. Pamatmat (1978) determined that this critical value for two intertidal species is 13.4 torr for the fiddler crab, *Uca pugnax,* and 1.3 torr for the snail, *Littorina irrorata.* A marked increase in the oxycalorific coefficient indicated the onset of anaerobic metabolism and the accumulation of metabolic end products.

Tube-dwelling or burrowing animals apparently are aerobic as long as water is being pumped through their tube or burrow (*Nereis virens,* a polychaete, Scott, 1976; burrowing crabs, Thompson and Pritchard, 1969; burrowing bivalves, Hammen, 1976; Taylor, 1976; an asteroid; Shick, 1976; snails, Kushins and Mangum, 1971; see Vernberg, 1972, for other examples). An excellent example of physiological adaptation to habitat variation is seen in the isopod, *Cirolana borealis,* a scavenger which burrows into the flesh of dead fish. During feeding this isopod becomes anaerobic when it enters the deeper anoxic regions of the dead fish (de Zwann and Skjoldal, 1979).

As the tide recedes, intertidal bivalve molluscs either close their shells tightly for long periods of time (oysters, Hammen, 1976) or partially close their shells (shell gapes). Recently, Widdows *et al.* (1979) determined the physiological response of four bivalve species of this latter type (*Mytilus edulis, M. galloprovincialis, Cardium edule,* and *Modiolus demissus*). In air, all four species consumed less oxygen than they did when immersed; in addition, *Cardium* and *Modiolus* were influenced less by the change of media. During aerial exposure, anaerobic end-products accumulated in the tissues of *Mytilus,* but not in *Cardium.* These investigators correlated this differential response with the degree of shell-gaping; i.e., *Cardium* gaps more. Since *Mytilus* closes its shell more, less oxygen is taken up, and anaerobic pathways are utilized, resulting in end-product accumulation. However, *Cardium* is known to have functional anaerobic pathways during anoxia. Even when the bivalve, *Mercenaria mercenaria,* is submerged in aerated water, Gordon and Carriker (1978) reported indications of anaerobiosis as a result of rhythmic shell closing. The pH of the extrapallial fluid of *Mytilus edulis* decreases during shell closure under aerobic or anaerobic conditions. A pH reduction during anaerobiosis has functional significance in that the enzyme activities of pyruvate kinase and phosphoenolpyruvate carboxykinase are pH dependent (Wijsman, 1975).

In order to analyze the comparative response of crabs to different anaerobic conditions, Burke (1979) exposed *Carcinus maenas* and *Pachygrapusus crassipes* to different conditions of reduced oxygen availability. Typically *Pachygrapsus* inhabits rocky shores which are not subjected to conditions of reduced oxygen availability. In contrast, *Carcinus* lives in a variety of habitats and frequently encounter hypoxic conditions. During periods of high locomotor activity, both species showed similar capacities for anaerobic energy production; however, the two species differed in their response to hypoxia. With decreasing oxygen content of the water, the metabolic rate of *Carcinus* was constant over the range of 160–50 torr. In the range of 50 to about 20 torr, oxygen consumption decreased. Below this tension *Carcinus* abandoned aerobic respiration and increased lactic production. In contrast, *Pachygrapsus* neither regulated oxygen consumption nor increased lactic production under conditions of reduced oxygen availability. Thus, the anaerobic response to hypoxic conditions is independent of metabolic response to activity in these species, and the metabolic responses of *Carcinus* appear to be adaptive to the environment in which this species lives.

A detailed understanding of the processes involved in changing between aerobic and anaeriobic metabolism in benthic animals has not been achieved (Hammen, 1976). However, Hammen (1969), de Zwaan and Wijsman (1976), de Zwann (1977), and Widdows *et al.* (1979) have dealt with this problem in molluscs. Since highly evolved aerobic metabolic pathways are more advanced evolutionarily, Hammen (1976) feels that the pressure to retain an ancestral anaerobic capability can be correlated with frequent exposure to hypoxic and/or anoxic environments, such as those encountered by some benthic animals. Mangum and Van Winkle (1973) also felt that there was a phylogenetic trend for increased aerobic metabolic regulation as animals acquire structures that effectively insulate respiring tissue from the habitat.

G. ENERGY BUDGETS

Not only is it important to understand how an organism adapts metabolically to stress, but respiratory data are of fundamental significance to determining the energetics of a population and how energy flows between different trophic levels. In recent years, an ever-increasing interest in energy budgets of benthic species is evidenced by the increased number of published papers.

Energy budgets for various levels of biological organization have been published.

(1) Organismic level; barnacles, Wu and Levings (1978), grass shrimp and a parasitic isopod, Anderson (1977), and general discussions by Kleiber (1961) and Vernberg and Vernberg (1972).

(2) Population level; the gastropod, *Littorina irrorata,* Odum and Smalley (1959), the bivalve, *Modiolus demissus,* Kuenzler (1961), the bivalve, *Scrobicularia plana,* Hughes (1970), three species of *Nerita,* a gastropod, Hughes (1971a), the limpet, *Fissurella barbadensis,* Hughes (1971b), the gastropod, *Tegula funebralis,* Paine (1971), the bivalve, *Tellina tenuis,* Trevallion (1971), the isopod, *Tylos punctatus,* Haynes (1974) the oyster, *Crassostrea virginica,* Dame (1976), the oyster, *Ostrea edulis,* Rodhouse (1979), and the isopod, *Idotea battica,* Strong and Daborn (1979).

(3) Community level; see reviews of Weigert (1976), Kinne (1977), and Dame (1979).

The equation commonly used to determine the energy budget as proposed by Petrusewiez (1967) is

$$C = P + R + F + U$$

where $P = P_r + P_g$ and $P_g = \Delta B + E$. Each component may be measured in kilocalories per annum. C is the energy content of the food consumed by the population; P, the total energy produced as flesh or gametes; P_g, the energy content of the tissue due to growth and recruitment; B, the net increase in energy content of standing stock; E, elimination, or energy content lost to the population through mortality; P_r, the energy content of the gametes liberated during spawning; R, the energy lost due to metabolism (respiration); F, the energy lost as feces; and U, the energy lost as urine or other exudates.

Energy Budget Variables

The equation used to compute an energy budget is complex and consists of a number of discrete biological functions, each of which is influenced by a number of intra- and extraorganismic factors.

a. Energy Intake. Since an energy budget assumes that the intake of energy is equal to the sum of the energy-using and energy-transforming functions, the estimation of the amount of energy uptake is of fundamental importance. Within a population, the food intake rate may vary with body size. During a period of rapid growth, which is typical of small-sized organisms, the rate of energy intake is higher than during periods of little growth, which is characteristic of adults. Hence, to estimate intake for an entire population, the rates for each of the various life history stages must be determined. One complicating factor is that a high rate of energy uptake may not reflect a period of rapid growth. For example, during the onset of heightened gametogenic activity, the rate of energy uptake may be very high, but most of the energy is used in gamete production, not in organismic growth.

In addition to internal processes, extraorganismic factors may also influence the uptake of energy. One example is the relationship of temperature and feeding rate. In most poikilothermic marine invertebrates the feeding rate decreases with a decrease in temperature and increases with increasing temperature. At a higher temperature the rate of energy intake is increased, but the energy demands of metabolic machinery may increase faster, resulting in starvation of the organism. Thus, a higher rate of energy intake results in a decrease in biomass (P), but an increase in respiration (R).

Determination of field feeding rates is difficult, and much of the data used to compute energy intake is based on analysis of stomach contents or laboratory feeding experiments. For example, Darnell and Wissing (1975) estimated the food relationships in a population of a benthic-associated estuarine fish, *Lagodon rhomboides,* based on field and laboratory observations. They stressed the difficulties of arriving at total population estimates because physiological processes may change dramatically during the life history of a given individual organism.

Not only is there a potential change in energy input with developmental stage and age of an individual, but also at any one stage the input may vary with time and must be considered in estimating feeding rates. This may be very pronounced in intertidal zone species where feeding activities may be correlated with the phase of the tidal cycle. For example, some species feed at low tide, such as fiddler crabs (genus *Uca*), whereas others, such as oysters, feed at high tide. The caloric intake of a population could be difficult to determine for a species occupying a broad band of the intertidal zone because a subpopulation occupying higher levels in the littoral zone may feed at a different rate and for a different period of time than those near the low tide mark. Earlier Southward (1964) demonstrated this phenomenon in the barnacle, *Chthamalus dalli*. In addition to tidal changes, seasonal changes in physical factors, such as temperature, could influence the food intake of intertidal animals. Endogenous factors can also influence the feeding periodicity of an organism. Relatively little is known about the physiological mechanism which controls the appetite of marine organisms, although stretch receptors in the stomach, hormones, and blood sugar levels have been suggested as possible control mechanisms.

Organisms may be opportunistic feeders, and their mode of feeding may change rapidly and on an irregular basis. For example, the ratio of food intake between particulate and dissolved organic matter may change significantly. If food consumption values are based solely on laboratory experiments without knowledge of what the organism does in nature, this value may not be representative. This point emphasized one of the basic tenets of physiological ecology: laboratory experiments should be designed so that responses of organisms to conditions that simulate their natural environment can be analyzed. If this essential blend of field and laboratory experimentation does not exist, important insights into how physiological systems function can be obtained, but the data may have limited applicability in interpreting the physiological ecology of an organism.

b. Production. The change in biomass and the production of gametes by a population over a specific time period can be estimated with relatively more ease than population energy uptake. However, persistent problems of sampling a population to determine changes in the total weight of a population exist. Pelagic species are more difficult to sample than sessile species, and within a species certain life history stages are easier to sample than others. With changes in population structure, is the same amount of energy required to produce different types of tissues? For example, the relative amount of exoskeleton or shell in a population may change with season as a greater number of older organisms are found.

Reproductive effort, in terms of gamete production and release, varies from species to species and within a species because of both exogenous and endogenous factors (Stancyk, 1979). Also, cycles of reproduction can vary geographically or yearly within a species (Sastry, 1975).

c. Fecal Production. Some investigators have estimated production of fecal material by some intertidal species (see review of Pandian, 1975). For example, Frankenberg *et al.* (1961) reported that a population of the ghost shrimp, *Callianassa major,* produces fecal pellets at a rate sufficiently high to provide 0.06 g organic carbon per square meter per day for coprophagous organisms. Analyses of the chemical composition and calorific content of feces has received some attention (Pandian, 1975). Little is known of the functional impact of modified environmental conditions on fecal production. However, to construct better energy budgets, detailed analyses of this factor for a wide size range of animals of one species and for a number of species are needed.

d. Urine and Exudate Production. Many of the same comments made relative to our understanding of fecal production are appropriate to urine and exudate production. Estimates of the rate of urine production and the chemical composition of urine have been made based principally on relatively few laboratory experiments. However, little is known of the calorific value of urine and how the qualitative and quantitative production of urine is altered by fluctuation in environmental factors throughout an annual cycle.

Exudates, including mucous and molted exoskeletal material, can be collected and analyzed. For example, mucus production can be estimated for certain species under certain laboratory conditions, but estimates of the rate of production throughout the year based on field observations are not very precise. Better estimates of energy loss from shed exoskeleton are possible, although molting frequency is unknown for many crustaceans.

e. Respiration. Estimates of respiration rates are more numerous than for any of the previously discussed factors in the energy budget. Also, it has typically the greatest value of any of the terms. Respiration can be influenced by sex, stage of life history, starvation, pollutants, locomotor activity, temperature, salinity, external oxygen concentration, acclimation, and numerous other variables acting either singly or in combination. One example will demonstrate the complexity of determining respiration rates and some recent advances.

Temperature is known to influence metabolic rate, especially in invertebrates. Many of the earlier studies were done using organisms exposed to constant thermal regimes. After using various temperatures, an acclimated metabolic–temperature (M–T) curve could be constructed. Recently it has been shown that acclimation to cyclic thermal environments influences the metabolic responses of intertidal marine invertebrates differently from those acclimated to constant temperatures (Widdows, 1976; Bayne *et al.*, 1977; Dame and Vernberg, 1978). In general, the standard metabolic rates were lower in cyclically acclimated animals than those of animals acclimated to constant thermal regimes. Since these intertidal zone animals experience cyclic environments in nature, any estimate of annual respiratory rates should consider this factor.

Rather than attempt a comprehensive review of this subject which would be beyond the scope of this chapter, a few summary comments are made. Since respiratory rates appear to be extremely important in an energy budget, careful consideration of the principal variables influencing this rate must be given or misleading values will be obtained. Sufficient understanding and definition of the variables exist to permit the time-consuming, complex analyses needed to provide better estimates of yearly metabolic rates of a population of a species. More detailed studies on characterizing the annual thermal regime of benthic animals are needed since our knowledge of microenvironmental conditions encountered by animals in dynamic environments is extremely limited.

V. Osmotic and Ionic Regulation

The ability to chemically regulate internal body fluid composition is of particular value to those benthic species that live in an environment which is chemically unstable. The most common examples of stressful benthic habitats are the intertidal zone and estuaries; however, the region of salt domes in deeper waters is also stressful but its influence has been little studied. Evidence presented in the resistance adaptation section demonstrated the existence of a good correlation between salinity regime and distribution of benthic animals; i.e., the number of oceanic species decreases with decreasing salinity. Those animals living in stressful salinity habitats exhibit a number of mechanisms for maintaining physiological viability including reduced permeability of body surface to salts and or water, active uptake or extrusion of salts, regulation of body water volume, conservation of water or salts by the excretory organs, and regulation of cellular osmotic concentration (for reviews, see Lockwood, 1962; Vernberg and Vernberg, 1972; Gilles, 1979). Although these various physiological ploys have been used by a number of species, others avoid environment fluctuations behaviorally. For example, bivalves and barnacles may withdraw within their impermeable body covering and remain unscathed by external salinity variations, while other estuarine benthic animals may burrow to salinity-stable sites.

In general, benthic crustaceans living in low salinity water or those invading higher regions of the intertidal zone are excellent osmoregulators, whereas molluscs are euryhalinic osmoconformers.

The ability to osmoregulate is influenced by various factors other than external osmotic concentrations. For example, the osmoregulatory capability of the isopod, *Idotea chelipes,* decreased with increasing temperature as well as with decreasing salinity (Vlasblom *et al.,* 1977). Earlier, Spaargaren (1971) reported a difference in the response of two species of shrimp; *Crangon crangon* migrates to coastal and inland waters in the spring and returns to the open water of the North Sea with the approach of winter, while *C. allmanni* remains in deeper waters throughout the year. *Crangon allmanni* exhibits a high degree of osmoconformity, and no blood concentration differences were observed in this animal over the thermal range of 5°–15°C. In contrast, *C. crangon* can osmoregulate, but this ability is temperature related, with lower salinities being better tolerated at higher temperatures. These responses would appear to be correlated with the distributional pattern of these two species. Recently Boddeke (1975), based on 14 years of laboratory and field work, reported that the migration of *C. crangon* was induced by marked thermal fluctuations of water temperatures rather than any absolute thermal level.

Unlike most estuarine or marine bivalves, *Rangia cuneata* is capable of significant hyperosmotic blood regulation at salinities below 10‰, although the mechanism is unclear (Anderson, 1975). With decreasing salinity, many

TABLE III

Water-Regulation Ability of Selected Worms[a]

Species	Source of population	Percent water content in 560 m*M* Cl⁻	β "Best Fit"	Percent regulation
I. Osmoregulating Species				
Nereis limmicola	Lake Merced, San Francisco	88.4	0.10	90
Nereis diversicolor	Gdynia, Baltic Sea	82.1	0.10	90
Nereis diversicolor	Kamyshovaya Bay, Black Sea	82.4	0.20	80
Namalycastis indica	Madras, India	83.1	0.20	80
Nereis diversicolor	River Wansbeck, England	82.8	0.25	75
Nereis succinea	Hayward, San Francisco Bay	81.7	0.30	70
II. Osmoconforming Species				
Nereis vexillosa	Pt. Richmond, San Francisco Bay	79.3	0.30	70
Themiste cyscritum	Boiler Bay, Oregon	78.1	0.60	40
Abarenicola pacifica	Coos Bay, Oregon	85.6	0.65	35

[a] From Oglesby, 1975.

molluscs release amino acids intact from their free amino acid pool to the environment (Lynch and Wood, 1966; Kasschau, 1975; Gilles, 1979). However, Anderson (1975) reported that at low salinity, isolated gills of *Rangia* accumulated glycine, which was converted to large, insoluble, and osmotically inactive compounds.

Some barnacles are found in low salinity water, although they are osmoconformers (Fyhn and Costlow, 1975). The ability of *Balanus improvisus* to exhibit marked euryhalinity is dependent on cell volume regulation. Amino acids and other amino compounds, especially proline, alanine, glycine, and taurine, participated in adjusting the intracellular osmolality.

The ability of worms to invade waters of reduced salinity is dependent on a number of factors, one of which is water regulations. Oglesby (1975) found that those species which had the best regulatory ability, expressed as B values (Table III), could penetrate furtherest up a salinity gradient into freshwater. They also exhibited better tolerances to low salinities in the laboratory and could better regulate volume after salinity transfers. Several mechanisms involved in volume regulation have been proposed: (1) lowered permeability to water; (2) expulsion of some or all of the excess water via the urine; and (3) loss of salts from the body fluids which would reduce the gradient for osmotic water influx.

VI. Reproduction

Reproduction phenomena exhibited by benthic organisms represent a wide diversity in patterns of adaptive mechanisms which can be correlated with the numerous habital types in which they appear. Moreover, since some species spend part of their life cycle away from the benthic region in the overlying water column, each stage must be adapted to a different set of environmental factors. Because of its obvious importance, physiological ecologists have investigated various aspects of reproduction ranging from the molecular mobilization of biochemical substances needed for egg production to the reproductive role of the various populations of a community in interpreting ecosystem dynamics. In general, sexual reproduction of benthic animals can be arbitrarily divided into three phases: gonad development, spawning and fertilization, and development and growth. These phases proceed in an orderly, coordinated process, but each is environmentally influenced. Both endogenous and exogenous factors interact to determine the delicate balance between comple interactive processes necessary to complete a life cycle successfully. A comprehensive review of this subject for various animal groups is found in the continuing series "Reproduction of Marine Invertebrates" edited by Giese and Pearse (1974, 1975, 1977, 1979). Recently, Stancyk (1979) edited proceedings of a symposium dealing with the reproductive ecology of marine invertebrates. In the present chapter, the physiological ecological implication of reproduction will be emphasized.

A. GONAD DEVELOPMENT

The gametogenic cycle has been subdivided into a number of stages, including activation, growth and gametogenesis, ripening of gametes, spawning, and a quiescent or resting period. These various stages may be timed so that most members of a population are in the same stage; i.e., synchronous breeding. However, some populations may exhibit an asynchronous cycle, resulting in a proportion of the population breeding at any time. Sastry (1979) states that the reproductive cycle of a species is a genetically controlled response to the environment, especially to temperature, salinity, light, and food. Thus, exogenous and endogenous factors interact to produce mature ova and sperm.

Temperature influences gamete production. In many species, gonad growth and gametogenesis occurs either with increasing spring and summer temperatures or with decreasing temperatures in the fall. However, Calabrese (1969) found that the bivalve, *Mulinia lateralis,* undergoes gametogenesis throughout the year, but at a slower rate during the winter. Species with a wide biogeographical range may exhibit a differential reproductive response to temperatures at various parts of its range. In *Argopecten irradians* from Massachusetts, the development of primary germ cells and gonial cells occurs in winter and early spring, gamete differentiation occurs in April, and maturity is reached in July. In contrast, a population of this species from North Carolina does not show development of primary cells and gonial cells until spring, with gamete differentiation taking place in May (Sastry, 1970a). Populations of tropical species extending into higher latitudes exhibit gametogenesis in warmer months, whereas in the more polar species that extend into warmer regions, this process occurs other than in summer months.

Different thermal thresholds have been reported for different phases of development. For example, scallops exposed to 15°C will develop oogonia, but oocyte growth will not occur unless the animal is exposed to a higher temperature (Sastry, 1970b). Egg production in the burrowing estuarine prawn, *Upogebia africana,* was higher in populations found in a region receiving a heated effluent from a thermal power station than in populations from adjacent estuaries (Hill, 1977). This increase probably was due to the presence of larger and more numerous females, since a positive relationship between female size and number of eggs exists. Although altering the sex ratio, the heated effluent did not influence the initiation or duration of the breeding season.

Salinity is known to influence gametogenesis. Oysters will not develop gametes if the salinity is less than 6‰ (Butler, 1949). Also, reduced salinity inhibits development and incubation rates of the tube-building amphipod, *Carophium trianenonyx* (Shyamasundari, 1976). Although adults tolerate a range between 0.6 and 59.8‰, the eggs develop only within a range of 7.5–37.5‰. Temperature interacts with salinity to alter developmental limits. For example, the lower salinity limit is 10‰ at temperatures between 15° and 35°C and only 15‰ at 9°C.

Gametogenesis is not inhibited in all species at reduced salinities; this process occurs in the bivalve, *Donax cuneatus* (Rao, 1967), and the shipworm, *Nausitoria hedleyi* (Nair and Saraswathy, 1970), when the salinity is low.

Although Gimazane (1971) found that gametogenesis in a bivalve was not effected by photoperiod, Sastry (1970b) reported a correlation between gonad growth and gametogenesis in a scallop. An increase in incident light appeared to be related to the initiation of oogenesis in the intertidal sponge, *Haliclona permollis,* whereas temperature appears to have a secondary role in reproductive behavior (Elvin, 1976).

Some benthic organisms deposit egg masses in the intertidal zone where they could be exposed to desiccation at low tide. Pechenik (1978) found that, contrary to expectations, no differences in desiccation tolerance of encapsulated embryos and rates of water loss from eggs capsules were found between the intertidal mud snail, *Ilyanassa obsoleta* (= *Nassarius obsoletus*), and the subtidal snail, *N. trivittatus*. He felt that the adult behavior of tending to deposit egg capsules in more aquatic microenvironments was more important.

Gonad development is very dependent on the availability of food. In nature, the presence or absence of various food substances and the ability of the adult to utilize food are closely coupled to combinations of various environmental factors, such as temperature, salinity, and light. Therefore, the need to understand the energetics of reproduction in a complex changing environment is fundamental.

In various molluscs, the periods of food abundance and of gonadal development are nearly coincident. However, Sastry (1970a) reported latitudinal differences in reproduction of the scallop (*Argopecten irradians*). In the northern population, the period of gonadal development does not coincide with peak phytoplankton production, unlike the relationship in North Carolina and Florida. The allocation of energy to the process of gametogenesis varies between species and between individuals of the same species. For example, Bayne (1976) compared this phenomenon in eight species of bivalves; *Mytilus* used between 8 and 94% of the total biomass production for gamete production, while the average for all of these species is 39%. Griffiths (1977b) calculated gamete production in a bivalve from South Africa rocky shores as 1.17 × standing crop expressed as dry flesh weight, or 1.33 × energy value of the standing crop per year.

Energy derived from food may be stored when food is abundant and be utilized for reproduction at a later date. Translocation of reserves from storage organs to the gonads has been well documented. However, the details of how the timing of the various biochemical events is coordinated varies with species (see reviews of Bayne 1976; Sastry, 1979).

The complex interaction of external environmental and endogenous factors was visually summarized by Sastry (1975) in Fig. 4. At every stage, various factors interact to influence the successful completion of the reproductive cycle.

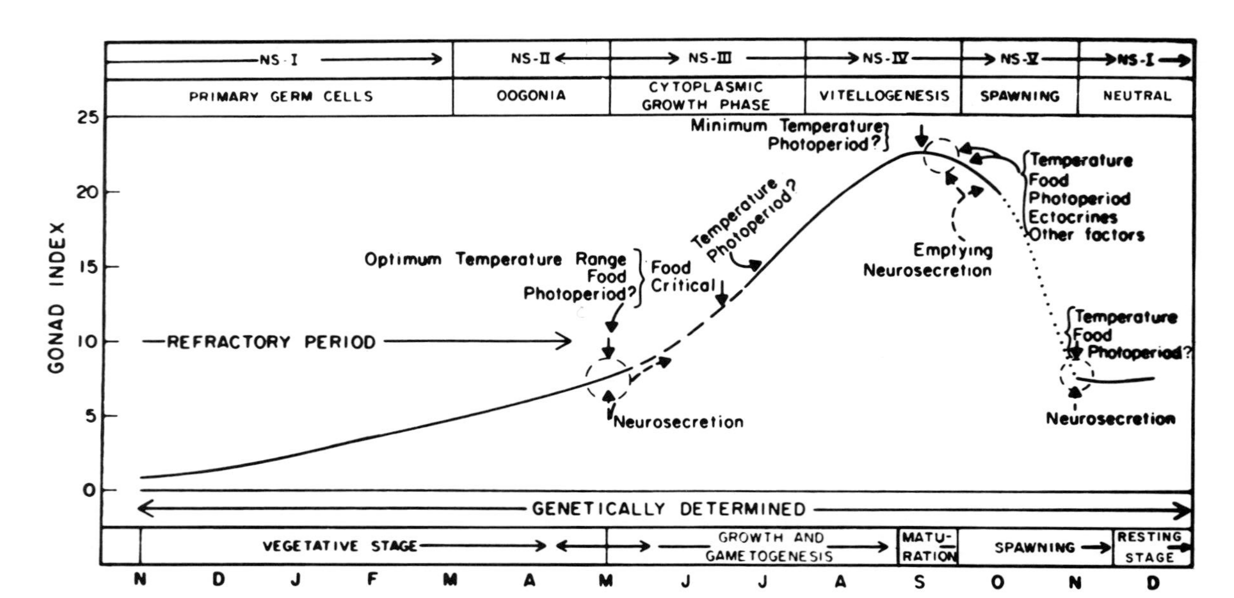

Fig. 4. Schematic drawing showing the interaction between exogenous and endogenous factors in the regulation of the annual reproductive cycle of *A. irradians*. (From Sastry, 1975.)

Seasonal differences in the reproduction of *Halichondria* were noted by Fell and Jacob (1979). In the Mystic Estuary, Connecticut, the reproductive period of postdormant specimens extends from May until July, while the reproductive period of postlarval specimens begins in July and continues at least into October. Since the reproductive periods of postdormant and postlarval specimens are separated, these authors suggest that reproduction is to a large extent under endogenous control.

The periodicity of a complete reproduction cycle varies in different latitudes (see review of Vernberg, 1962) and between species in one area. Monthly reproduction cycles in three species of tropical fiddler crabs were reported by Zucker (1978). Courtship behavior appeared to be synchronized and concentrated during the full moon period. A population of the holothurian, *Leptosynapta tenuis,* from North Carolina has two discrete breeding seasons (spring and autumn). Based on the midsummer absence of mature gametes, Green (1978) suggested that the high seawater temperatures depressed the gametogenic cycle. Two species of bivalves, *Macoma secta* and *M. nasuta,* are geographically sympatric species in California waters. Rae (1978) found that the *M. secta* population was synchronous in its progression through the gametogenic cycle and spawned in August; in contrast, *M. masuta* was asynchronous and spawned in May and from September through November. Since no physical barriers exist between these species, differential reproductive period could prevent interbreeding.

B. LARVAL DFVELOPMENT

Numerous abiotic and biotic factors affect larval development. Normal development appears to occur within a range of environmental conditions characteristic of a species, although populations of one species experiencing distinctly different environmental regions may prefer different ranges. This phenomenon is well demonstrated by the differential response of larvae of the polychaete, *Nereis diversicolor,* from two geographically separated areas in the Baltic Sea (Fig. 5). Also, in its northernmost limit, the embryos of a population of the amphipod, *Gammarus palustris,* develop relatively fast at temperatures above 10°C when compared to other gammarids (Gable and Croker, 1977). The series on reproduction of marine invertebrates edited by Giese and Pearse (1974, 1975, 1977, 1979) summarizes much of the literature on this important topic. More recently, emphasis has been placed on the influence of pollution on the survival and growth of larvae of benthic species (see resistance section). Another research area receiving additional attention deals with the genetics of benthic animals. Based on studies of larval survival and growth of the bivalve, *Mytilus edulis,* in different salinities, Innes and Haley (1977) postulated the presence of genes that influence larval growth, which depend on salinity for their expression and may be related to past selection influence of a fluctuating environment.

Until recently, because a number of factors, including inadequate rearing

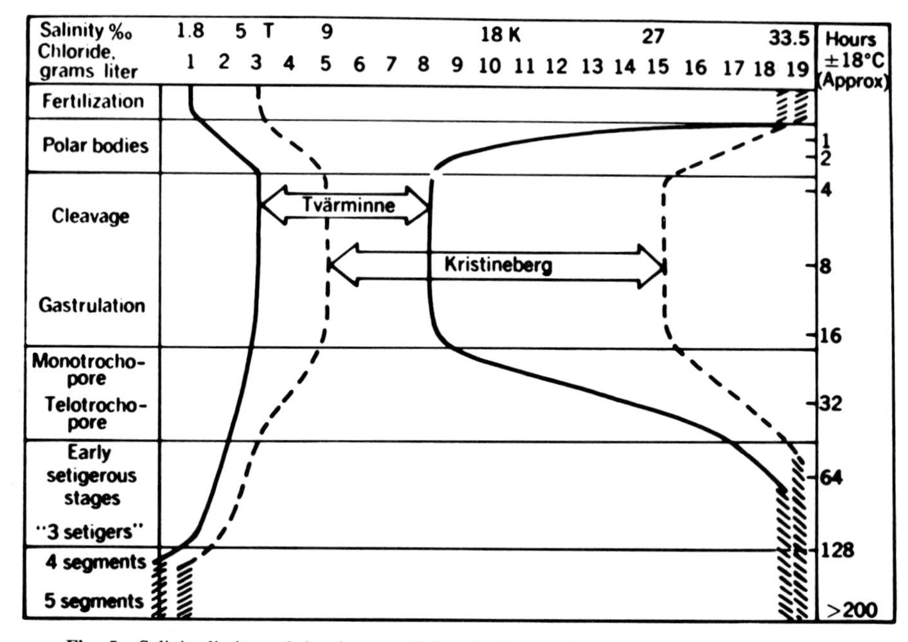

Fig. 5 Salinity limits and developmental time for larvae of *Nereis diversicolor,* a polychaete, from two geographically separated areas—Tvarminne, Finland, and Kristineberg, Sweden. T and K represent approximate surface salinity of these areas. The stages and times are approximations at about 18°C. (From Smith, 1964, in Vernberg and Vernberg, 1970.)

techniques and a lack of sensitive techniques to analyze small amounts of biological material, relatively little data existed on the physiology and biochemistry of developmental stages of benthic invertebrates. A few examples of current studies are cited. Changes in the biochemical composition (carbohydrate, protein, lipid, and total RNA and DNA) of developing eggs of three species of barnacles showed species variation and changes in composition with stage of development (Achituv and Barnes, 1978). One species, *Tetraclita squamosa,* had large amounts of reserves which could serve as metabolic substrates during the development of the planktonic larvae under nutrient-poor conditions. In another study, the correlation between zinc levels and zinc-metalloenzyme activity during development of four molluscan species was investigated (George and Coombs, 1975). Zinc-dependent digestive hydrolase enzymic activities were correlated with the dietary components, and malate dehydrogenase activity increased in oyster larvae when stored fat was metabolized.

C. SPAWNING

The release of gametes has been shown to be influenced by a variety of abiotic and biotic factors (see the excellent review of Giese and Pearse, 1974). As would be expected, great variation between the numerous species existing in the benthic environment is noted. For example, some species spawn as the environmental temperature increases with the approach of spring, others spawn then the maximum temperatures of summer are reached, and still others release gametes as temperatures decrease in autumn. Populations of the same species from geographically separated latitudes may spawn under different thermal regimes, or in some species, males may have a different thermal threshold than females (Sastry, 1979). Chemical substances will induce spawning. Some species produce compounds which stimulate spawning in other members of the same species, and man-produced substances, such as Kraft mill effluents, have also been shown to influence spawning.

Larval Settlement

A critical stage in the life history of a benthic species is when the free-living larva leaves the water column to develop into a benthic adult. If an unfavorable benthic habitat is selected, the adult stage might perish, especially in the case of a sessile species. In the water column, three physical clues that have vector properties are light direction, gravity, and water flow. Other environmental factors, such as temperatures, salinity, and dissolved chemical substances, are scalar quantities which are useful when variations in space can be detected.

Light controls the behavior of the larval stage of many benthic animals. In general, larvae of most species that have been studied are photopositive when first released, starting life by swimming toward light, and becoming photonegative with age. Thus, the larvae are brought to the phytoplankton-rich waters where they grow and develop. When they are ready to settle, they become photonegative and retreat from light to the bottom (Thorson, 1964). Light response can be correlated with habitat preferences of the adults, as illustrated by the two species of intertidal sponges, *Mycale macilenta* and *Haliclina* sp. As typical of intertidal sponges, larvae of these two species are liberated at an advanced stage and have a free-swimming period lasting only a few hours. Adult *M. macilenta* are cryptic species occurring under stones, and their larvae are photonegative throughout the swimming period. On the other hand, adults of *Haliclona* sp. live in a higher tidal region in more exposed conditions, and the larvae swim actively for 9–10 hr after they have been released. Their strong photopositive response keeps them swimming near the surface of the water until they are ready to settle (Berquist and Sinclair, 1968). Response to light may also vary from stage to stage in the life cycle. For example, the trochophore stages of *Mytilus edulis* show no response to light, the young veligers are photonegative, the straight-hinge veliger stages will concentrate toward light, the veloncha larvae do not respond to light, and the eyed-veliger stages are positively phototactic. Then at the time of settlement, the larvae become photonegative. The adaptive value of light response can be illustrated by these mussels; for example, young larvae do not become photopositive until the swimming and defensive mechanisms, including the ability to retract the velum between the shell valves, have become well developed (Bayne, 1964b). Larvae appear to generally prefer a diffuse, not too strong light, and they are more responsive to blue and green wavelenths (450–510 mu) that penetrate most deeply in the sea (Clarke and Oster, 1934).

Environmental factors, such as high temperature and reduced salinity, can modify the response to light, and they tend to reduce the photopositive response, as illustrated by developmental stages of *Mytilus edulis* (Bayne, 1964a). A similar response in adult marine animals has been explained by the fact that the respiration rate of these organisms is conditioned in part by temperature and in part by light. The exchange of gases between animal and environment, which must be in balance, will shift in the same direction with increasing temperature and light. Thus, if respiration is to continue optimally, an animal must avoid light when water temperature rises. Larval responses to light are thought to have the same basis. If surface layers of the water were to become too warm, a reversal of the photopositive response would tend to remove the larvae to deeper, cooler water where they can better adapt metabolically. Reduced salinity caused by heavy run off of freshwater from rivers or by prolonged periods of rain also tend to lessen the photopositive response in larvae. During these periods the water will become stratified, with lower salinity water staying near the surface. To avoid reduced salinity, the larvae become photonegative, seeking out higher salinity waters at greater depths. Larvae of intertidal zone animals, however, are much less affected by low salinity waters than are larvae of pelagic animals.

As larvae approach metamorphosis and settlement, their response to light, gravity, and hydrostatic pressure often changes. They tend to become photonegative and geopositive and they no longer respond to increased pressure by swimming. This combination of responses effectively takes them out of the surface water and places them in contact with the substratum.

Invertebrate larvae are sensitive to small changes in hydrostatic pressure. Generally an increase in pressure results in larvae swimming upward, whereas the response to a decreased pressure is to move downward, either by active swimming or by passive sinking [Knight-Jones and Morgan (1966) and Crisp (1974) reviewed this subject].

Upon reaching the settling stage in their development, larvae must choose the proper substratum. Although photonegative at the time of settlement, this photonegativity must not encourage them to settle during the night because they might settle in a position directly exposed to direct sunshine, which is often injurious to intertidal zone animals. Because total darkness will postpone or even prevent settling in many animals, it appears that there is a direct correlation between response to light at the time of settling and the position the organism maintains in the intertidal zone, as demonstrated by three species of barnacles. Larvae of the barnacle, *Chthamalus stellatus,* living in the upper part of the tidal zone, settled most abundantly in direct sunlight; larvae of *Balanus amphitrite,* which lives lower in the intertidal zone, settled abundantly in bright but not direct sunlight; and larvae of *Balanus tintinnabulum,* inhabiting a still lower part of the intertidal zone, settled most abundantly during dusk and at daybreak (Daniel, 1957). At the time of settling, larvae frequently exhibit a searching response by alternately crawling and swimming. If a suitable surface is not found, settlement and subsequent metamorphosis to an adult body form may be delayed and in extreme cases, the larvae may lose its ability to metamorphose altogether.

D. ADULT GROWTH

As found in larval growth studies, various environmental factors affect the growth of adult organisms. Although not all macrobenthic species have been examined, some commercially important species have been studied for years. One research trend has been to determine growth patterns in order to determine secondary production in benthic ecosystems. For example, Conan and Shafee (1978) analyzed growth and recruitment of the black scallop (*Chlamys varia*) over a 5-year period because it is an important component in the food web of the Bay of Brest, France. Breteler (1976) published results of a similar 4-year study on the shore crab, *Carcinus maenas,* in the Dutch Wadden Sea. Comparative studies on growth and population studies of closely related or common species have been made to understand inter- and intraspecific dynamics in niche diversification.

Seed and Brown (1975) attempted to explain the markedly different population structures of three common bivalve molluscs from Strangford Lough, Northern Ireland, in terms of their annual reproductive cycles, patterns of recruitment, growth, and mortality. *Cerastoderma edule* and *Mytilus edulis* are abundant intertidally, whereas *Modiolus modiolus* forms extensive sublittoral reefs. *Cerastoderma* has a pronounced seasonal cycle, spawning over a relatively restricted period between late June and September, but no marked spawning period is detectable in *Modiolus,* and *Mytilus* spawns from early spring until the autumn with correspondingly extended, and often very erratic, periods of settlement. The apparently stable *Modiolus* population could be explained in terms of slow but almost continual recruitment, moderate growth, and a relatively low mortality once beyond a critical size range. *Mytilus* populations with a persistently high proportion of small individuals exhibit heavy but irregular periods of recruitment, variable individual growth rates coupled with exceedingly slow growth of the majority of individuals, and locally intense predation. The generally well defined polymodal distribution in *Cerastoderma* is attributed largely to the restricted period of recruitment during the nongrowing season, together with relatively uniform growth rates of all individuals within each year class.

Pollution effects on growth have been investigated. For example, Edwards

(1978) reported that chronic exposure to water-soluble crude oil fractions reduced the growth rate and respiration of the shrimp, *Crangon crangon*. In addition, the net carbon turnover was lessened.

The development of growth models is important to the better management fisheries and to the understanding of ecological processes, such as population dynamics and energy budgets. A few example are McCaughran and Powell (1977), a growth model for the Alaska king crab; Botsford and Wickham (1978), age–species, density-dependent models for the California Dungeness crab; Dame (1972), intertidal oysters; and McDonald (1977) intertidal crabs.

E. ASEXUAL REPRODUCTION

Although more emphasis has been put on sexual reproduction of benthic macrofauna, the dynamics and energetics of asexual reproduction has recently been reviewed. Sebens (1979) presented a model which defines three optimal sizes for indeterminately growing species, especially anthozoans; (1) optimal size for a solitary individual, (2) optimal size of a polyp within a colony, and (3) optimal size at asexual division. Based on field translocation studies of sea anemones, Shick *et al.* (1979) suggested that asexual reproduction can lead to amplification of particularly successful genotypes. In *Metridium senile* subjected to areas of low tidal current velocity, there is reduced vegetative proliferation and individuals are larger in body size; smaller sized individuals are produced asexually.

Earlier, Kinne (1958) demonstrated that the growth form of sections of the colonial hydroid, *Cordylophora caspia,* could be differentially altered by salinity and temperature. Since these sections came from one colony and were assumed to be genetically similar, different structural responses were environmentally induced. For example, the colonies grown in freshwater and compared with sections grown at 15 and 30‰ were relatively low, the hydrocauli were short and unramified, and the total length of stolons was long. Important changes in the hydranths also occurred; i.e., they were short, their breadth was maximum, the number of tentacles was reduced, and the diameter of a tentacle was greatest. In freshwater, structure changes at the cellular level included cells that were more numerous, higher, and narrower, and in possession of larger nuclei. All of these changes appeared to be adaptations to a hypotonic environment, effectively reducing the cellular free surface and increasing cell number and metabolism.

References

Achituv, Y., and Barnes, H. (1978). Studies in the biochemistry of cirripede eggs. VI. Changes in the general biochemical composition during development of *Tetraclita squamosa rufotincta* Pilsbry, *Balanus perforatus* Brug., and *Pollicipes cornucopia*. *Mar. Biol.* **32,** 171–176.

Achituv, Y., and Borut, A. (1975). Temperature and water relations in *Tetraclita squamosa rufotincta* Pilsbry cirripedia from the Gulf of Elat (Red Sea). *In* "Ninth European Marine Biology Symposium" (H. Barnes, ed.) pp. 95–108. University Press, Aberdeen, Scotland.
Ahsanullah, M., and Newell, R. C. (1977). The effects of humidity and temperature of water loss in *Carcinus maenas* (L) and *Portunus marmoreus* (Leach). *Comp. Biochem. Physiol.* **56,** 593-601.
Alderdice, D. F. (1972). Factor combination. Responses of marine poikelotherms to environmental factors acting in concert. *In* "Marine Ecology" (O. Kinne, ed.), Vol. 1, pp. 1659–1722. Wiley (Interscience), New York.
Aldrich, J. C. (1975). On the oxygen consumption of the crabs *Cancer pagurus* and *Maia squinado. Comp. Biochem. Physiol.* **50,** 223–228.
Alee, W. C., Emerson, A. E., Park, 0., Park, T., and Schmidt, K. P. (1949). "Principles of Animal Ecology." Saunders, Philadelphia, Pennsylvania.
Anderson, G. (1977). The effects of parasitism on energy flow through laboratory shrimp population. *Mar. Biol.* **42,** 239–251.
Anderson J. W. (1975). The uptake and incorporation of glycine by the gills of Rangia cuneata (Mollusca: Bivalvia) in response to variations in salinity and sodium. *In* "Physiological Ecology of Estuarine Organisms" (F. J. Vernberg, ed.), pp. 239–258. Univ. of South Carolina Press, Columbia.
Andronikov, B. (1975). Heat resistance of gametes of marine invertebrates in relation to temperature conditions under which the species exist. *Mar. Biol.* **30,** 1–11.
Ansell, A. D., and Trevallion, A. (1969). Behavioral adaptations of intertidal molluscs from a tropical sandy beach. *J. Exp. Mar. Biol. Ecol.* **4,** 9–35.
Bakus, G. J. (1969). Energetics and feeding in shallow marine waters. *Int. Rev. Gen. Exp. Zool.* **4,** 275–369.
Barker, P. L., and Gibson, R. (1978). Observations on the structure of the mouthparts, histology of the alimentary tract, and digestive physiology of the mud crab *Scylla serrata. J. Exp. Mar. Biol. Ecol.* **32,** 177–196.
Barnes, H., Barnes, M., and Finlayson, D. M. (1963). The metabolism during starvation in *Balanus balanoides. J. Mar. Biol. Assoc. U.K.* **43,** 213–233.
Barnes, R. S. K., and Greenwood, J. G. (1978). The response of the intertidal gastropod *Hydrobia ulvae* to sediments of differing particle size. *J. Exp. Mar. Biol. Ecol.* **31,** 43–54.
Bayne, B. L. (1964a). The responses of the larvae of *Mytilus edulis* L. to light and to gravity. *Oikos* **15,** 162–74.
Bayne, B. L. (1964b). Primary and secondary settlement in *Mytilus edulis* L. (Mollusca). *J. Anim. Ecol.* **33,** 513–23.
Bayne, B. L. (1976). Aspects of reproduction in bivalve molluscs. In "Estuarine Processes" (M. Wiley, ed.), Vol. 1, pp. 432–448. Academic Press, New York.
Bayne B. L. Gabbott, P. A., and Widdows, J. (1975). Some effects of stress in the adult on the eggs and larvae of *Mytilus edulis* L. *J. Mar. Biol., U.K.* **55,** 675–689.
Bayne, B. L., Bayne, C. J., Carefoot, T. C., and Thompson, R. J. (1976). The Physiological ecology of *Mytilus californianus* Conrad. 1. Metabolism and energy balance. *Oecologia* **22,** 211–228.
Bayne, B. L., and Scullard, C. (1978). Rates of feeding by *Thais lapillus* (L.) *J. Exp. Mar. Biol. Ecol.* **32,** 113–129.
Bayne, B. L., Widdows, J., and Worral, C. (1977). Some temperature relationships in the physiology of two ecologically distinct bivalve populations. *In* "Physiological Responses of Marine Biota to Pollutants" (F. J. Vernberg, A. Calabrese, F. Thurberg, and W. Vernberg, eds.), pp. 379–400. Academic Press, New York.
Berquist, P. R., and Sinclair, M. E. (1968). The morphology and behavior of larvae of some intertidal sponges. *N.Z.J. Mar. Freshwater Res.* **2,** 426–37.
Bliss, D. E. (1978). Transition from water to land in decapod crustaceans *Am. Zool.* **8,** 355–392.

Boddeke, R. (1975). Autumn migration and vertical distribution of the brown shrimp Crangon crangon L. in relations to environmental conditions. *In* "Ninth European Marine Biology Symposium" (H. Barnes, ed.), pp. 483–494. University Press, Aberdeen, Scotland.
Botsfor, L., and Wickham, D. E. (1978). Behavior of age-specific, density-dependent models and the northern California dungeness crab (*Cancer magister*) fishery. *J. Fish. Res. Board Can.*, **35,** 833–843.
Brance, G. M. and Newell, R. C. (1978). A comparative study of metabolic energy expenditure in the limpets *Patella cochlear, P. oculus* and *P. granatina. Mar. Biol.* **49,** 351–361.
Breteler, W. C. M. K. (1976). Settlement, growth, and production of the shore crab, *Carcinus maenas,* on tidal flats in the Dutch Wadden Sea. *Neth. J. Sea Res.* **10,** 354–376.
Broekhuysen, C. J. (1940). A preliminary investigation of the importance of desiccation, temperature and salinity as factors controlling the vertical distribution of certain marine gastropods in False Bay, South Africa. *Trans. R. Soc. S. Afr.* **28,** 255–292.
Burke, E. M. (1979). Aerobic and anaerobic metabolism activity and hypoxia in two species of intertidal crabs. *Biol Bull.* (*Woods Hole, Mass.*) **156,** 157–168.
Butler, P. A. (1949). Gametogenesis in the oyster under conditions of depresses salinity. *Biol. Bull.* (*Woods Hole, Mass.*) **96,** 263–269.
Calabrese, A. (1969). Reproductive cycle of the coot clam, *Mulinia lateralis* (Say), in Long Island Sound. *Veliger* **12,** 265–269.
Calow, P. (1977). Conversion efficiencies in heterotrophic organisms. *Biol. Rev. Cambridge Philos. Soc.* **52,** 385–409.
Carr, W. E. S. (1967a). Chemoreception in the mud snail *Nassarius obsoletus.* I. Properties of stimulatory substances extracted, from shrimp. *Nassarius obsoletus. Biol. Bull.* (*Woods Hole, Mass.*) **132,** 70–105.
Carr, W. E. S. (1967b). Chemoreception in the mud snail Nassarius obsoletus. II. Identification of stimulatory substances. *Biol. Bull.* (*Woods Hole, Mass.*) **132,** 106–127.
Clarke, G. L., and Oster, R. H. (1934). The penetration of the blue and red components of daylight into Atlantic coastal waters and its relation to phytoplankton metabolism. *Biol. Bull.* (*Woods Hole, Mass.*) **67,** 59–75.
Coles, S. L., and Jokiel, P. L. (1978). Synergistic effects of temperature, salinity, and light on the hermatypic coral, *Montipora verrucosa. Mar. Biol.* **49,** 187–195.
Conan, G. and Shafee, M. (1978). Growth and biannual recruitment of the black scallop *Chlamys varia* (L.) in Lanveoc Area, Bay of Brest. *J. Exp. Mar. Biol. Ecol.* **35,** 59–71.
Cook, P. A., and Gabbott, P. A. (1972). Seasonal changes in the biochemical composition of the adult barnacle, *Balanus balanoides,* and the possible relationships between biochemical composition and cold-tolerance. *J. Mar. Biol. Assoc. U.K.* **52,** 805–825.
Coull, B. C., ed. (1977). "Ecology of Marine Benthos," Vol. 6. Univ. South Carolina Press, Columbia, South Carolina.
Crisp, D. J., ed. (1964). "An Assessment of Plankton Grazing by Barnacles," pp. 251–264. Blackwell, Oxford.
Crisp, D. J. (1974). Factors influencing the settlement of marine invertebrate larvae. *In* "Chemoreception in Marine Organisms" (P. T. Grant and A. M. Mackie, eds.), pp. 177–265. Academic Press, New York.
Crisp, D. J., Davenport, J., and Gabbott, P. A. (1977). Freezing tolerance in *Balanus balanoides. Comp. Biochem. Physiol. A* **57,** 359–361.
Dame, R. F. (1972). The ecological energies of growth, respiration, and assimilation in the intertidal american oyster *Crassostrea virginica. Mar. Biol.* **17,** 243–250.
Dame, R. F. (1976). Energy flow in an intertidal oyster population. *Estuarine Coastal Mar. Sci.* **4,** 243–253.
Dame, R. F., ed. (1979). "Marine-Estuarine Systems Simulation." Univ. of South Carolina Press, Columbia.
Dame, R. F., and Vernberg, F. J. (1978). The influence of constant and cyclic acclimation tempera-

ture on the metabolic rate of *Panopeus herbstii* and *Uca pugilator*. *Biol. Bull. (Woods Hole, Mass.)* **154,** 188–197.

Daniel, A. (1957). Illumination and its effect on the settlement of barnacle cyprids. *Proc. Zool. Soc. London* **129,** 305–313.

Darnell, R. M., and Wissing, T. E. (1975). Nitrogen turnover and food relationships of the pinfish *Lagodon rhombiodes* in a North Carolina estuary. *In* "Physiological Ecology of Estuarine Organisms" (F. J. Vernberg, ed.), pp. 81–110. Univ. of South Carolina Press, Columbia.

Davies, P. S. (1969). Physiological ecology of *Patella*. III. Desiccation effects. *J. Mar. Biol Assoc. U.K.* **49,** 291–304.

Dehnel, P. A. (1958). Effect of photoperiod on the oxygen consumption of two species of intertidal crabs. *Nature (London)* **181,** 1415–1417.

DeVries, A. L. (1974). Survival of freezing temperatures. *In* "Biochemical and Biophysical Perspectives in Marine Biology" D. C. Malins, and J. R. Sargent, eds.), Vol. 1, pp. 290–330. Academic Press, New York.

de Zwann, A. (1977). Anaerobic energy metabolism in bivalve molluscs. *Oceanogr. Mar. Biol.* **15,** 103–187.

de Zwann, A. and Skjoldal, H. R. (1979). Anaerobic energy metabolism of the scavenging isopod *Cirolana borealis* (Lilljeborg). *J. Comp. Physiol.* **129,** 327–331.

de Zwann, A., and Wijsman, T. C. M. (1976). Anaerobic metabolism in Bivalvia (Mollusca). Characteristics of anaerobic metabolism. *Comp. Biochem. Physiol. B* **54,** 313–324.

Dimock, R. V., Jr. (1977). Effects of evisceration on oxygen consumption by *Stichopus parvimensis*. *Mar. Biol. Ecol.* **28,** 125–132.

Dorgelo, J. (1976). Salt tolerance in Crustacea and the influence of temperature upon it. *Biol. Rev. Cambridge Philos. Soc.* **51,** 255–290.

Edwards, R. R. C. (1978). Effects of water-soluble oil fractions on metabolism, growth, and carbon budget of the shrimp *Crangon crangon*. *Mar. Biol.* **46,** 259–265.

Elner, R. W., and Hughes, R. N. (1978). Energy maximization in the diet of the shore crab, *Carcinus maenas*. *J. Anim. Ecol.* **47,** 103–116.

Elvin, D. W. (1976). Seasonal growth and reproduction on or of an intertidal sponge, *Haliclona permollis* (Bowerbank). *Biol. Bull.* (*Woods Hole, Mass.*) **151,** 108–125.

Engel, D. W., and Eggert, L. D. (1974). The effect of salinity and sex on the respiration rates of excised gills of the blue crab, *Callinectes sapidus*. *Comp. Biochem. Physiol.* **47,** 1005–1011.

Fell, E. E., and Jacob, W. F. (1979). Reproduction and development of *Halichondria* sp. in the Mystic Estuary, Connecticut. *Biol. Bull.* (*Woods Hole, Mass.*) **156,** 62–75.

Ferguson, J. C. (1971). Uptake and release of free amino acids by starfishes. *Biol. Bull.* (*Woods Hole, Mass.*) **141,** 122–129.

Foster, B. A. (1971). Desiccation as a factor in the intertidal zonation of barnacles. *Mar. Biol.* **8,** 12–29.

Fraenkel, F. (1966). The heart resistance of intertidal snails at Shirahama, Wakeyamaken, Japan. *Seto Mar. Biol. Lab.* **14,** 185–195.

Frankenberg, D., Coles, S. L., and Johannes, R. E. (1967). The potential trophic significance of *Callianassa major* fecal pellets. *Limnol. Oceanogr.* **12,** 113–120.

Friedrich, H. (1969). "Marine Biology." Univ. of Washington Press, Seattle.

Fyhn, H. J., and Costlow, J. D. (1975). Cellular adjustments to salinity variations in an estuarine barnacle, *Balanus improvisus*. *In* "Physiological Ecology of Estuarine Organisms" (F. J. Vernberg, ed.), pp. 227–238. Univ. of South Carolina Press, Columbia.

Gable, M., and Croker, R. A. (1977). The salt marsh amphipod, *Gammarus palustris* Bousfield, 1969 at the northern limit of its distribution. I. Ecology and life cycle. *Estuarine Coastal Mar. Sci.* **5,** 123–134.

Gable, M., and Croker, R. (1978). The salt marsh amphipod, *Gammarus palustris* Bousfield, 1969 at the northern limit of its distribution. II. Temperature-salinity tolerances. *Estuarine Coastal Mar. Sci.* **6,** 225–230.

Gamble, J. C. (1970). Effect of low dissolved oxygen concentrations on the ventilation rhythm of three tubicolous crustaceans, with special reference to the phenomenon of intermittent ventilation. *Mar. Biol.* **6,** 121–127.

George, S., and Coombs, T. (1975). A comparison of trace-metal and metalloenzyme profiles in different molluscs and during development of the oyster. *In* "Ninth European Marine Biology Symposium" (H. Barnes, ed.), pp. 433–449. University Press, Aberdeen, Scotland.

Giese, A. C., and Pearse, J. S., eds. (1974). "Reproduction of Marine Invertebrates," Vol. 1. Academic Press, New York.

Giese, A. C., and Pearse, J. S., eds. (1975). "Reproduction of Marine Invertebrates," Vols. 2 and 3. Academic Press, New York.

Giese, A. C., and Pearse, J. S., eds. (1977). "Reproduction of Marine Invertebrates," Vol. 4. Academic Press, New York.

Giese, A. C., and Pearse, J. S., eds. (1979). "Reproduction of Marine Invertebrates," Vol. 5. Academic Press, New York.

Gilles, R., ed. (1979). "Mechanisms of Osmoregulation in Animals." Wiley, New York.

Gimazane, J. P. (1971). Introduction a l'étude expérimentale du cycle sexual d'un mollusque bivalve *Cardium edule* L. Analyse des populations, évolution de la gonade et action de quelques facteurs: Nutrition, température, photopériode. Doctoral Thesis, Université de Caen.

Gordon, J., and Carriker, M. R. (1978). Growth lines in a bivalve mollusk: Subdaily patterns and dissolution of the shell. *Science* **202,** 519–521.

Grainger, J. N. R. (1975). Mechanism of death at high temperatures in *Helix* and *Patella. Therm. Biol.* **1,** 11–13.

Green, J. D. (1978). The annual reproductive cycle of an Apoduous Holothurian, *Leptosynapta tenuis,* a biomodal breeding season. *Biol. Bull.* (*Woods Hole, Mass.*) **154,** 68–78.

Griffiths, R. J. (1977a). Temperature acclimation in *Actinia equina* L. (Anthozoa). *J. Exp. Mar. Biol. Ecol.* **28,** 285–292.

Griffiths, R. J. (1977b). Reproductive cycles in littoral populations of *Choromytilus meridionalis* and *Aulacomya ater* with a quantitative assessment of gamete production in the former. *J. Exp. Mar. Biol. Ecol.* **30,** 53–71.

Gunter, G. (1945). Studies on marine fishes of Texas. *Publ. Inst. Mar. Sci., Univ. Tex.* **1,** 1–190.

Hamby, R. J. (1975). Heat effects on a marine snail. *Biol. Bull.* (*Woods Hole, Hass.*) **149,** 331–347.

Hammen, C. S. (1969). Lactate and succinate oxidoreductases in marine invertebrates. *Mar. Biol.* **4,** 233–238.

Hammen, C. S. (1976). Respiratory adaptations: Invertebrates. *In* "Estuarine Processes" (M. Wiley, ed.), Vol. 1, pp. 347–355. Academic Press, New York.

Hargens, A. R., and Shabica, S. V. (1973). Protection against lethal freezing temperatures by mucus in an antarctic limpet. *Cryobiology* **10,** 331–337.

Haynes, W. B. (1974). Sand-beach energetics: Importance of the isopod *Tylos punctatus. Ecology* **55,** 838–847.

Hazlett, B. A. (1968). Stimuli involved in the feeding behavior of the hermit crab *Clibanarius vittatus* (Decapoda, Paguridea). *Crustaceana* **15,**305–311

Hedgpeth, J. W. (1957). Classification of marine environments. *Mem., Geol. Soc. Am.* **67,** 17–28.

Herreid, C. F., II (1969). Integument permeability of crabs and adaptation to land. *Comp. Biochem. Physiol.* **29,** 423–429.

Hill, B. J. (1977). The effect of heated effluent on egg production in the estuarine prawn *Upogebia africana* (Ortmann). *J. Exp. Mar. Biol. Ecol.* **29,** 291–302.

Hughes, R. N. (1970). An energy budget for a tide-flat population of the bivalve *Scrobicularia plana* (Da Cota). *J. Anim. Ecol.* **39,** 357–381.

Hughes, R. N. (1971a). Ecological energetics of *Nerita* (Archaeogastropoda: Neritacea) populations on Barbados, West Indies. *Mar. Biol.* **11,** 12–22.

Hughes, R. N. (1971b). Ecological energetics of the keyhole limpet *Fissurella barbadensis* Gmelin. *J. Exp. Mar. Biol. Ecol.* **6,** 167–168.

Hunter, J., and Arthur, D. R. (1978). Some aspects of the ecology of *Peloscolex benedeni* Udekem (Oligochaeta: Tubificidae) in the Thames Estuary. *Estuarine Coastal Mar. Sci.* **6,** 197–208.
Innes, D. J., and Haley, L. E. (1977). Genetic aspects of larval growth under reduced salinity in *Mytilus edulis Biol. Bull.* (*Woods Hole, Mass.*) **153,** 312–321.
Jennings J. B. (1965). "Feeding, Digestion, and Assimilation in Animals." Pergamon, Oxford.
Jorgensen, C. B. (1966). "Biology of Suspension Feeding." Permagon, Oxford.
Jorgensen, C. B. (1976). August Putter, August Krogh, and modern ideas on the use of dissolved organic matter in aquatic environments. *Biol. Rev. Cambridge Philos. Soc.* **51,** 292–328.
Kangas, P., and Skoog, G. (1978). Salinity tolerance of *Theodoxus fluviatilis* (Mollusca, Gastropoda) from freshwater and from different salinity regimes in the Baltic Sea. *Estuarine Coastal Mar. Sci.* **6,** 409–416.
Kanwisher, J. W. (1966). Freezing in intertidal animals. *In* "Cryobiology". (H. T. Meryman, ed.) pp. 487–494. Academic Press, New York.
Kasschau, M. (1975). Changes in concentrations of free amino acids in larval stages of trematode *Himasthla quissetensis* in its intermediate host *Nassarius obsoletus. Comp. Biochem. Physiol. B* **51,** 273–280.
Keegan, B. F., Ceidigh, P. O., and Boaden, P. J. S. (1977). "Biology of Benthic Organisms." Pergamon, Oxford.
Kennedey, V. S., and Mihursky, J. A. (1971). Upper temperature tolerances of some estuarine bivalves. *Chesapeake Sci.* **12,** 193–204.
Khlebovich, V. V. (1969). Aspects of animal evolution related to critical salinity and internal state. *Mar. Biol.* **2,** 338–345.
Kinne, O. (1958). Adaptation to salinity variations—some facts and problems. *In* "Physiological Adaptation" (C. L. Prosser, ed.), pp. 92–106. Am. Physiol. Soc., Washington, D. C.
Kinne O. (1964). The effects of temperature and salinity on marine and brackish water animals. II. Salinity and temperature salinity combinations. *Oceanogr. Mar. Biol.* **2,** 281–339.
Kinne, O. (1971). Salinity-invertebrate animals. *In* "Marine Ecology," (O. Kinne, ed.), pp. 821–966. Wiley (Interscience), New York.
Kinne, O., ed. (1977). Ecosystems research *Helgol. Wiss. Meeresunters.* **30,** 1–735.
Kirby-Smith, W. W. (1976). The detritus problem and the feeding and digestion of an estuarine organism. *In* "Estuarine Processes" (M. Wiley, ed.), Vol. 1, pp. 469–480. Academic Press, New York.
Kleiber, M. (1961). "The Fire of Life: An Introduction to Animal Energetics." Wiley, New York.
Knight-Jones, E. W., and Morgan, E. (1966). Responses of marine animals to changes in hydrostatic pressure. *Oceanogr. Mar. Biol.* **4,** 267–99.
Kudenov, J. D. (1978). The feeding ecology of *Axiothella rubrocincta. J. Exp. Mar. Biol. Ecol.* **31,** 209–221.
Kuenzler, E. J. (1961). Structure and energy flow of a mussel population in a Georgia salt marsh. *Limnol. Oceanogr.* **6,** 191–204.
Kushins, L. J., and Mangum, C. P. (1971). Responses to low oxygen conditions in two species of the mud snail *Nassarius. Comp. Biochem. Physiol.A* **39,** 421–435.
Laird, C. E., and Haefner, P. A., Jr. (1976). Effects of intrinsic and environmental factors on oxygen consumption in the blue crab, *Callinectes sapidus* Rathbun. *J. Exp. Mar. Biol. Ecol.* **22,** 171–178.
Laverack, M. S. (1963). Aspects of chemoreception in Crustacea. *Proc. Int. Congr. Zool., 16th, 1963,* pp. 72–73
Lawrence, J. M., and Hughes-Games, L. (1972). The diurnal rhythm of feeding and passage of food through the gut of *Diadema setosum. Isr. J. Zool.* **21,** 13–16.
Lindstedt, K. J. (1971). Chemical control of feeding behavior. *Comp. Biochem. Physiol.* **39,** 553–581.
Lockwood, A. P. M. (1962). The osmoregulation of Crustacea. *Biol. Rev. Cambridge Philos. Soc.* **37,** 257–305.

Loosanoff, V. L., and Tommers, F. D. (1948). Effect of suspended silt and other substances on rate of feeding of oysters. *Science* **107,** 69–70.

Lynch, M. P., and Wood, L. (1966). Effect of environmental salinity on free amino acids of *Crassostrea virginica* Gmelin. *Comp. Biochem. Physiol.* **19,** 783–90.

McCaughran, D. A., and Powell, G. C. (1977). Growth model for Alaska King Crab (*Paralithodes camtschatica*). *J. Fish. Res. Board Can.* **34,** 989–995.

McDonald, J. (1977). The comparative intertidal ecology and niche relations of the sympatric mud crabs *Panopeus herbstii* Milne-Edwards and *Eurypanopeus depressus* (Smith) at North Inlet, South Carolina, USA. (Decapoda: Brachyura: Xanthidae). Ph.D., University of South Carolina, Columbia.

McLeese, D. W. (1956). Effects of temperature, salinity and oxygen on the survival of the American Lobster. *J. Fish. Res. oar Can.* **13,** 247–272.

Mangum, C. (1969). Low temperature blockage of the feeding response in boreal and temperate zone polychaetes. *Chesapeake Sci.* **10,** 64–65.

Mangum, C. P., and Cox, D. O. (1971). Analysis of feeding responses in the onuphid polychaete *Diopatra cuprea* (Bosc.). *Biol. Bull.* (*Woods Hole, Mass.*) **140,** 215–229.

Mangum, C. P., and Miyamoto, D. M. (1970). The relation between spontaneous activity cycles and diurnal rhythms of metabolism in the polychaetous annelid, *Glycera dibranchiata. Mar. Biol.* **7,** 7–10.

Mangum, C., and Van Winkle, W. (1973). Responses of aquatic invertebrates to declining oxygen conditions. *Am. Zool.* **13,** 529–541.

Murphy, D., and Pierce, S. (1975). The physiological basis for changes in the freezing tolerance of intertidal molluscs. *J. Exp. Zool.* **193,** 313–322.

Nair, N. B., and Saraswathy, M. (1970). Some recent studies on the shipworms of India. *J. Mar. Biol. Assoc. India, Symp. Ser.* **3,** Part 3, 718–729.

Newell, R. C. (1970). "The Biology of Intertidal Animals." Am. Elsevier, New York.

Newell, R. C. (1976). Adaptations to intertidal life. *In* "Adaptation to Environment: Essays on the Physiology of Marine Animals" (R. C. Newell, ed.), pp. 1–82. Butterworth, London.

Newkirk, G. F., Waugh, D. L., and Haley, L. E. (1977). Genetics of larval tolerance to reduced salinities in Two populations of oysters, *Crassostrea virginica. J. Fish. Res. Board Can.* **32,** 384–387.

Nicolaisen, W., and Kanneworff, E. (1969). On the borrowing and feeding habits of the amphipods *Bathyporeia pilosa* Lindstrom and *Bathyporeia sarsi* Watkin. *Ophelia* **6,** 231–250.

Norse, E. A., and Estevez, M. (1977). Studies on portunid crabs from the Eastern Pacific. I. Zonation along environmental stress gradients from the coast of Colombia. *Mar. Biol.* **40,** 365–373.

Odum, E. P., and Smalley, A. E. (1959). Comparison of population energy flow of a herbivorous and a deposit-feeding invertebrate in a salt marsh ecosystem. *Proc. Natl. Acad. Sci. U.S.A.* **45,** 617–622.

Oglesby, L. C. (1975). An analysis of water-content regulation in selected worms. *In* "Physiological Ecology of Estuarine Animals" (F. J. Vernberg, ed.), pp. 181–204. Univ. of South Carolina Press, Columbia.

Paine, R. T. (1971). Energy flow in a natural population of a herbivorous gastropod *Tegula funebralis. Limnol. Oceanogr.* **16,** 86–98.

Palmer, R. (1979). A histological and histochemical study ot digestion in the bivalve *Arctica islandica* L. *Biol. Bull.* (*Woods Hole, Mass.*) **156,** 15–129.

Pamatmat, M. M. (1978). Oxygen uptake and heat production in a metabolic conformer (*Littorina irrorata*) and a metabolic regulator (*Uca pugnax*). *Mar. Biol.* **48,** 317–326.

Pandian, T. J. (1975). Mechanisms of heterotrophy. *In* "Marine Ecology" (O. Kinne, ed.), Vol. 2, pp. 61–249. Wiley, New York.

Patel, B., and Crisp, D. J. (1960). Rates of development of the embryos of several species of barnacles. *Physiol. Zool.* **33,** 104–119.

Pechenik, J. A. (1978). Adaptations to intertidal development: Studies on *Nassarius obsoletus*. *Biol. Bull.* (*Woods Hole, Mass.*) **154,** 282–291.

Pequignat, E. (1972). Some new data on skin-digestion and absorption in urchins and sea stars. *Mar. Biol.* **12,** 28–41.

Pequignat, E. (1973). A kinetic and autoradiographic study of the direct assimilation of amino acids and glucose of organs of the mussel *Mytilus edulis*. *Mar. Biol.* **19,** 227–244.

Petrusewiez, K. (1967). Suggested list of more important concepts in productivity studies (definitions and symbols). *In* "Secondary Productivity of Terrestrial Ecosystems" (K. Petrusewiez, ed.), Panstwowe Wydawn. Naukowe, Warszawa, pp. 51–82.

Precht, H. (1958). Concepts of the temperature adaptation of unchanging reaction systems of cold-blooded animals. *In* "Physiological Adaptation" (C. L. Prosser, ed.), Am. Physiol. Soc., Washington, D.C., pp. 50–77.

Prosser, C. L. (1958). The nature of physiological adaptation. *In* "Physiological Adaptation" (C. L. Prosser, ed.), Am. Physiol. Soc., Washington, D.C., pp. 167–180.

Rae, J. G. (1978). Reproduction in two sympatric species of *Macoma* (Bivalvia). *Biol. Bull.* (*Woods Hole, Mass.*) **155,** 207–219.

Rakusa-Suszczewski, S., and McWhinnie, M. A. (1976). Resistance to freezing by antarctic fauna: supercooling and osmoregulation. *Comp. Biochem. Physiol.* **54,** 292–300.

Rao, K. S. (1967). Annual reproductive cycle of the wedge clam, *Donax cuneatus* Linnaeus. *J. Mar. Biol. Assoc. India* **9,** 141–146.

Rasmont, R. (1968). Nutrition and digestion. *In* "Chemical Zoology" (M. Florkin and B. T. Scheer, eds.) Vol. 2, pp. 43–51. Academic Press, New York.

Reimer, A. A. (1973). Feeding behaviour in the sea anemone *Calliactis polypus*. *Comp. Biochem. Physiol.* **44,** 1289–1301.

Remane, A. (1933). Verteilung und Organisation der benthonischen Mikrofauna der Kieler Bucht. *Wiss. Meeresunters., Kiel* **21,** 161–221.

Remane, A., and Schlieper, C. (1971). "Biology of Brackish Water." Wiley, New York.

Roberts, D. (1979). Deposit-feeding mechanisms and resource partitioning in tropical holothurians. *J. Exp. Mar. Biol. Ecol.* **37,** 43–56.

Rodhouse, P. G. (1979). A note on the energy budget for an oyster population in a temperate estuary. *J. Exp. Mar. Biol. Ecol.* **37,** 205–212.

Rohde, K., and Sandland, R. (1975). Factors influencing clustering in the snail *Cerithium moniliferum*. *Mar. Biol.* **30,** 203–215.

Roland, W., and Ring, R. A. (1977). Cold, freezing, and desiccation tolerance of the limpet *digitalis* (Eschoscholtz). *Cryobiology* **14,** 228–235.

Sameoto, D. D. (1969). Comparative ecology, life histories, and behaviour of intertidal and sand-borrowing amphipods (Crustabea: Haustoriidae) at Cape Cod. *J. Fish. Res. Board Can.* **26,** 361–388.

Sander, F. (1973). A comparative study of respiration in two tropical marine polychaetes. *Comp. Biochem. Physiol.* **46,** 311–323.

Sastry, A. N. (1970a). Reproductive physiological variation in latitudinally separated populations of the bay scallop, *Aequipecten irradians* Lamarck. *Physiol. Zool.* **41,** 44–53.

Sastry, A. N. (1970b). Environmental Regulation of oocyte growth in the bay scallop, *Aequipecten irradians* Lamarck. *Experientia* **26,** 1371–1372.

Sastry, A. N. (1975). Physiology and ecology of reproduction in marine invertebrates. *In* "Physiological Ecology of Estuarine Organisms" (F. J. Vernberg, ed.), pp. 279–299. Univ. of South Carolina Press, Columbia.

Sastry, A. N. (1979). Pelecypoda (excluding Ostreidae). *In* "Reproduction of Marine Invertebrates" (A. C. Giese and J. S. Pearse, eds.), Vol. 5, pp. 113–292. Academic Press, New York.

Scott, D. M. (1976). Circadian rhythm of anaerobiosis in an polychaete annelid. *Nature* (*London*) **262,** 811–813.

Sebens, K. P. (1979). The energetics of asexual reproduction and colony formation in benthic marine invertebrates. *Am. Zool.* **19,** 683–698.

Seed, R., and Brown, R. A. (1975). The influence of reproductive cycle, growth, and mortality on population structure in *Modiolus modiolus* (L.), *Cerastoderma edule* (L.), and *Mytilus edulis* L., (Mollusca: Bivalvia). *Proc. 9th Europ. mar. biol. Symp.* (H. Barnes, ed.), Aberdeen Univ. Press, Aberdeen, Scotland, pp. 257–274.

Shick, J. M. (1976). Physiological and behavioral responses to hypoxia and hydrogen sulfide in the infaunal asteroid *Ctenodiscus crispatus. Mar. Biol.* **37,** 279–289.

Shick, J. M., Hoffman, R. J., and Lamb, A. N. (1979). Asexual reproduction, population structure, and genotype-environment interactions in sea anemones *Am. Zool.* **19,** 699–713.

Shirley, T. C., and Findley, A. M. (1978). Circadian rhythm of oxygen consumption in the marsh periwinkle *Littorina irrorata* (Say, 1822). *Comp. Biochem. Physiol.* **59,** 339–342.

Shyamasundari, K. (1976). Effects of salinity and temperature on the development of eggs in the tube building amphipod *Corophium triaenonyx* Stebbing. *Biol. bull.* (*Woods Hole, Mass.*) **150,** 286–293.

Simpson, R. D. (1976). Physical and biotic factors limiting the distribution and abundance of *Littoral molluscs* on Macquarie Island (Sub-Antarctic). *J. Exp. Biol. Ecol.* **21,** 11–49.

Skoog, C. (1976). Effects of acclimatization and physiological state on the tolerance to high temperatures and reactions to desiccation of *Theodoxus fluviatilis* and *Lymnea peregra. Oikos* **27,** 50–56.

Smith, R. I. (1964). On the early development of *Nereis diversicolor* in different salinities. *J. Morphol.* **114,** 437–464.

Southward, A. J. (1964). The relationship between temperature and rhythmic cirral activity in some Cirripedia considered in connection with their geographical distribution. *Helgol. Wiss. Meeresunters.* **10,** 391–403.

Spaargaren, D. H. (1971). Aspects of the osmotic regulation in the shrimps *Crangon crangon* and *Crangon allmanni. Neth. J. Sea Res.* **5,** 275–335.

Stancyk, S. E., ed. (1979), "Reproductive Ecology of Marine Invertebrates," Belle W. Baruch Library in Marine Science, No. 9. Univ. of South Carolina Press, Columbia.

Stancyk, S. E., and Shaffer, P. L. (1977). The salinity tolerance of *Ophiothrix angulata* (Say) Echinodermata: Ophiuroidea in latitudinally separate populations. *J. Exp. Mar. Biol. Ecol.* **29,** 35–43.

Stefansson, U., and Atkinson, L. P. (1967). "Physical and Chemical Properties of the Shelf and Slope Waters off North Carolina," Tech. Rep., pp. 1–71. Duke University Marine Laboratory, Beaufort, North Carolina.

Stephens, G. C. (1972). Amino acid accumulation and assimilation in marine organisms. *In* "Nitrogen Metabolism and the Environment" (J. W. Campbell and L. Goldstein), eds.), pp. 155–84. Academic Press, New York.

Stephens, G. C., and Schinske, R. A. (1957). Uptake of amino acids from sea water by ciliary-mucoid filter feeding animals. *Biol. Bull.* (*Woods Hole, Mass.*) **113,** 356–7.

Stephens, G. C., and Schinske, R. A. (1961). Uptake of amino acids by marine invertebrates. *Limn. Oceanogr.* **6,** 175–181.

Stephenson, T. A., and Stephenson, A. (1972). "Life Between Tidemarks on Rocky Shores." Freeman, San Francisco, California.

Strong, K. W., and Daborn, G. R. (1979). Growth and energy utilisation of the intertidal isopod *Idotea baltica* (Pallas) (Crustacea: Isopoda). *J. Exp. Mar. Biol. Ecol.* **41,** 101–124.

Swedmark, B. (1964). The interstitial fauna of marine sand. *Biol. Rev. Cambridge Philos. Soc.* **39,** 1–42.

Taylor, A. C. (1976). Burrowing behavior and anaerobiosis in the bivalve *Arctica islandica* (L.) *J. Mar. Biol. Assoc. U.K.* **56,** 95–109.

Teal, J. M., and Carey, F. G. (1967). The matabolism of marsh crabs under conditions of reduced oxygen pressure. *Physiol. Zool.* **40,** 83–91.

Thayer, G. W., Schaaf, W. E., and Angelovic, J. W. (1973). Caloric measurements of some estuarine organisms. *Fish. Bull.* **71,** 289–296.

Theede, H. (1975). Aspects of individual adaptation to salinity in marine invertebrates. *In* "Physiological Ecology of Estuarine Organisms" (J. Vernberg, ed.), pp. 213–226. Univ. of South Carolina Press, Columbia.

Theede, H., Ponat, A., Hiroki, K., and Schlieper, C. (1969). Studies on the resistance of marine bottom invertebrates to oxygen-deficiency and hydrogen sulphide. *Mar. Biol.* **2,** 325–337.

Thompson, R. K., and Pritchard, A. W. (1969). Respiratory adaptations of two burrowing crustaceans, *Callianassa californiensis* and *Upogebia pugettensis* (Decapoda, Thalassinidae). *Biol. Bull.* (*Woods Hole, Mass.*) **136,** 274–287.

Thorp, J. H., and Hoss, D. E. (1975). Effects of salinity and cyclic temperature on survival of two sympatric species of grass shrimp (Palaemonetes), and their relationship to natural distributions. *J. Exp. Mar. Ecol.* **18,** 19–28.

Thorson, G. (1957). Bottom communities (sublittoral or shallow shelf). *Mem., Geol. Soc. Am.* **67,** 461–534.

Thorson, G. (1964). Light as an ecological factor in the dispersal and settlement of larvae of marine bottom invertebrates. *Ophelia* **1,** 167–208.

Trevallion, A. (1971). Studies on *Tellina tenuis* Da Costa. III. Aspects of general biology and energy flow. *J. Exp. Mar. Biol. Ecol.* **7,** 95–122.

Trott, T. J., and Dimock, R. V. (1978). Intraspecific trail following by the mud snail *Ilyanassa obsoleta. Mar. Behav. Physiol.* **5,** 91–101.

van Winkle, W., and Mangum, C. (1975). Oxyconformers and oxyregulators. A quantitative index. *J. Exp. Mar. Biol. Ecol.* **17,** 103–110.

Vargo, S. L., and Sastry, A. N. (1977). Acute temperature and low dissolved oxygen tolerances of brachyuran crab (Cancer irroratus) Larvae. *Mar. Biol.* **40,** 165–171.

Vernberg, F. J. (1956). Study of the oxygen consumption of exised tissues of certain marine decapod Crustacea in relation to habitat. *Physiol. Zool.* **29,** 227–234.

Vernberg, F. J. (1962). Latitudinal effects on physiological properties of animal populations. *Annu. Rev. Physiol.* **24,** 517–546.

Vernberg, F. J. (1969). Acclimation of intertidal crabs. *Am. Zool.* **9,** 333–341.

Vernberg, F. J. (1972). Dissolved gases—animals. *In* "Marine Ecology" (O. Kinne, ed.), Vol. 1, pp. 1491–1526. Wiley (Interscience), New York.

Vernberg, F. J. (1979). Multiple factor and synergetics stresses in aquatic systems. *In* "Energy and Environmental Stress in Aquatic Systems" (H. Thorp and J. Gibbons, eds.), pp. 726–748. Technical Information Center, U.S. Department of Energy, Washington, D.C.

Vernberg, F. J., and Vernberg, W. B. (1970). "The Animal and the Environment." Holt, New York.

Vernberg, F. J., and Vernberg, W. B., eds. (1974). "Pollution and Physiology of Marine Organisms." Academic Press, New York.

Vernberg, F. J., and Vernberg, W. B. (1975). Adaptations to extreme environments. *In* "Physiological Ecology of Estuarine Animals" (F. J. Vernberg, ed.), pp. 165–180. Univ. of South Carolina Press, Columbia.

Vernberg, F. J., Schlieper, C., and Schneider, D. E. (1963). The influence of temperature and salinity on ciliary activity of excised gill tissue of molluscs from North Carolina. *Comp. Biochem. Physiol.* **8,** 271–285.

Vernberg, F. J., Calabrese, A., Thurberg, F. P., and Vernberg, W. B., eds. (1977). "Physiological Responses of Marine Biota to Pollutants." Academic Press, New York.

Vernberg, W. B. (1975). Multiple factor effects on animals. *In* "Physiological Adaptation to the Environment (F. J. Vernberg, ed.), pp. 521–538. Crowell-Collier, New York.

Vernberg, W. B., and Vernberg F. J. (1968). Studies on the physiological variation between tropical and temperate zone fiddler crabs of the genus *Uca*. X. The influence of temperature on cytochrome-c oxidase activity. *Comp. Biochem. Physiol.* **26,** 499–508.

Vernberg, W. B., and Vernberg, F. J. (1972). "Environmental Physiology of Marine Animals." Springer-Verlag, Berlin and New York.

Vernberg, W., and Vernberg, F. J. (1975). "The Physiological Ecology of Larval *Nassarius obsoletus*," pp. 179–190. University Press, Aberdeen, Scotland.

Vernberg, W. B., DeCoursey, P. J., and O'Hara, J. (1974). Multiple environmental factor effects on physiology and behavior of the fiddler crab. *Uca pugilator*. *In* "Pollution and Physiology of Marine Organisms" (F. J. Vernberg and W. B. Vernberg, eds.), pp. 381–425. Academic Press, New York.

Vernberg, W. B., Calabrese, A., Thurberg, F. P., and Vernberg, F. J., eds. (1979). "Marine Pollution: Functional Responses." Academic Press, New York.

vlasblom, A. G., Graafsma, S. J., and Verhoeven, J. T. (1977). Survival osmoregulation ability, and respiration of idotea chelipes (Crustacea, Ispoda) from Lake Veere in different salinities and temperatures. *Hydrobiologia* **52,** 33–38.

Wallis, R. L. (1975). Thermal tolerance of *Mytilus edulis* of eastern Australia. *Mar. Biol.* **30,** 183–191.

Wallis, R. L. (1976). Aspects of the thermal tolerance of the tropical mussel *Trichomya hirsuta* L.—A multivariable approach. *Aust. J. Mar. Freshwater Res.* **27,** 475–86.

Walshe-Maetz, B. M. (1956). Controle respiratoire et métabolisme chez les Crustaces. *Vie Milieu* **7,** 523–543.

Weigert, R. G. (1976). "Ecological Energetics." Dowden, Hutchinson, & Ross, Inc., Stroudsburg, Pennsylvania.

Wells, H. W. (1961). The fauna of oyster beds, with special reference to the salinity factor. *Ecol. Monogr.* **31,** 239–266.

Widdows, J. (1976). Physiological adaptation of *Mytilus edulis* to cyclic temperatures. *J. Comp. Physiol. B* **105,** 115–128.

Widdows, S. J., Bayne, B. L., Livingstone, D. R., Newell, R., and Donkin, P. (1979). Physiological and biochemical responses of bivalve molluscs to exposure to air. *Comp. Biochem. Physiol. A.* **62,** 301–302.

Wijsman, T. C. M. (1975). pH fluctuations in *Mytilus edulis* L. in relation to shell movements under aerobic and anaerobic conditions. *In* "Ninth European Marine Biology Symposium" (H. Barnes, ed.), pp. 139–149. Aberdeen Univ. Press.

Wilson, J. G. (1978). Upper temperature tolerances of *Tellina tenuis* and *T. fabula*. *Mar. Biol.* **45,** 123–128.

Wolcott, T. G. (1973). Physiological ecology and intertidal zonation in limpets (Acmaea): A critical look at "limiting factors." *Biol. Bull.* (*Woods Hole, Mass.*) **145,** 389–422.

Wu, R. S. S., and Levings, C. D. (1978). An energy budget for individual barnacles (*Balanus gladula*). *Mar. Biol.* **45,** 225–235.

Young, C. M. (1928). Feeding mechanisms in invertebrates. *Biol. Rev.* **3,** 21–79.

Young, C. M. (1978). Desiccation tolerances for three hermit crab species *Clibanarius vittatus* (Bosc), *Pagurus pollicaris Say* and *P. longicarpus* Say (Decapoda, Anomura) in the North Inlet Estuary, South Carolina, USA. *Estuarine Coastal Mar. Sci.* **6,** 117–122.

Zann, L. P. (1973). Relationships between intertidal zonation and circatidal rhythmicity in littoral gastropods. *Mar. Biol.* **18,** 243–250.

Zipser, E., and Vermeij, G. J. (1978). Crushing behavior of tropical and temperate crabs. *J. Exp. Mar. Biol. Ecol.* **31,** 155–172.

Zucker, N. (1978). Monthly reproductive cycles in three sympatric hood-building tropical fiddler crabs (Genus *Uca*). *Biol. Bull.* (*Woods Hole, Mass.*) **155,** 410–424.

Pelagic Macrofauna

7

Malcolm S. Gordon and Bruce W. Belman

I. Introduction

The macrofaunal inhabitants of the open-sea and midwater environments of the world's oceans are a varied and diverse assemblage of organisms belonging to many phyla. Defined solely on the basis of physical size, the grouping includes representatives of the coelenterates, ctenophores, platyhelminths, molluscs, arthropods, protochordates, and vertebrates. For purposes of this chapter, in order to avoid undue duplication of materials included in other chapters (especially Chapters 4 and 8), we emphasize those forms which are nonplanktonic inhabitants of the epi-, meso-, or bathypelagic environments at some normal stage of their life histories. This definition narrows the group to the molluscs, arthropods, and vertebrates.

Our definition still leaves us with a heterogeneous group, both phylogenetically and in terms of what might be called "degree of pelagicness." It seems unreasonable to us to limit the group to only those animals living their entire lives in one or more of the three pelagic environments. Doing so would eliminate such undeniably pelagic forms as the sea turtles and birds, among others. Accordingly, we have used a relatively nonrestrictive criterion for "degree of pelagicness" in considering which kinds of animals to discuss.

The epi-, meso-, and bathypelagic environments have a number of special

FUNCTIONAL ADAPTATIONS OF MARINE ORGANISMS

ISBN 0-12-718280-2

characteristics which have led to the evolution of a wide range of special functional adaptations in their animal inhabitants. Some of the most important of these characteristics and associated adaptations include (a) The physical size of the environment, which has led to the development of rapid, sustained swimming, sensory methods for location of distant objects (food, mates, predators), and geographical localization; (b) The large scale spatial uniformity of the environment (primarily in the horizontal dimension, but especially for the deeper living meso- and bathypelagic forms, in the vertical dimension as well) which has led to development of spatial orientation and local position finding; (c) the lack of edges or solid objects (away from the sea surface), which has led to adaptations for buoyancy control and for visual transparency; (d) the widespread rarity of food (especially in the meso- and bathypelagic regions), which has led to reductions in metabolic needs and probably to increased metabolic efficiency; (e) The properties of the deep sea (especially uniform low temperatures, high hydrostatic pressures, low ambient light levels, and, in some regions, low oxygen levels), which has led to a range of biochemical, physiological, reproductive, and behavioral adaptations.

Epipelagic and deep-sea organisms, particularly the larger and more active forms, are often difficult to capture and maintain in good condition. Experimental studies of the functional adaptations of these forms are, as a result, relatively few and incomplete. Despite this, there is a substantial literature which is far too large, diverse, and complex to cover completely here. Accordingly, we have selected for discussion those features of our subject animals that seem primarily to be adaptations to the special characteristics of the open sea. We also emphasize the properties of intact organisms and their organ systems. Most biochemical research has limited relevance to the special features of the open sea.

Recent general discussions or reviews including information on aspects other than those covered here include Anctil (1975), Andersen (1969), Arnold (1974), Ashmole (1971), Berger and Hart (1974), Brauer (1972), Brett (1971, 1972), Cheng (1975), Dunson (1975), Emlen (1975), Harrison (1972, 1974), Hasler (1971), Hochachka (1971, 1975), Locket (1975), Macdonald (1975), McFarland (1971), McFarland and Munz (1975), Malins (1975), Menzies *et al.* (1973), Munz and McFarland (1973, 1975), Norris (1968, 1975), Packard (1972), Peaker and Linzell (1975), Pennycuick (1975), Randall (1970), Renaud and Popper (1975), Sleigh and Macdonald (1972), and Steen (1970).

II. Invertebrates

The invertebrate components of the three pelagic faunas are diverse assemblages of organisms including representatives of most major phyla. In terms of diversity, the most important are the Crustacea (about 1200 species), the Coelenterata (about 350 species), and the Mollusca (about 200 species). Less diverse but still important in terms of biomass are the Urochordata, the Chaetognatha, and the Ctenophora. Far less common in either biomass or diversity are the Platyhelminthes and the Annelida (Hardy, 1956; Sverdrup *et al.*, 1942). The vast majority of pelagic invertebrates are tiny and are generally considered "zooplankton." Here, to the extent that it is possible, we emphasize "macrofauna" or those pelagic species not clearly defined as zooplankton.

The pelagic macrofauna is immensely varied in terms of morphology, physiology, and behavior. Few of the pelagic species have been the subjects of experimental studies of any kind. Fewer still have been studied physiologically. There are two categories of reasons for this situation. The first involves the nature of the questions asked by workers interested in pelagic organisms and the approaches taken in trying to answer them. Childress (1976) points out that to work effectively in the general area of "pelagic biology," a physiologist must either obtain considerable knowledge of the methods and findings of oceanographers or else work closely with oceanographers. Physiologists often are unable to design appropriate experiments and apparatus since the necessary understanding of the environments and natural histories of the organisms to be examined is lacking. Oceanographers, on the other hand, tend to have limited physiological backgrounds and often approach problems in physiology from a population or community point of view. This "separation of disciplines" has resulted in numerous physiologically unsophisticated studies, with resultant data often being relatively meaningless in terms of the normal environmental conditions.

The second reason for physiological studies of pelagic macrofauna being rare and uneven in content and approach involves the extreme difficulties inherent in the capture and laboratory maintenance of most of these organisms in conditions suitable for meaningful physiological studies. This is particularly true of deep-living species. Recent advances in the technology of capture and maintenance equipment have begun to allow significant physiological experimentation on a variety of pelagic organisms (Brauer, 1972; Clarke, 1969; Gordon *et al.*, 1976; Hochachka *et al.*, 1975; Hopkins *et al.*, 1973; Macdonald, 1975).

A. FEEDING

The pelagic zones of the oceans contain limitless food supplies for individual organisms, but the problems of obtaining the food are formidable, since the supply is thinly dispersed. Pelagic organisms must, therefore, either have mechanisms for concentrating small food items or they must possess adaptations for capturing and consuming large prey. Zooplankton organisms generally utilize the first strategy. The morphological and physiological adaptations which are involved in the feeding of zooplankton are well described by Conover (1968), Jørgensen (1966), and Marshall (1973). Large pelagic invertebrates use feeding strategies involving capture of both small prey (suspension feeding) and large food items. The mechanisms for feeding (both capture and consumption) seen in pelagic forms are generally similar to those occurring in related intertidal and subtidal forms. This statement is based primarily on morphological studies of mouth parts and feeding appendages and many analyses of stomach contents.

A recent review of the biology of pelagic shrimps describes what is known about their feeding mechanisms (Omori, 1974). Decapod crustacea, whether pelagic or not, are generally omnivorous or predatory. Feeding in pelagic shrimps does not appear to be particularly specialized, although variations in the structures of the feeding appendages suggest some marked differences in the size ranges of food items. There may also be a correlation between sizes of food items in the guts of many pelagic shrimps and the general forms of the feeding appendages. Tchnidonova (1959) observed that while several species, such as *Hymenodora glacialis,* consume small prey whole, other species grind the food prior to ingesting it. Renfro and Pearcy (1966) indicate that some species (e.g., *Pasiphaea pacifica*) have adaptations for manipulating large prey and lack the structures to deal with small food items, whereas others (e.g., *Sergestes similis*) can deal with food of a wide size range.

There are few laboratory observations of feeding by pelagic invertebrates. Omori (1971) observed feeding patterns of the shrimp, *Sergia lucens,* under laboratory conditions. In this species feeding may follow either of two distinct patterns. Quoting from Omori (1974):

> When the shrimp is swimming forward, the third maxilliped and the first to third pereiopods are usually stretched slightly forward on either side so that, with the long setae of the fringes, they can form a scoop-like net as a catching device. In the first process, an *Artemia* nauplius which entered this device and touched the setae was brought into a position where it could be seized by the second maxillipeds. The mandibles then macerated the prey and the food passed to the mouth. Unless physical contact with the prey was made within this device, the encounter was unsuccessful. The second process is an active hunting of living prey. When the shrimp encountered the prey at a distance of about twice its body length, it suddenly circled around the prey several times so that the prey was brought into the center of a strong vortex caused by the body and widely open antennal flagellae. The prey was then easily held by the second maxillipeds.

As conditions improve for capture and maintenance of open ocean pelagic organisms, more studies of this sort may be attempted. In general, the morphologies of feeding appendages in pelagic crustaceans and molluscs suggest that we may find few startling or unique adaptations. The basic problem would seem to be more one of finding and capturing suitable prey in a sparsely populated environment than of the actual manipulation and ingestion of that prey (see section II-D).

B. METABOLIC AND RESPIRATORY ADAPTATIONS

1. General Comments

A vast literature has accumulated which is directed toward developing an understanding of the metabolic responses of organisms to the major physicochemical and biological characteristics of their environments. Similarly, large amounts of information are available which attempt to relate metabolic properties to specific organismal factors, such as body surface area, body weight, activity level, reproductive state, and age. Older reviews that consider important aspects of the metabolism and metabolic adaptations of the two major groups of pelagic invertebrates, the Crustacea (Wolvekamp and Waterman, 1960) and the Mollusca (Ghiretti, 1966) should be consulted for general background.

Since metabolism often accounts for the largest fraction of an organism's energy use, metabolic rates are of considerable ecological interest. Ecologists generally are interested in estimating metabolic rates in the field, where they cannot be readily measured. Laboratory results are, therefore, sometimes uncritically extrapolated to produce estimates of field values. A frequent result is that effects due to many of the variables which affect metabolic rates are not clearly separated. For pelagic organisms in particular, efforts like these have often produced studies with quite contradictory results (Childress, 1976; and Marshall, 1973, consider these problems).

The interests of physiologists in metabolic rates often focus on developing an understanding of which factors determine such rates and how these factors operate at the organismal, tissue, or biochemical levels. As Childress (1976) points out, an important feature of the physiological approach to organismic metabolic rates is a careful separation of the variables so that the metabolic influences of each variable can be quantified. The physiological approach has a disadvantage in that an immense amount of work is often needed to produce predictive ability in hypotheses. This predictive ability may, however, be very powerful when compared to more rapid ecological approaches. With this in mind, we now discuss the factors involved in determining metabolic patterns in pelagic invertebrates, emphasizing a quantitative, physiological point of view.

2. *Measurement of Metabolic Rates*

Metabolic rates in pelagic organisms are usually measured in terms of oxygen consumption, because this parameter can be measured easily, rapidly, continuously, and non-destructively by a wide variety of methods. Other types of measurement have been used in some studies. Rates of waste nitrogen excretion have been used as measures of rates of nitrogen metabolism. This approach has limited value, however, since nitrogen is excreted in several forms, since most studies measure only 80–90% of the total excretory rate, and since nitrogen excretion cannot be measured continuously. CO_2 production in marine animals is difficult to measure due to the high solubility of CO_2 in seawater and to the presence of large amounts of carbonate and bicarbonate. This method is seldom used in metabolic studies of marine organisms. The least ambiguous method for measuring metabolic rate, heat production, is simply too difficult to use with the small organisms found in pelagic environments. Detailed analyses of the use of oxygen consumption in determining metabolic rates, of the patterns of measurement, and of the expression of results are provided by Childress (1976). In this paper, oxygen consumption rate will be used synonymously with metabolic rate.

3. *Factors Determining Metabolic Rate*

a. Activity. For most pelagic invertebrates, the level of physical activity is probably the single most important factor producing an observed metabolic rate. Activity is, however, the least studied of all factors affecting metabolic rates. An analysis of the definitions of metabolic rates which correspond to particular activity levels in pelagic invertebrates is presented by Childress (1976). Childress concludes that in many previously published studies in which activity was not in some way quantified it is not possible to satisfactorily analyze the metabolic effects of factors other than activity. It is especially important, in metabolic studies of pelagic invertebrates, to control activity. Childress (1976) proposes this be done by (a) minimizing handling effects and agitation in experimental situations; (b) making quantitative measurements or observations of behavior and correlating these with the measured metabolic rates; and (c) analyzing continuous records of oxygen consumption rates for sustained maxima and minima as well as for mean values. With respect to quantitative studies of the relationships between activity and metabolism, Childress points out that utilizing the methodologies and concepts of workers with fishes (Brett, 1964; Fry, 1971) is probably not practical. Pelagic invertebrates usually do not orient into currents, do not swim rapidly except for short bursts, and cannot hover in confined spaces.

In general, the metabolic and activity rates of invertebrates increase with increasing temperature (Kinne, 1970). Variations in metabolic rate often modify the scope for activity, but activity rates and metabolic rates do not necessarily parallel each other. Sudden bursts of activity may be followed by slow repayment of metabolic debts. Likewise, temporary declines in activity may not be accompanied by comparable declines in metabolic rates. In constantly swimming species, such as some copepods, activity per se may not be the major regulating factor in metabolism. Since a direct relationship between temperature and activity level has been demonstrated in many crustacea (Wolvekamp and Waterman, 1960), the principle factor is probably temperature. For neutrally buoyant species which do not swim continuously, locomotor activity is probably limited much of the time. This lethargic behavior is interrupted by short periods of considerable activity when catching prey or avoiding predators. There are few direct estimates of metabolic scope for organisms like these. The oxygen consumption of the bathypelagic mysid, *Gnathophausia ingens,* at maximum sustainable activity is about 9.5 times that at minimum activity (Childress, 1971a). It may be that many deep-living pelagic species utilize only minimum levels of activity under natural conditions. There is accumulating evidence that oxygen-minimum layer organisms can only obtain sufficient oxygen for their needs if they have quite low activity levels (Childress, 1971a,b, 1977). In further support of the hypothesis of lethargy at depth are observations from submersible vehicles (Barham, 1971).

Considerable work must be done to develop an understanding of how metabolism and activity are related in pelagic organisms, especially in deep-living ones. The emphasis in such work should be on developing methods for making measurements of metabolic rates that occur during short bursts of intense activity as well as during sustained, extremely low activity levels just sufficient for an animal to maintain its position in the water column.

b. Temperature. The open-ocean environment includes a wide range of different thermal conditions. Surface water temperatures may be as high as 30°C in some regions, whereas deeper water temperatures are usually 4°–7°C below the photic zone, <4°C below about 1500m, and as low as −2°C in polar regions. Pelagic species may, therefore, occur either over wide temperature ranges both geographically (populations) and daily (individuals), or may live in what are often completely thermally stable environments. Patterns of physiological responses to temperature and to temperature changes are subjects of considerable interest and have been extensively studied. The effects of temperature on metabolism and activity in invertebrates are reviewed in detail by Kinne (1970).

There are relatively few definitive studies of the effects of temperature on the metabolic rates of oceanic invertebrates. For pelagic crustacea living in midwater environments, the available data suggest that oxygen consumption rates increase with temperature, with Q_{10} values of 1.5–3.0 (Childress, 1977; Marshall, 1973; Mauchline and Fisher, 1969; Teal, 1971). However, many factors interact with temperature in determining particular metabolic rates. For pelagic invertebrates these factors include ambient levels of dissolved oxygen and hydrostatic pressure. Such interactions are complex and largely unstudied (see below). Invertebrate species occurring in thermally unstable environments often show remarkable abilities to hold their rates of metabolism relatively constant (Bullock, 1955; Prosser, 1967). Other species are decidedly stenothermal and are significantly affected by small temperature changes. Evaluations of the biochemical bases of the compensatory responses involved in metabolic regulation in ectotherms have been made by Hochachka and Somero (1971, 1973) and Hochachka (1974).

There have been few studies of temperature acclimation in pelagic species. In one such study, Halcrow (1963) indicated a considerable capacity in the copepod, *Calanus finmarchicus,* to acclimate to different temperatures. This description of temperature acclimation characteristics of a pelagic species has interesting implications for defining energetic aspects of vertical migrations. Are such species as euphausiids, for example, acclimated to shallow, deep, or intermediate depth temperatures?

It is obvious from the small amount of available information dealing with the effects of temperature on metabolism in pelagic invertebrates that considerably more work must be done before we can begin to understand how temperature may act independently of other factors in determining metabolic patterns. The interactions of activity, available oxygen, and hydrostatic pressure will need to be carefully evaluated in such work.

c. Oxygen. Kinne (1970) describes invertebrate examples of both metabolic conformity and metabolic regulation as functions of available oxygen. He also discusses the relationships between ambient oxygen levels, temperature, and metabolism in invertebrates generally. Our coverage here is limited to open-ocean invertebrates and how these species deal physiologically with low oxygen levels.

Oceanic dissolved oxygen concentrations are almost always adequate to support routine metabolic rates of animals. However, there are—particularly in the Pacific Ocean—characteristic regions of minimum oxygen found at intermediate depths. Dissolved oxygen levels in these regions are often considerably less than 0.5 ml/liter (Sewell and Fage, 1948). The metazoan animals occurring in these oxygen minimum layers (either all or part of the time) include a diverse assemblage of invertebrate species (Banse, 1964). Several studies have shown that many of these species are capable of respiring totally aerobically under such conditions (Childress, 1968, 1969, 1971a,b, 1975, 1977; Quetin and Childress, 1976; Teal and Carey, 1967a).

In one species of pelagic crustacean, the lophogastrid mysid, *Gnathophausia ingens,* the physiological mechanisms that permit aerobic metabolism under extremely low oxygen conditions have been examined in detail. *Gnathophausia ingens* occurs continuously in oxygen minimum layer conditions of from 0.20 to 1.25 ml O_2/liter along the coast of California. Childress (1968) showed this species to live largely aerobically at oxygen concentrations as low as 0.20 ml/liter, with a relatively constant rate of oxygen consumption of 0.8 ml O_2/kg/min at 5°C. The ability of this species to maintain aerobic metabolism appears to result from a combination of several physiological and morphological factors. These include (1) large gill surface areas (between 5 and 15 cm^3/g); (2) low diffusion distances across the gills (1.5–2.5 μm); (3) high ventilation volumes (up to 8 body volumes/min); (4) high utilization of the available oxygen in the respiratory streams (50–80%); (5) rapid blood flows through the gills (turnover time, 2–3 min); and (5) relatively low routine metabolic rates (Belman and Childress, 1976; Childress, 1968, 1971a). Comparable detailed studies are not yet available for other oxygen minimum layer organisms.

Studies of the relationships existing between low oxygen concentrations and the rates of oxygen consumption in a variety of midwater crustaceans have shown that most species can regulate their consumption down to the lowest oxygen concentrations at which they are found (as little as 0.1 ml O_2/liter); they appear to have generally limited or no anaerobic metabolic abilities (Childress, 1975). It has been concluded by Childress (1976) that the ability to regulate oxygen consumption rates at low oxygen tensions depends strongly on the maintenance of extremely low activity levels (lethargy at depth, discussed in section IIB-3a). Midwater species that cannot regulate their oxygen consumption rates at low ambient oxygen levels generally have considerable anaerobic metabolic abilities. Since anaerobic metabolism produces low energy per unit of food material, species relying on anaerobiosis face severe problems in the nutrient-poor deep sea. Childress (1975) noted that only three of twenty-eight species he examined exhibited substantial anaerobic metabolic abilities. Two of these are reported to be parasites on jellyfish and would therefore seem to have an abundant food supply. The third, a large (to 1 cm) copepod, *Gaussia princeps,* is a strong vertical migrator and presumably could metabolize anaerobic end products following ascent into more oxygen-rich water above the oxygen minimum layer.

Many significant questions regarding adaptive strategies of organisms residing, either all or part of the time, in low-oxygen environments remain to be answered. Examination of the physiological mechanisms for uptake and transport of oxygen across low oxygen gradients, such as those occurring in the mysid, *G. ingens,* may offer significant new insights. Studies of the characteristics of the respiratory pigments (if they exist) in deep-living crustaceans would likewise prove valuable.

d. Hydrostatic Pressure. An abundant literature exists on the diverse effects of hydrostatic pressure on molecular, biochemical, and behavioral phenomena in invertebrates. However, relatively few studies have been directed toward determining pressure effects on metabolism at the whole organism level. Macdonald (1975) provides the most recent review of this field. From the available data it appears that shallow-living pelagic species show no greater pressure tolerance than do littoral species; both tend to be relatively pressure sensitive. Mesopelagic and bathypelagic species are relatively pressure tolerant by comparison. In the more shallow-living vertically migratory species temperature appears to be the major determinant of metabolic rate, whereas pressure has little effect on metabolism over the normal vertical depth ranges of individual species (Childress, 1977; Pearcy and Small, 1968; Quetin and Childress, 1976; Smith and Teal, 1973; Teal and Carey, 1967b). In deeper-living species of crustaceans which migrate to within 100m of the surface, Teal (1971) reported an increased rate of metabolism with increased hydrostatic pressure. Pressure appears therefore to oppose temperature effects and to enable these species to maintain near constant metabolic rates with changes in depth. Recent work with the pelagic squid, *Heteroteuthis heteropsis,* indicates that this species, like the deeper-living crustaceans, is essentially pressure insensitive over the normal range of pressures to which it is exposed (Belman, 1978).

The interactions in organisms of hydrostatic pressure with other factors determining metabolic rates under given sets of conditions are poorly understood. A study of these problems in the large copepod, *Gaussia princeps* (Childress, 1977), illustrates aspects of the nature of these interactions. *Gaussia princeps* is a diurnal migrator which spends its days below 400m in the oxygen-minimum layer off southern California and migrates to 200–300m at night. Childress examined the effects of hydrostatic pressure on oxygen consumption rates in this species at three temperatures and under different levels of available oxygen. Hydrostatic pressure had a significant stimulatory effect on oxygen consumption rate in this species at pressures as low as 28 atm, but higher pressures (61–181 atm) had little or no effect. Childress interprets these data to mean that at shallow night-time depths, pressure is the dominant factor determining the rate of oxygen consumption, whereas temperature is the major determining factor at the deeper, oxygen-poor, daytime depths. The significance of these findings for *G. princeps* is probably that a lowered metabolic rate at greater depths is advantageous in terms of energy conservation. This species is unable to regulate its oxygen consumption rate at the levels of available oxygen occurring in the minimum layer and appears to rely in part on anaerobic metabolic pathways. Decreased metabolic rates at greater depths would therefore be a significant factor in dealing with the problem of obtaining sufficient energy.

e. Other Factors. In addition to the factors just outlined, a number of other physical, chemical, and biological conditions probably have effects on metabolic patterns in pelagic species. Salinity, while of considerable importance in estuarine and intertidal environments, usually varies so little in oceanic environments that measurable effects on metabolism are not likely. Possible exceptions to this are very small organisms (e.g., eggs, larvae) living in the few surface mm. The effects of food availability on metabolism, though clearly of importance in understanding energy relationships, have been little studied. The changes in metabolic rate associated with the amounts and types of stored food and with the initial consumption of the food material are likely to be productive areas for future research.

What Childress (1976) has described as "composite factors" deal with studies of how observed changes in metabolism are related to the complex of primary causative factors which co-vary. The examples that most clearly illustrate this approach are variation in metabolism with season in individual species (Conover, 1968) and with depth distribution among species (Childress, 1971a, 1975).

C. REPRODUCTION

The reproductive physiology of strictly pelagic invertebrates is broadly similar to that in other invertebrate species. Reproductive adaptations related to life in the pelagic environment have not been examined to any great extent, so reference must be made to more general information provided in the reviews of Charniaux-Cotton (1960), Fretter and Graham (1964), and Giese and Pearse (1974a). Pelagic invertebrates like other successful organisms, have adapted their reproductive habits to the particular conditions of their environment. The environment, particularly in deep water below approximately 1000m, poses several problems for pelagic organisms. These include the absence of solar light, lack of seasonal change, and high spatial separation. Therefore mate location, fertilization, and survival of eggs and larvae are difficult processes. The solutions to these problems are largely unstudied. In shallow-water environments, including the pelagic, most marine organisms reproduce in a cyclic manner, seemingly in response to changing external conditions (Giese and Pearse, 1974b). Very little information is available on breeding in bathypelagic organisms. It is not known if such organisms have defined breeding seasons.

There is a predominance of females in populations of pelagic mysids, euphausiids, and copepods (Mauchline, 1971, 1972; Mednikov, 1961) and an apparent absence of synchronous annual cycles in deep-sea benthic forms (Rokop, 1974). These facts suggest that for mysids, which mate only when ovarian eggs are ripe and ready for laying, the final stages of egg development may be dependent on social aggregations. The factors involved in cueing for the formation of such aggregations are largely unknown, but vision, mechanoreception, and chemoreception have been suggested as important factors.

The attraction of mates in shallow pelagic habitats may involve visual stimuli and/or cues from changes in solar radiation. A particularly good example of visual cues in reproduction is found in the squid, *Loligo pealei,* where sexual behavior is stimulated by the appearance of the egg mass.

There have been several laboratory observations of copulatory behavior of the pelagic mysid, *Metamysidopsis elongata* (Clutter, 1969). Copulation in this species occurs only at night, within 2–3 min following the molting of the female. Clutter concluded that a pheromone is extruded to attract adult males of the species. The sensitivity of response to such compounds is unstudied in pelagic species, although the work of Fuzzessery and Childress (1976), which demonstrates extremely high chemosensitivity to amino acids in a bathypelagic mysid, may offer fruitful ground for future studies of pelagic reproductive strategies.

D. SENSORY PHYSIOLOGY

1. General Comments

The immense volume of the open ocean and the sparse spatial distribution of pelagic organisms suggest that such animals face unique problems of communication. Obviously communication does in fact occur and a good deal is known about several aspects of the processes involved. Macdonald (1975) summarized the problems and some of the solutions, primarily for fishes. Information on communication in pelagic invertebrates is sparse. Reviews by Barber (1961), Bullock and Horridge (1965), Harvey (1961), and Waterman (1961) offer good summaries of the physiological background.

2. *Bioluminescence*

A wide variety of invertebrate species belonging to most major phyla are known to emit light. Among pelagic forms many coelenterates are known to be luminous, as are most ctenophores. Several marine polychaetes are luminous, the most striking examples being members of the pelagic family Tomopteridae. There are numerous pelagic crustaceans that possess light organs, often arranged in very complex patterns. In the deep-sea shrimp, *Acanthephyra purpurea,* for example, not only are the light organs highly differentiated with both reflecting layers and lenses, but this species possesses glands from which luminescent substances may be forcefully ejected. Bioluminescent light is of obvious importance to pelagic species, particularly to those which reside below the photic zone. Such light is often utilized in camouflage, either by blinding potential predators with a luminous cloud, as in the case of the squid, *Heteroteuthis* (Clarke, 1963), or in counter-shading, an important adaptation in several crustacea that possess ventrally located photophores (Dennell, 1955). Luminescence is also probably involved in a variety of ways in communication, as in the identification of potential mates or potential prey. Transmission of information by the nature of the light flashes and their spatial distribution suggests that, in euphausiids for example, light signals are used to help the species remain in shoals (Hardy, 1962; Nicol, 1962). The functional roles of bioluminescence in deep-sea invertebrates have been reviewed by Clarke (1963) and Nicol (1968).

Various studies of the physiology of bioluminescence have been conducted with both pelagic and nonpelagic organisms. While such studies have clarified many of the morphological and physiological details, much remains to be done before an understanding of the basic mechanisms is complete. The review by Brown (1973) discusses current studies of the biochemistry, physiology, and morphology of luminescent systems and should be consulted for detailed information.

3. *Vision*

A large and expanding literature exists dealing with aspects of vision and vision-related phenomena in invertebrates. A substantial portion of this work is involved with structural details of the photoreceptor organs (see Eakin, 1972, for a summary of photoreceptor structures). In general, the photoreceptors of invertebrates from the open ocean environment do not appear to be significantly different structurally from those of related benthic or littoral forms. This is in contrast to the observations of Munk (1966), who found that, compared to shallow-living fishes, the eyes of deep-sea fishes are usually larger and often possess more transparent lenses placed so that maximum amounts of light are transmitted to the retina. The reflecting tapetum, composed of guanine crystals, located behind the retina in deep-sea elasmobranchs and other fishes (Denton and Nicol, 1964), is apparently aligned to reflect light back through the retina and thus to increase its absorption. While such structures are known from several invertebrate eyes (e.g., the copepod, *Macrocyclops albidus;* Fahrenback, 1964) their presence in deep-living pelagic invertebrates has not been noted. These factors suggest that "pelagicness" per se may not be a strong evolutionary force in the structural development of invertebrate photoreceptors.

Adaptations of the visual systems of pelagic organisms appear to be significantly related to the photic environment. A most productive field of study involves the adaptation of the visual pigments of pelagic species to the spectral transmission of the water in which they live (see Lythgoe, 1972, for a consideration of this problem and a summary of the majority of the work, primarily done with fishes). Spectral properties of the visual pigments of many invertebrates (Goldsmith, 1972) have been studied, but only a few of these species fall under the strictly pelagic definition. It is apparent from the pelagic crustacean species so far examined (six euphausiids and three decapods) that the peak spectral absorbance (λ_{max}) of the visual pigments is quite similar to the maximum transmittance of the open ocean water that they inhabit. This is in close keeping with the findings of Denton and Warren (1957), Munz (1958, 1964), and others for mesopelagic fishes. The conclusion that the visual pigments of deep-sea organisms have maximum absorbance corresponding to wavelengths of maximal transmission of deep-sea water would appear to be applicable to crustaceans as well. The data are very limited, however, and this would seem to be an area for fruitful future study.

One interesting aspect of the study of vision in deep-sea organisms has been an evaluation of the relationship of the spectral sensitivity of visual pigments to spectral emission curves for bioluminescent light. Such studies for pelagic invertebrates are limited to the work of Kampa (1955) and Kampa and Boden (1957), who found that the peak of the emission spectrum of the bioluminescence of *Euphausia pacifica* was located at 476 nm, while the rhodopsin in the eyes showed a λ_{max} of 462 nm. A similar study of the mesopelagic fish, *Pachystomias,* which has large red-tissue covered photophores located near the eyes (emitting only red light), indicated that maximal retinal absorbance was near 575 nm. (Denton *et al.,* 1970). The conclusion that this fish is particularly sensitive to bioluminescence from others of the same species seems justified. An examination of these relationships in a number of pelagic crustaceans should prove quite interesting and may aid in understanding basic questions of communication in the deep-sea environment.

The structure of the cephalopod eye has been extensively examined, as have several properties of the visual pigments. These studies are reviewed by Hara and Hara (1972) and Wells (1966), who suggest that pelagic cephalopods, while possessing remarkably elaborate visual systems, do not differ markedly in terms of visual adaptations from primarily benthic or littoral species.

4. *Chemoreception*

The role of chemoreception in sensing food, the presence of members of the same species, and potential predators has been examined to only a limited extent in pelagic invertebrates. The only relevant study we know of is the recent work of Fuzzessery and Childress (1976) in which the role of amino acid chemosensitivity in feeding activity was compared in littoral and bathypelagic crustacea. The pelagic species examined were *Gnathophausia ingens* (lophogastrid mysid) and *Pleuroncodes planipes* (galatheid crab). Feeding responses were elicited by equimolar solutions of L-glutamic acid, taurine, and DL-α-amino-n-butyric acid. The greatest behavioral responses in the pelagic species occurred with thresholds between 10^{-10} and 10^{-12} M. This is particularly striking when compared to threshold values of 10^{-8} to 10^{-6} M for two of the littoral species examined by these authors and values of 10^{-4} to 10^{-6} M recorded for other littoral species (Levandowsky and Hodgson, 1965; McLeese, 1970). Fuzzessery and Childress also carried out electrophysiological studies to examine the sensitivity of dactyl receptors and antennal receptors in the same group of animals. The dactyl receptor threshold for *G. ingens* was 10^{-8} M amino acid mixture, the lowest threshold yet reported in crustacean receptors. Antennal receptor thresholds in this species were 10^{-7} M, similar to that reported in the subtidal lobster, *Homarus americanus* (Ache, 1972). The available data suggest that aquatic crustacea, in general, rely heavily on chemical control over feeding behavior. The finding, by Fuzzessery and Childress, of extremely high chemoreceptor abilities in the bathypelagic mysid, *G. ingens,* suggests that this capacity is of essential importance in locating food in the aphotic, nutritionally dilute environment of the deep sea in which this species occurs.

E. BUOYANCY

1. *General Comments*

The maintenance of neutral or near-neutral buoyancy has long been considered to be of primary importance in energy conservation by pelagic organisms. A second, oft-cited advantage is that of allowing pelagic animals to remain inconspicuous by not having to swim. The energetics of neutral buoyancy have been elegantly explored for fish by Alexander (1972), but it has been little considered in pelagic invertebrates other than zooplankton (Vlymen, 1970). Pelagic organisms generally employ one or more of several mechanisms to achieve neutral or near-neutral buoyancy.

2. Gas-Filled Floats

Gas bubbles are the simplest flotation mechanisms used by pelagic animals. They are widely used by the coelenterate siphonophores. The composition and chemical origins of the gas in the float of the siphonophore Portuguese man-of-war, *Physalia,* were studied by Wittenberg (1960). The evidence indicates that the gas produced by the gas gland to inflate the float is carbon monoxide, derived from the metabolism of the amino acid L-serine. The gas remaining in the float after some time contains 0.5–13% carbon monoxide, 15–20% oxygen, negligible carbon dioxide, and the rest nitrogen. These later gas compositions apparently result from diffusional exchanges of gases across the float walls, plus metabolic consumption of oxygen.

Several pelagic cephalopods, including the genera *Nautilus* and *Spirula,* achieve neutral buoyancy by means of gas-filled, rigid shells. The mechanism responsible for filling the shells with gas involves the osmotic removal of water as a result of active transport of sodium from the shell chambers closest to the animals. As water is withdrawn, gas (primarily nitrogen) diffuses into the chambers (Denton and Gilpin-Brown, 1966, 1971).

Another similar mechanism occurs in the cuttlebone of *Sepia officinalis,* a common relatively shallow-living cephalopod. In this species, lift is produced by entrapment of gas (mostly nitrogen and some oxygen) in a rigid laminar structure of calcium carbonate reinforced by chitin. The mechanism by which the cuttlebone is able to maintain the gas-filled state is similar to that of *Nautilus* (Denton and Gilpin-Brown, 1959, 1961).

3. Replacement of Dissolved Ions by Lighter Ions

In the cranchid squids, the large, fluid-filled coelomic cavity makes up about two-thirds of the volume of the animal. Fluid in this cavity has a specific gravity of 1.010 to 1.012, is isosmotic with sea water, and has a pH of about 5.2. The concentration of ammonium ion is exceptionally high, with the principle anion being chloride. It appears that these squids are excluding the heavy SO_4^{2-} ion and accumulating metabolically produced ammonia. While this is an effective method of reducing total weight, these organisms face the disadvantage of carrying about a volume of fluid nearly twice as large as that of the living tissue (Denton *et al.,* 1969). A number of other pelagic invertebrates use the mechanism of excluding SO_4^{2-} ions to reduce weight. The ostracod, *Gigantocypris mulleri,* is one example (Macdonald, 1975), and there are several others (Denton and Shaw, 1962).

4. Lipids

Many planktonic organisms contain substantial amounts of lipids, both as a form of energy storage and for use in achieving neutral or near-neutral buoyancy. While the utilization of lipids as buoyancy mechanisms in fishes has been examined in some instances (Alexander, 1972; Capen, 1967), the use of this mechanism in pelagic invertebrates is almost unstudied. Most pelagic crustacea probably use a balance of lipid and protein to achieve relatively neutral buoyancy. In more dense (less buoyant) shallow-water crustaceans, elevated lipid levels are most often used to provide the necessary lift. Triglycerides, such as occur in *Euphausia,* have a specific gravity of 0.91–0.92 (average sea water at 0°C has a specific gravity of 1.028), and they provide an up-thrust of about 0.106 g/ml (Lewis, 1970; Macdonald, 1975). Many species of marine zooplankton accumulate wax esters rather than triglycerides (Lee *et al.*, 1971; Lee and Hirota, 1973). The buoyancy effects of the two types of lipids are similar.

A revealing study of the chemical composition and relative buoyancy of sixteen species of midwater crustaceans as a function of their minimum depths of occurrence (depths below which 90% of the population occurs) was conducted by Childress and Nygaard (1974). These authors measured relative buoyancy by weighing live animals in seawater and air. Relative buoyancy ranged from −0.7 to 40 mg/g, with a general trend of decreasing relative buoyancy with increasing depth of occurrence. While most species examined had compositional ranges of water, ash, C, N, lipid, carbohydrate, chitin, and protein close to the levels expected for marine animals in general, very high lipid levels (up to 20% of the wet weight) occurred in some. Thus it appeared that relative buoyancy in these crustaceans was accounted for by a balance between lipid and protein, neither replacing the other, but rather both displacing neutrally buoyant material. Childress and Nygaard (1974) concluded that the major mechanisms for approaching neutral buoyancy in midwater crustacea are the displacement of buoyantly neutral body fluids by positively buoyant lipids, with only slight displacement of negatively buoyant materials, and an overall reduction of the organic content of the animal's body. A particularly intriguing aspect of this study was the finding that reduced relative buoyancy in species occurring in the 400–700 m range is not a determining factor in the general trend of a reduced respiratory rate with increasing minimum depth of occurrence (Childress, 1971b). Since the reduction in relative buoyancy with increasing depth is generally considered to be an energy conserving mechanism in pelagic species, this finding suggests that overall activity, and therefore metabolic rate, may be independent of relative buoyancy. Childress and Nygaard (1974) suggest that it may be more appropriate to consider that density of midwater crustaceans has evolved to be compatible with activity levels, so that support in the water column will require only a small fraction of the overall metabolism. These authors argue that the primary value in achieving neutral buoyancy may not lie in energy conservation, but instead in allowing pelagic organisms to remain nearly motionless, thereby avoiding drawing attention to themselves.

An unusual aspect of the work of Childress and Nygaard (1974) is the finding that four species of deep-living crustacea have positive relative buoyancies, that is, are less dense than sea water, and therefore, tend to float. While one species, *Systellapsis cristata* (depth range 650–1100 m) became positively buoyant when subjected to hydrostatic pressure of 700 psi; two others *Hymenodora frontalis* (400–1100 m) and *Paracallisoma coecus* (500–1100 m) still floated at 1000 psi. This would seem to indicate that these species may have to swim and therefore utilize a considerable fraction of their metabolism to remain at their natural depths.

III. Vertebrates

The vertebrate members of the pelagic macrofauna include a wide variety of bony and elasmobranch fishes, the sea turtles, several species of sea snakes, many oceanic birds, and many mammalian pinnipeds and cetaceans. With the exception of the reptilian taxa, both of which contain only a few species, each of these groups is highly diversified in terms of morphologies, life histories, physiologies, behaviors, and ecologies.

Only a few forms from each group have been the subjects of experimental functional studies. A large proportion of these few studies have been directed at features of the animals not primarily related to how they cope with the special characteristics of the open-sea and deep-sea environments. As a result, the following discussion is incomplete and uneven. Our hope is that these aspects of the chapter will help to stimulate additional work in the future on more of these interesting and specialized creatures.

A. RESPIRATORY ADAPTATIONS

Epipelagic fishes vary in their body forms and locomotor activity patterns from the lethargic, vertically compressed, and almost disc-shaped ocean sunfishes, to the continuously active, highly streamlined dolphins, mackerels, tunas, marlins and their relations, plus several kinds of sharks. Many of the specialized functional morphological features of the body forms of these fishes are discussed by Fierstine and Walters (1968), Alexander (1970), and Ameyaw-Akumfi (1975).

Although most species are unstudied, it seems probable that the majority of the less active epipelagic fishes respire by mechanisms similar to those used by most other fishes (Randall, 1970; Steen, 1971). Many of the high speed forms, however, appear to be specialized in this respect. The relative surface areas of the gills are substantially larger than in most other fishes (Hughes and Morgan, 1973). The average gill area for a large number of fish species representing a variety of systematic groups, locomotor activity patterns, body sizes, and other parameters is 4.9 cm^2/g body weight (Randall, 1970). A few species of mackerel and tunas have been studied in this regard. For Atlantic mackerel (*Scomber*) Gray (1954) estimated 11.6 cm^2/g. Muir and Hughes (1969) carried out a detailed study on three species of tunas, which included careful consideration of the effects of body weights on the area relationships. From comparisons of the regression equations for gill area against body weight for various species, they concluded that tunas have larger gill areas per unit body weight than any other fish investigated. At a body weight of 1 kg the tunas' areas were 3–5 times larger than those of the average fishes studied by Gray (1954). The bases for these large areas are proportionately longer gill filaments and very close spacing of secondary lamellae.

The gill structure of many active, high speed oceanic fishes is termed "reticulate." Instead of having the paired gill filaments (the primary lamellae) on each gill arch largely free from physical connections with other filaments, both within each pair and immediately above and below, the filaments in these forms are more or less fused with each other. These fusions become more extensive and well developed with increasing age of the animals. The hemibranchs of the gills of larger specimens often become stiff, solid sheets of tissue, penetrated by rows of small perforations in the positions where the free secondary lamellae on the vertically adjacent filaments would approach each other in more ordinary fishes.

The bases for these reticulate structures are fusions of two kinds between the involved gill elements (Muir and Kendall, 1968; Hughes and Morgan, 1973). Filamentar fusion involves extensive elaboration of the mucosal epithelia of the leading and following edges of the filaments in each hemibranch, producing fusion between vertically adjacent filaments. The secondary lamellae are not involved, although their ends may be embedded in the fused masses of surface epithelia, thus holding them in fairly fixed positions. Filamentar fusions occur at least in some tunas, the wahoos (*Acanthocybium* spp.), the marlins (*Tetrapterus* spp.), and the swordfish (*Xiphias gladius*). Lamellar fusion involves epithelial and connective tissue elaboration along the edges of the secondary lamellae, resulting in fusions of parts of the edges of the lamellae on vertically adjacent filaments. There is little or no involvement of the mucosal epithelia on the leading and following edges of the filaments. Lamellar fusions appear to be limited in occurrence to some marlins and to the tunas; they are not known to occur in the closely related mackerels.

Details of the nature and degrees of development of both types of fusions vary between species, as well as with age within a species. The overall result of the fusions is to substantially reduce the surface areas open for water flow, which can then reach the secondary lamellae. The restriction of these open areas in older and larger fishes to narrow slits or small pores produces reductions in surface areas open to water flow of up to 90% as compared with the areas open for water flow between the secondary lamellae.

Muir and Kendall (1968) give the following data for a typical big eye tuna

(*Thunnus obesus*): Midway along the length of the gill filaments there is an average of one pair of inflow and outflow pores per mm of filament length. The pores are located directly above what would in other fishes be the channels between vertically adjacent filaments. The area of an average pore on the inflow surface is 0.09 mm^2 and on the outflow surface is 0.07 mm^2. Between successive pairs of pores there are 21 secondary lamellae per mm of filament length. The areas of the interlamellar spaces per mm of filament length add up to about 0.58 mm^2. Thus the pore areas are 16% on the inflow side and 13% on the outflow side of the interlamellar channel areas in the adjacent regions.

There has been considerable speculation in the literature as to the possible functional significance of these fusions. Muir and Kendall (1968) and Brown and Muir (1970) review the various theories and suggest that at least one of the major functions is that of restricting the speed and volume of water passing through the interlamellar spaces. The cruising swimming speeds of the fishes having these structures are generally high. Thus, it is probable that water flow velocities over their gill surfaces are well above the velocities occurring in most other fishes (Walters and Fierstine, 1964; Magnuson and Prescott, 1966; Yuen, 1966; Magnuson, 1970). These higher velocities could be structurally damaging to the delicate secondary lamellae, and they could impair the efficiency of gas exchanges across the lamellar membranes. The reticulate structure could assist in the avoidance of both problems.

A further structural specialization of the respiratory apparatus of the high speed fishes is that the musculature used by most fishes for pumping water through the gills is greatly reduced in size or sometimes almost completely absent (Kishinouye, 1923; Gibbs and Collette, 1967). Gill irrigation is therefore largely the result of continuous swimming with the mouth and opercula more or less open (Hall, 1930; Magnuson and Prescott, 1966). This pattern of respiration is called "ram" perfusion or ventilation and occurs widely in fishes when they swim at higher speeds (Brown and Muir, 1970; Randall, 1970; Roberts, 1975). It may serve as an energy conservation mechanism, since respiratory pumping probably involves a significant fraction of energy needs in most fishes (Roberts, 1974, 1975).

We know of only one experimental study devoted to the operational features of the respiratory mechanisms in these active fishes. Stevens (1972) studied Pacific skipjack (*Katsuwonus pelamis*) and Kawakawa tuna (*Euthynnus affinis*) with respect to several major respiratory variables. He used restrained and lightly anaesthetized fish in an experimental holding chamber and free-swimming fish in a larger tank. His results did not contribute directly to understanding of the possible functions of the reticulate gill structures described earlier. However, the results did indicate that respiratory values for tunas are well above those of all other fishes studied so far; the parameters measured included efficiency of removal of oxygen from the respiratory water stream (90%, as compared with

10–30% for other fishes), ease of transfer of oxygen from water to blood (transfer factor 4–10× greater than for other fishes), and calculated cardiac output (weight specific rate as high as that of resting man). Thus tunas appear to be very effective at taking up oxygen from rapidly moving water.

Respiratory adaptations of epipelagic reptiles, birds, and mammals have received substantial experimental attention, especially the mammals. However, unlike the fishes just discussed, the adaptations appear to be primarily related to a feature of their life histories and behaviors which is by no means unique to the epipelagic environment, namely diving. We will not, therefore describe these adaptations, but refer interested readers to several relevant papers and reviews: Andersen (1966, 1969), Berkson (1966, 1967), Graham *et al.* (1975), Heatwole and Seymour (1975), Kooyman (1972, 1975), and Wahrenbrock *et al.* (1974).

B. METABOLIC ADAPTATIONS

1. Epipelagic Forms

The energy needs of animals are affected by such a multiplicity of factors that, with the exceptions of man and a relatively few other species, only fragmentary pictures are available of the metabolic effects of natural environmental conditions. Data permitting even rough estimates of overall energy budgets in nature are rare. To our knowledge, there has been only one attempt made, using experimental methods, to develop an overall energy budget for an epipelagic vertebrate under natural conditions. Smith (1973) attempted this for an unusual and highly specialized pelagic fish, the sargassum fish, *Histrio histrio*. A primary goal of Smith's work was to try to determine whether or not *Histrio* possesses any special metabolic adaptations for the restricted food supply levels characteristic of at least its primary habitat in the North Atlantic Ocean, the Sargasso Sea. Smith's (1973) study involved the assumption that excretion and secretion from the fish in soluble, rather than filterable particulate form was not energetically significant. He also made several informed guesses as to caloric equivalences of certain measured quantities (e.g., an "arbitrary constant" of 3.95 Kcal/g carbon for egested carbon). He performed relatively short term experiments (7–26 days) at only two times of year (October, May) and in only one temperature interval (21°–24°C). Three size classes of *Histrio* were studied. Energetic components estimated were ingestion (I), egestion (E), growth (G), respiration (R) partitioned into standard (maintenance) metabolism and metabolism of digestion and assimilation (specific dynamic action), and production (P) including reproductive products, mucus, soluble excretions, and secretions. The basic relationship between these is

$$I = E + G + R + P$$

The primary results (Table I) indicated that small *Histrio* expend a larger proportion of their energy intake on growth and a smaller proportion on respiration. Larger *Histrio* not unexpectedly reverse this pattern. Efficiences of gross growth (G/I) and net growth (G/A, $A = I - E$) for the two smaller size classes were higher than literature values for nonpelagic carnivorous fishes; values for these quantities in the largest size class were similar to literature values. Assimilation efficiencies (A/I) were high in comparison with other carnivorous fishes. Standard metabolic rates (*Histrio* almost never swims) in comparison with fishes from more food-rich environments were low for small sargassum fish and high for large ones. The overall conclusion was that *Histrio* is adapted to the food-limited Sargasso Sea environment primarily in terms of assimilation and growth efficiencies. There is no overall metabolic rate adaptation.

Smith also made a preliminary estimate of the availability in its natural environment of sufficient food to supply the energy needs of *Histrio*. On the basis of dietary information indicating a specialized pattern of feeding on certain types of shrimps which also live in the *Sargassum* weed, he estimated that available food falls short of indicated need by about 40%. He lists four possible reasons for this impossible result: (1) his experiments did not adequately simulate natural conditions; (2) the fish are not as specialized in their diet as he assumed; (3) his data on food abundance and turnover were inaccurate; and (4) seasonal variations in energy needs occur which were not taken into account.

A few studies have attempted to determine the metabolic bases for the sustained high levels of activity shown by the usually continuously-swimming, high-speed epipelagic fishes (all of which are "stayer" or consistent swimmer

TABLE I

Relative Partitioning of Energy Budgets of North Atlantic Sargassum Fish *(Histrio histrio)*[a,b,c]

Size class	I	E	A	G	R_{std}	R_{sda}	P
Length (mm)							
10–29	100	28	72	35	18	15	4
37–51	100	26	74	24	22	20	8
81–86	100	18	82	16	19	36	11
Weight (g)							
1 ± 0.3			100	48	25	21	6
13 ± 2.4			100	32	30	27	11
28 ± 0.7			100	19	23	44	14

[a] Based on Smith (1973).

[b] Symbols: I, ingestion; E, egestion; A, assimilation; G, growth; R, respiration (std, standard or maintenance metabolism; sda, specific dynamic action metabolism); P, production.

[c] All figures given in % of I or A.

species in the activity pattern classification of Boddeke *et al.*, 1959). These studies involve biochemical, tissue metabolic, and whole animal physiological approaches. For general background for the discussion which follows see Bilinski (1974), Brett (1972, 1973), Brett and Glass (1973), Fry (1971), and Morris (1974).

Attempts at measurements of the swimming speeds of medium-sized tunas and wahoos in the open sea have led to estimates of short-lived bursts of speed reaching 77 km/hr, and of sustained cruising speeds on the order of 15 km/hr (Walters and Fierstine, 1964; Magnuson and Prescott, 1966; Yuen, 1966; Anonymous, 1969; Magnuson, 1970, 1973). In relative terms, these burst speeds were near 20 body lengths (B.L.) per sec, and the sustained cruising speeds were 3-6 B.L./sec. Aquarium observations on two tuna species yielded lower cruising speeds of 1-2 B.L./sec (Magnuson and Prescott, 1966; Magnuson, 1970). Literature values (Blaxter, 1969) for maximal sustained swimming speeds for a few nonpelagic "stayer" fish species (primarily freshwater salmonids and cyprinids, but including the Atlantic herring, *Clupea harengus*) are in the range 3-4 B.L./sec. More routine cruising speeds for these other species are 2-3 B.L./sec.

The most fully and carefully studied epipelagic fish species with respect to both swimming performance and activity metabolism is the eastern North Pacific jack mackerel, *Trachurus symmetricus*. Hunter (1971) carried out a detailed investigation of a large group of medium-sized (9-18 cm total lengths) jack mackerel with respect to endurance at a range of swimming speeds at a single temperature (18°C) in the normal ecological temperature range. Using 6 hr duration tests, Hunter found that 10% of 10-12 cm total length fish fatigued at speeds near 7 B.L./sec, while 50% of this size range fatigued near 8 B.L./sec. Over the entire size range of fish he found that 50% fatigued at an average swimming speed of $22.4(L^{0.6})$ cm/sec, where L = total length. Extrapolating slightly using this latter relationship, this would mean a relative speed near 7 B.L./sec for a 20 cm total length fish. Maximal burst swimming speeds, sustainable for no more than 3-5 min, were in the range 10-12 B.L./sec.

Hunter (1971) also determined the shape of the relationship relating duration of swimming period (based on times to fatigue for 50% of fish tested at each speed) to swimming speed. This curve was exponential in form for the first 22 min and linear after that. Pritchard *et al.* (1971) provided a possible partial explanation for the segmentation of this speed-endurance curve on the basis of measurements of glyocogen, lactic acid, and fat concentrations in the red and white muscles and glycogen in the livers of the fatigued fish (Table II). The exponential portion of the curve, in which fish collapsed from fatigue after bursts of rapid swimming, was correlated with depletion of glycogen in the white muscles and liver, but not in the red muscles. The linear portion, in which fish fatigued after various longer periods of sustained swimming, was correlated with

TABLE II

Biochemical Changes Relating to Energy Metabolism in the Compositions of the White Muscles, Red Muscles, and Livers of California Jack Mackerel (*Trachurus symmetricus*) Subjected to Various Swimming Speeds and Durations[a,b]

Tissue component	Controls	Burst metabolism fatigued	Sustained swimming fatigued
Glycogen			
Red muscle (mg/100 g)	750	380	83[c]
White muscle (mg/100 g)	160	13[c]	23[c]
Liver (%)	15	6[c]	5[c]
Lactic acid (mg/100 g)			
Red muscle	77	140[c]	86
White muscle	430	650[c]	520
Lipids (% dry wt)			
Red muscle	24	23	27
White muscle	1.1	0.8	1.4

[a] Based on Pritchard *et al.* (1971)
[b] All figures are mean values.
[c] Statistically significantly different from control levels at $P \leq 0.05$

depletion of glycogen in all three tissues. Tissue lactic acid concentrations were high in all situations, and showed no consistent relationship with fatigue. Lipid changes were small and unrelated to fatigue. These biochemical results suggest that jack mackerel derive most of their energy for swimming from carbohydrate metabolism. Anaerobic processes probably predominate in the white muscle during burst swimming, and aerobic processes probably predominate in both white and red muscles otherwise. This is in partial contrast to the evidence for most nonpelagic fishes, which appear to use anaerobic glycolysis in white muscle for burst metabolism, but otherwise use red muscle aerobic lipid metabolism (Bilinski, 1974; Bone, 1966; Hudson, 1973).

Pritchard *et al.* (1971) speculate that this metabolic difference may represent a specialization in the jack mackerel for high speed swimming., They suggest that other species of scombroid and carangid fishes may function similarly. They also suggest the possibility that jack mackerel, when swimming at slower speeds than were used in their experiments, may be more like other fish species and use only their red muscles for propulsion, with aerobic lipid metabolism for energy.

Studies of rates of oxygen consumption by tissue preparations support the general features of this description. Wittenberger (1968), working at 23°C with tissue slices from the Black Sea counterpart of the jack mackeral (*Trachurus mediterraneus ponticus*), measured red muscle oxygen consumption rates 2.2

times higher than white muscle rates in "normal" fish. Gordon (1972b) reported ratios of oxygen consumption rates for substrate saturated minces of red and white muscle from California *T. symmetricus* of 12.8:1 at 5°C, 3.1:1 at 15°C, and 2.7:1 at 25°C. The changes in value of this ratio at different temperatures reflect a greater metabolic sensitivity to temperature (larger Q_{10} values) on the part of white muscle as compared to red muscle preparations.

Several recent papers by Russian workers describe results of biochemical and metabolic studies on the Black Sea jack mackerel (*Trachurus mediterraneus ponticus*), which complement and supplement the work with the California species. Shul'man *et al.* (1973), in an incomplete study, indicated that lipid utilization by both white and red muscles may be a major energy source for swimming in this other species—in contrast to the results of Pritchard *et al.* (1971). Trusevich (1974) measured changes in white muscle concentrations of energy-rich phosphate compounds (creatine phosphate, adenosine mono-, di-, and triphosphates) in this second species. Fish were subjected to varying durations of swimming up to fatigue at fairly high water velocities in three temperature ranges from 11° to 22°C. The endurance of the fish was less at lower temperatures, and the patterns of changes in the concentrations of the phosphate compounds varied with temperature. Shchepkin *et al.* (1974) measured total lipid contents and the abundances of major lipid types in red and white somatic, cardiac, and gastric muscles of *I. mediterraneus ponticus*. Total lipid content was highest in red muscle and lowest in white. Triglycerides were the most abundant fraction in the somatic muscles. Cardiac muscle had high phospholipid and cholesterol contents, while cholesterol esters were most abundant in gastric muscle. The total lipid content of the jack mackerel was substantially higher than that of a sedentary species of scorpion fish (*Scorpaena porcus*) from the same region.

Gordon and Chow (1974 and personal communication) have studied the effects of both swimming speed and temperature on the oxygen consumption rates of intact California jack mackerel. Fish studied were maintained at 18°C, the seawater temperature at which they had been caught. Oxygen uptake rates were measured at relative swimming speeds of 0.8–3.8 B.L./sec at 10°, 15°, and 20°C. This temperature range covers the range of seawater temperatures in which this jack mackerel normally occurs. The major results are summarized in Tables III and IV. The most significant feature is the decreasing thermal sensitivity of the oxygen uptake rate with increasing swimming speed. The activity metabolism of this jack mackerel, expressed either as Q_{10} for oxygen consumption rates at given activity levels or as scope for metabolic activity, apparently becomes independent of temperature at swimming speeds above about 3 B.L./sec.

The absolute levels of jack mackerel metabolic rates are not unusual in comparison with maximal active metabolic rates for a variety of nonepipelagic fishes (Brett, 1972). Note, however, that the highest swimming speed used in these

TABLE III

Rates of Oxygen Consumption of California Jack Mackerel (*Trachurus symmetricus*) Swimming at Various Relative Speeds at Three Temperatures[a,b]

Temperature (°C)	Relative speed (B.L./sec)	Oxygen uptake rate [ml (STP) kg/hr]
10	0.8	70
10	2.5	110
10	3.7	290
15	0.8	60
15	2.4	110
15	3.8	300
20	0.8	120
20	2.5	140
20	3.6	290

[a] Based on Gordon and Chow (1974 and personal communication).
[b] All figures are averages. All fish in size range 22–24 cm standard length.

experiments was only about half the normal cruising speed for the species (Hunter, 1971).

Several of the physical parameters of swimming (e.g., tail beat frequencies and amplitudes) have been studied in the California jack mackerel by Hunter and Zweifel (1971) and Gordon and Chow (1974). Hunter and Zweifel (1971) also summarize the limited amount of comparable physical data available on other epipelagic fishes, plus various freshwater fishes. The high speed epipelagic species seem to be essentially like other "stayer" species.

TABLE IV

Thermal Sensitivity (Q_{10}) of Rates of Oxygen Consumption of California Jack Mackerel (*Trachurus symmetricus*) at Various Swimming Speeds[a]

Relative speed (B.L./sec)	Q_{10} over temperature intervals (°C)		
	10–15	15–20	10–20
0.0[b]	0.74	4.45	1.80
0.8	0.76	4.08	1.75
2.5	1.14	1.51	1.31
3.7	1.04	0.96	1.00

[a] Based on Gordon and Chow (1974 and personal communication).
[b] These figures are based on calculated values for standard metabolic rates.

Gordon (1975) presents data on the effects of elevated hydrostatic pressures on the oxygen uptake of *in vitro* preparations of white muscles from the California jack mackerel. Within the ecologically normal range of pressure–temperature combinations for the species, pressure had no significant metabolic effects.

A number of other aspects of the physiology of the Mediterranean and Black Sea jack mackerel (*Trachurus mediterraneus*) were studied over a period of years by Pora and co-workers in Rumania. These studies generally do not relate directly to the adaptations of this species to the epipelagic environment (e.g., Pora *et al.*, 1959, 1967).

Metabolic information on other species of active epipelagic fishes is limited and, because of the types of experiments involved, difficult to relate to specific levels of activity. Stevens (1972) measured oxygen consumption rates on restrained skipjack tuna (*Katsuwonus pelamis*) and Kawakawa tuna (*Euthynnus affinis*). The fishes were lightly anesthetized and confined to a box in which their gills were perfused with water at controlled flow rates and pressures. Average rates of oxygen uptake at 23°–25°C were 480 ml kg/hr (690 mg kg/hr) for six skipjack (weights 1.47–1.74 kg), and 350 ml kg/hr (500 mg kg/hr) for one Kawakawa. Stevens points out that these were not resting rates, but were associated with the high water perfusion rates needed to keep the fishes alive. It seems probable that they were also not maximal rates. They fall within the upper ranges of previously measured active metabolic rates for nonepipelagic fishes (Brett, 1972). The data of Pora *et al.* (1955) and of Muravskaya (1972) for Black Sea jack mackerel (*Trachurus mediterraneus*) are also based on restrained fishes or fishes otherwise unable to swim freely at controllable speeds. Lipskaya (1974) presents some limited metabolic data on young of several species of tropical epipelagic fishes.

The properties of the energy metabolism of *in vitro* animal tissue preparations rarely resemble those of the energy metabolism of the intact animals. There are multiple causes for the differences. Some of the more important causes include the removal, for the *in vitro* preparations, of most normal control mechanisms, the more or less severe disruption of structures and spatial relationships during tissue isolation and handling, and the usually large changes in substrate concentrations involved in the use of artificial nutrient media. These reasons are sufficient to serve as warnings against seeking a great deal of ecological significance in the results of *in vitro* tissue metabolic studies. However, in view of the difficulties involved in studying the metabolism of most kinds of open-sea and deep-sea animals as intact organisms, tissue studies are worthwhile as possible indications of major features. The following discussion should be viewed in this context.

As was mentioned in the discussion of the jack mackerel (*Trachurus symmetricus*), the rates of oxygen consumption of the lateral red muscles in high speed epipelagic fishes are much higher than the oxygen uptake rates of white muscles in the trunk musculatures. More extensive documentation of this point is provided by Gordon (1968, 1972b). Table V lists the ratios of measured oxygen consumption rates for substrate saturated minces of the two muscle types from eight species of epipelagic fishes inhabiting three different oceanic regions. The absolute values of the red muscle rates at 25°C were all close to or substantially higher than comparably measured oxygen uptake rates for mammalian (white rat) thigh muscle (Gordon, 1968). These observations support the idea that the red muscles are the primary energy sources for the high speed swimming of these forms. Greer-Walker and Pull (1975) summarize data on the relative abundances of red and white muscles in 84 fish species having a wide range of activity patterns.

Oxygen uptake rates for substrate saturated minces of white muscles from twelve species of Californian and Hawaiian Pacific epipelagic fishes were significantly higher than rates for similar preparations from 21 species of littoral and benthic species from the same areas (Gordon, 1972a). These observations may indicate a greater aerobic capacity in the white muscles of the epipelagic forms than in the white muscles of the non-epipelagic species. An interesting minor feature of these data (Gordon, 1972a) is the fact that the highest oxygen uptake rates for white muscles measured were for preparations deriving from the "prejuvenile" epipelagic stages of a goatfish (*Parupeneus pleurostigma*) and a squirrelfish (*Holocentrus lacteoguttatus*), both from Hawaii. Prejuvenile stages are part of the life histories of at least three tropical reef-dwelling fish families; the goatfishes (Mullidae), the squirrelfishes (Holocentridae), and the surgeonfishes (Acanthuridae). The conversion of these fishes from epipelagic prejuveniles to

TABLE V

Ratios of Average Oxygen Consumption Rates of Minces of Red and White Muscles from Active Epipelagic Fishes[a,b]

Region	Species	Ratios		
		5°C	15°C	25°C
Southern California	*Decapterus hypodus*	3.3	3.5	5.8
	Trachurus symmetricus	12.8	3.1	2.7
Galapagos	*Sardinops sajax*	2.2	3.6	2.7
	Fodiator acutus	3.2	5.1	5.8
	Scomber japonicus	2.2	2.9	3.5
Hawaii	*Parexocoetus brachypterus*	4.1	5.1	7.4
	Decapterus pinnulatus	2.6	4.2	4.6
	Katsuwonus pelamis	6.1	6.5	7.9

[a] Based on Gordon (1968, 1972b).
[b] Ratios are values of the fraction $\dot{V}_{O_2}$ red muscle/$\dot{V}_{O_2}$ white muscle.

(usually) littoral juveniles occurs rapidly—often within 24 hours—and is associated with changes in morphology, behavior, pigmentation, and possibly metabolism.

The differences in the relative sensitivities of oxygen uptake rates to changes in temperature in the two muscle types may have functional significance. Six of the eight species listed in Table V are indicated to have greater thermal sensitivity in their red (versus white) muscles; the tabulated ratios consistently increase in value with increase in temperature. This pattern could indicate an ability to carry out high speed bursts of activity at lower water temperatures, despite an overall decline in cruising speeds. Food getting and predator evasion capacities could be preserved in this fashion. The two exceptional species, *Trachurus symmetricus* and *Sardinops sajax,* both normally live in water temperatures of 10°–20°C. Reasons for their deviating from the pattern shown by the other fishes are not apparent. The rapid fall in white muscle rates at lower temperatures in the jack mackerel is particularly puzzling.

The skipjack tuna (*Katsuwonus*) is unusual in another respect, namely that it is partially endothermic and, for a fish, an excellent thermoregulator (see section III,C of this chapter). For this reason, Gordon (1968) carried his study of *in vitro* muscle metabolic rates in this species up to the environmentally unrealistic temperature of 35°C. Red muscle oxygen consumption rates were high and constant between 25° and 35° while white muscle rates were low and constant. Gordon interpreted the thermostability of the red muscles as assuring that the energy supply for high cruising speed swimming would not be affected by even fairly substantial variations in body temperatures. White muscle thermostability (which extended down to 5°C) was considered a possible indication of constant availability of extra energy for burst swimming efforts.

Wittenberger (1968, 1972) has investigated several aspects of energy metabolism related to carbohydrate utilization in the Black Sea jack mackerel (*Trachurus mediterraneus*) and the Atlantic mackerel (*Scomber scombrus*). The experiments were designed primarily to contribute information to an ongoing discussion of the functions of the red and white muscles of fishes in general, rather than to further ecological understanding of these particular species. The jack mackerel work (Wittenberger, 1968) involved measurements of concentrations of pyruvic and lactic acids and glycogen in red and white muscles and livers of fish subjected to 8 min periods of electrical stimulation designed to make them exercise vigorously. The results apparently were variable, as few statistically significant changes in tissue compositions were found. The statistically significant changes in composition do not seem to us to permit development of an internally consistent model of events. The Atlantic mackerel work (Wittenberger, 1972) involved estimates of rates of glycogen glucose turnover in red and white muscles and in the liver, in intact, confined fishes injected with ^{14}C-labelled glucose. The radioactive label was incorporated into red muscle glycogen about 40 times more rapidly than into the glycogen of either white muscle or liver.

Modigh and Tota (1975) have studied various properties of mitochondrial respiration in juvenile bluefin tuna (*Thunnus thynnus*) red and white skeletal and cardiac muscles. Tuna mitochondria appear to be generally similar to other mitochondria.

Dando (1969) studied several features of lactic acid metabolism in 36 species of fish, including the Atlantic mackerel (*Scomber scombrus*) and a jack mackerel (*Trachurus trachurus*). Lactate dehydrogenase activities in the livers of both species were relatively low, placing them well within a large group of species having enzyme activity levels below 100 mU/mg protein (a second group of species had activity levels above 1000 mU/mg protein). Blood and white muscle concentrations of lactic acid and pyruvic acid were measured on Atlantic mackerel captured in trawl nets, after several hours of trawling. As would be expected in stressed and exhausted fish, the lactate levels in both tissues were the highest measured of the four trawled species studied. Blood pyruvate levels were also somewhat elevated in comparison with other trawled species, but muscle pyruvate levels were relatively low.

Another aspect of the possible bases for high speed swimming was touched on by Sather and Rogers (1967) in a study of the major inorganic electrolytes in the blood and in the white and red muscles of skipjack tuna (*Katsuwonus pelamis*). Table VI summarizes the results. Plasma electrolyte composition was little different from that found in most teleost fishes. The red muscles had higher sodium contents than did the white, while the white muscles had higher contents of water, potassium, and magnesium. Extracellular volumes (inulin spaces) were somewhat larger in the red than in the white muscles. In comparison with the

TABLE VI

Concentrations of Major Inorganic Electrolytes in the Blood Plasma, and the Red and White Muscles of the Skipjack Tuna, *Katsuwonus pelamis*[a]

Ion	Plasma[b] (mEq/liter)	White muscle[c] (mEq/kg wet wt)	Red muscle[c] (mEq/kg wet wt)
Na^+	204	11.5	20.9
K^+	6.8	106	80
Cl^-	177	38	40
Ca^{2+}	7.6	4.9	1.9
Mg^{2+}	2.2	25	17

[a] Based on Sather and Rogers (1967). All values are averages.

[b] Plasma osmotic concentration: 415 mOs*M*/liter.

[c] Muscle water contents: 74% for white, 73% for red. Inulin spaces: 19% for white, 24% for red.

muscles of several less active, nonepipelagic fishes skipjack red muscles were relatively low in sodium and potassium concentrations and relatively high in chloride. Skipjack white muscles compared similarly were even lower in sodium but high in both potassium and chloride. Both types of skipjack muscle were lower in calcium content and higher in magnesium content than eel muscles (Sather and Rogers, 1967).

Estimates of intracellular concentrations based upon the data just summarized were (all in mEq/liter cell water) potassium: 189 for white, 161 for red; calcium: 6.3 for white, 0.1 for red; magnesium: 44 for white, 34 for red. Intracellular sodium and chloride contents were low and not reliably calculable, because of the high concentrations of these ions in the extracellular fluids. These intracellular concentrations would be consistent with lower resting membrane potentials in red muscle than in white; hence, the greater excitability on the part of red muscle. The calcium and magnesium values could indicate greater strength of contraction and also more rapid energy release from adenosine triphosphate (ATP) in the white muscles than in the red. Meaningful interspecific comparisons with less active, nonepipelagic marine teleosts are not presently possible. We know of no fully comparable set of data for any species.

Metabolic information relevant to the environmental factors being considered in this chapter is almost nonexistent for epipelagic reptiles, birds, and mammals. There are a few recent exceptions. Prange (1976) studied swimming speed, oxygen consumption rates, and several aspects of hydrodynamics in medium sized (250–900 g) green turtles (*Chelonia mydas*). Over a range of swimming speeds varying from about 0.0–2.0 B.L./sec, oxygen consumption rates increased by about four times. The calculated oxygen cost of swimming for the turtles, in comparison with other vertebrates of comparable body mass, was less than that for flying birds, but greater than that for nonepipelagic fish. The estimated aerobic efficiency of swimming was between 1 and 10%. Prange estimated that 21% of body mass could be the amount of fat which might be used up by an adult turtle during one cycle of its breeding migration in the South Atlantic Ocean.

The metabolic cost of flight in the laughing gull (*Larus atricilla*), (a medium sized, usually coastal sea gull) which is of comparable size, shape, and general behavior to a number of species of epipelagic gulls, was studied by Tucker (1972). Bernstein *et al.* (1973) compared Tucker's results with a similar set of data for a terrestrial bird, the fish crow (*Corvus ossifragus*). A major result was the finding that the energy cost for horizontal flight was about one-third higher for the crow than for the gull. Over the ranges of air speeds tested (6–12 m/sec), energy costs for horizontal flight were independent of air speed in both species. These costs averaged about seven times resting metabolic rates.

Baudinette and Schmidt-Nielsen (1974) measured energy costs for prolonged gliding flight in another coastal gull species, the herring gull (*Larus argentatus*).

Results from two individual well-trained birds indicated that gliding flight requires energy expenditure at rates only about double resting rates. Gliding, therefore, clearly conserves energy in comparison with horizontal flapping flight. This observation is clearly relevant to consideration of energy budgets for most long-winged oceanic birds, the great majority of which glide as often and as long as they can.

Hochachka *et al.* (1975) have made a preliminary search for alternate end products of anaerobic metabolism in four diving vertebrates including the green turtle, a seal (*Phoca vitulina*), a sea lion (*Zalophus californianus*), and a porpoise (*Tursiops gilli*). The specific goal was to determine whether or not amino acid metabolism may be an important anaerobic energy source in addition to glycolysis. As pointed out earlier, we do not consider diving per se to be a special adaptation for the open sea. However, anaerobic metabolism may well be important in high speed swimming of open-sea mammals such as some seals and porpoises (Lang and Norris, 1965; Lang and Pryor, 1966). Both the seal and the porpoise studied showed after-dive changes in blood concentrations of succinic acid and alanine suggesting that anaerobic amino acid metabolism may be a significant source of energy.

Wahrenbrock *et al.* (1974) present the first data ever obtained on growth, metabolism, and aspects of respiratory mechanisms in a captive, healthy baleen whale. They studied a growing juvenile female California gray whale (*Eschrichtius robustus*) over a period of a year, from its capture at about 10 weeks of age. During this period the whale apparently grew normally, from an initial length of 5.5 m to a length at time of release of 8.3 m, with associated body weights at these two times of 2000 and 6350 kg. Length increased linearly with time, weight exponentially. During the final 2 months of captivity the whale was eating 800 kg/day of squid and increasing in weight at the rate of 1.5 kg/hr. Respiratory parameters measured were resting lung volume, tidal volume, respiratory rate, minute ventilation, respiratory "dead space", oxygen consumption, arterial blood pH, and oxygen and carbon dioxide tensions. In comparison with terrestrial mammals, growth related changes in relative body weight, length, lung volume, minute ventilation, metabolic rate, and growth efficiency were all within usual ranges. Lung volumes and metabolic rates were relatively high, based on size scaling extrapolations, probably due to the immaturity of the animal.

2. *Deep-Sea Forms*

The multiplicity and complexity of the technical problems associated with the capture, handling, and maintenance of intact, living deep-sea fishes are such (Brauer, 1972) that the great majority of the metabolic investigations on these animals have been done at the biochemical level. Various enzymes and their substrates are more resistant to the stresses of capture than are the fishes in which they function (Hochachka, 1971, 1975; Macdonald, 1975). The literature on ecologically relevant, non-biochemical metabolic studies of deep-sea fishes is sparse, often inferential, and fragmentary. This is especially the case for the papers relating to buoyancy control and vertical migrations. Estimates of the energy requirements for these processes have been primarily based on measurements of tissue compositions and densities, swim bladder gas compositions, and various physical parameters (Alexander, 1972; Baldridge, 1972; Blaxter *et al.*, 1971; Marshall, 1972).

Childress and Nygaard (1973) made indirect estimates of the metabolic rates of 37 species of Californian midwater fishes on the basis of their protein contents and various estimates of energy equivalences of relevant parameters. The main result was a prediction that metabolic rates should decline rapidly with increasing minimum depths of occurrence. Rates for fishes having minimum depths of occurrence between 400 and 1200 m are predicted to be only 5–10% of rates for species which reach the surface during their nightly vertical migrations. The limited amount of direct metabolic information available supports this prediction.

We know of only one estimate of oxygen consumption rates of deep-sea fishes in the deep sea. Smith and Hessler (1974), in a technological *tour-de-force*, measured the routine metabolic rates of one individual each of two species of benthopelagic fishes; a hagfish (*Eptatretus deani*) and a rat-tail (*Coryphaenoides acrolepis*). These fishes were separately trapped in a closed plastic box fitted with a recording oxygen electrode, and their oxygen uptake rates measured over periods of 13 and 3½ hr, respectively, all at a depth of 1230 m and a temperature of 3.5°C. When compared with oxygen consumption rates at similar temperatures of shallow-water fishes that are somewhat related to the deep-sea forms, the hagfish was found to have a metabolic rate about one-fourth that of another hagfish, while the rat-tail had a rate about 5% that of a codfish. Combined with observations of very low or zero levels of physical activity in the deep-sea forms during much of their lives (Barham, 1971), the indications are that these deep-sea fishes have metabolic rates adapted to the low levels of food availability in their environment.

Oxygen uptake rates of intact fishes have been measured in only three additional deep-sea species, all mesopelagic forms from depths of 400–1000 m in the northeastern Pacific Ocean. The three are the fangtooth, *Anoplogaster cornuta* (Meek and Childress, 1973; Gordon *et al.*, 1976); an eelpout, *Melanostigma pammelas* (Childress, 1971; Belman and Gordon, 1979); and a liparid, *Nectoliparis pelagicus* (Childress, 1971b). The liparid will not be discussed further, since Childress (1971b) published only pooled data in which measurements on this species were mixed with results from six other species from a similar depth range. The eelpout has been the subject of an extensive metabolic investigation by Belman and Gordon, who have studied the effects on oxygen consumption rates of ecologically relevant ranges of temperature, hydrostatic pressure, and oxygen partial pressure. These results were recently published and will not be discussed here. Childress (1971b) presented his data on this eelpout in the same form as that for the liparid.

The fangtooth is a medium-sized (up to about 15 cm standard length and 80 g fresh weight) black mesopelagic fish without a swimbladder. It is circumtropical in its distribution and occurs regularly in small numbers in midwater at depths of 500–900 m off southern California. This depth range places much of its habitat well within the regional oxygen minimum layer. It is resilient enough so that the majority of trawled individuals, especially if severe thermal shocks have been avoided, will survive in surface aquaria kept at near ecologically normal temperatures (5°–7°C) for periods of several days to several weeks. It is usually quite lethargic in aquaria; an individual in the best condition will generally rest virtually immobile against a wall of the aquarium or swim slowly in straight lines by means of regular beats of its pectoral fins. It does not respond actively to sudden changes in light intensity. Meek and Childress (1973) measured oxygen consumption rates of five individual fangtooth fish at 5°C. They studied the effects on routine metabolic rates of hydrostatic pressure (at 1 and 68 atm) and oxygen partial pressure. They also made a preliminary study of the relationship between metabolic rate and body weight.

Gordon *et al.* (1976) also measured rates of oxygen uptake. They worked with ten fish, tested only at 1 atm hydrostatic pressure and studied the effects on near routine metabolic rates of temperature (3°, 7°, and 10°C) and oxygen partial pressure. These authors found that (1) ecologically relevant hydrostatic pressures have no significant metabolic effects; (2) the fangtooth is metabolically sensitive to temperature in the range 3°–7°C ($Q_{10} = 2.5$), but temperature insensitive in the range 7°–10°C (average $Q_{10} = 1.3$); (3) the measured metabolic rates at 3°–5°C are close to the rates predicted for the species by Childress and Nygaard (1973); and (4) the fangtooth can take up oxygen at easily measurable rates at partial pressures all the way down to zero; thus, it has the possibility of functioning aerobically even in the most oxygen-deficient regions of the oceanic oxygen minimum layer. The critical oxygen tension [(P_c) the partial pressure below which an organism can no longer regulate its oxygen uptake rate at a constant level] for the fangtooth varies in proportion to the routine metabolic rate (Table VII); thus, for the species to operate aerobically in the most oxygen-deficient regions of the oxygen minimum layer, it must limit its locomotor activity while there.

Even less information is available concerning tissue metabolic rates in deep-sea fishes. Gordon (1975) has reported results of oxygen consumption rates measured at 1 atm pressure and at three temperatures (5°, 15°, 25°C) for substrate saturated white muscle minces from four species of mesopelagic fishes from the eastern North Pacific. Three lantern fishes (family Myctophidae) and one hatchetfish (family Sternoptychidae) were involved, including forms which have

TABLE VII

Weight-Specific Rates of Oxygen Consumption and Critical Oxygen Tensions (P_c) for the Mesopelagic Fangtooth Fish, *Anoplogaster cornuta*[a,b]

	Fish wet weight (g)	Oxygen uptake rate (ml kg/hr)	P_c (mm Hg)
	26	20	35
	75	22	18
	36	24	35
	34	31	22
	38	32	23
	36	35	44
	34	44	35
	39	46	45
$\bar{X}$	39	33	35

[a] Based upon Gordon *et al.* (1976).
[b] Least squares regression equation for all points: $P_c = 0.74$ (oxygen consumption) + 8.6. All measurements are made at the single temperature of 7°C, since this is the usual environmental temperature for the species

minimum depths of occurrence ranging from the surface to about 300 m. The lanternfishes are all vertically migratory, whereas the hatchetfish is considered to be relatively sedentary.

The results indicated that the muscles of all species were thermostable ($Q_{10} \simeq 1.0$) over the interval of 5°–15°C, and that only the deepest living lanternfish, which rarely comes closer to the surface than about 200 m depth, showed a statistically significantly higher rate at 25°C. These patterns would be consistent with the fishes' maintaining nearly constant swimming capacities at all normally occurring temperatures in their vertical ranges. The single group showing a higher rate was well outside the ecological range of temperatures for the species. The overall range of muscle metabolic rates for the mesopelagic fishes fell within the ranges of rates measured on a number of species of cool-water shallow-water fishes, studied by identical methods, also by Gordon (1972a). The indications are that the muscles of the mesopelagic species studied are not significantly more cold-adapted than are the muscles of cool-water shallow-water species.

Gordon (1975) also reported on the combined effects of temperature and hydrostatic pressure on the oxygen uptake of substrate saturated minces of white muscle from the mesopelagic scorpionfish, *Ectreposebastes imus*. This species is little known in behavioral terms, but appears to have a minimum depth of occurrence in the Galápagos Islands (where the study was carried out) of about 150–200 m. It is probably at least somewhat vertically migratory. This scorpion fish is indicated to have muscle metabolism which is moderately sensitive to temperature in the 5°–15°C range ($Q_{10} \simeq 2$), but is insensitive in the 15°–25°C range ($Q_{10} \simeq 1$). Over the pressure range from 1–1000 atm, there are no statistically significant variations in muscle metabolic rates at 5° or 25°C; however, at 15° (near the probable ecological thermal maximum for the species), pressures up to 400 atm seem to stimulate muscle metabolism. It is possible that over the ecologically normal ranges of temperature and pressure for this species, the metabolic effects of these two variables on the white muscles may act to compensate for each other, resulting in an overall muscle metabolic rate that is independent of depth.

An interesting finding relating to metabolic rates is the fact that a variety of vertically migratory mesopelagic fishes have both low blood oxygen carrying capacities and low blood hemoglobin oxygen affinities (Douglas *et al.,* 1976). It had been considered probable that many of these fishes had low carrying capacities, since they have low blood hematocrits (Blaxter *et al.,* 1971). Douglas *et al.* (1976) determined oxygen capacities in the bloods of nine species of mesopelagic fishes from the eastern North Pacific; all were in the low range of 1.5–3.6 vol %. The oxygen loading curves for two species were measured at 0°C. They were quite similar to each other and showed 50% oxygen loading at high oxygen partial pressures near 80 mm Hg. Most of the fishes studied spend at least parts of their lives in well developed oceanic oxygen minimum layers. Under the conditions existing in these layers, it is probable that the blood hemoglobins in these fishes are only about 1% saturated at their gills. The suggestion is made by Douglas *et al.* that these forms may well be facultative anaerobes.

C. TEMPERATURE ADAPTATIONS

The patterns, mechanisms, and controls relating to the physiological effects of temperature on the few pelagic vertebrates that have been studied appear to be generally similar to those existing in most nonpelagic vertebrates belonging to the same systematic categories. Accordingly, we restrict ourselves here to listing some recent relevant articles; Boersma (1975), Brett (1971), Calder and King (1974), Dunson (1975), Elsner *et al.* (1974), Fry, (1971), Hochachka and Somero (1973), and Irving (1969).

The sole significant exception to this general situation occurs in the high speed epipelagic fishes, many of which have evolved vascular structures at different places in their bodies to function as countercurrent heat exchangers. These exchangers permit the fish to operate as regional endotherms capable of substantial amounts of thermoregulation in large portions of their trunk musculatures (Carey *et al.,* 1971; Carey, 1973; Graham, 1975). The energy which produces such high body temperatures in the fishes possessing these exchangers derives from the high metabolic rates of the swimming muscles (see Section III,B) and in turn assists the swimming muscles in the generation of even more power. This latter aspect seems to be of greatest importance in the generation of power for high speed bursts of swimming. The thermoregulatory abilities of these fishes, even though they are imperfect, probably also permit them to make rapid vertical and horizontal migrations involving substantial changes of environmental temperatures without suffering severe metabolic consequences, or needing time for thermal acclimation.

Carey *et al.* (1971), Carey (1973), and Graham (1975) review what is known about the phylogenetic distribution in the fishes of regional endothermy and its major properties, including its various anatomical bases. We will restate some of the main points.

(1) Regional endothermy occurs to significant degree only in various of the mackerel sharks (Isuridae) and in some of the evolutionarily more advanced mackerels, bonitos, and tunas (Scombridae).

(2) There are two basic categories of vascular heat exchangers, lateral and central. Lateral exchangers arise from subcutaneous arteries and veins which branch intersegmentally to form *retia mirabilia* that penetrate the red muscle masses of the trunk. Central exchangers are located in the hemal arches of the vertebrae and consist of vertical "retes" formed from branches of the dorsal aorta and the posterior cardinal vein.

(3) The magnitudes of the temperature differentials that different species maintain between their red muscle masses and the outside water and the shapes of the thermal profiles within their trunk musculatures vary, depending on the degrees of development of the two types of exchangers. Species with well developed central but poorly developed lateral exchangers can generate high core temperatures, but can cool rapidly radially. Species with small central but well developed lateral exchangers have lower core temperatures, but may be both warmer and more uniformly warm radially. Species with well developed exchangers of both types have both higher core temperatures and more uniformly warm radial temperature distributions.

Graham (1973) and Stevens *et al.* (1974) have considered some other possible implications of the existence of these exchangers. Graham (1973) points out that the hemoglobins of warm-bodied scombrids show lower temperature effects on oxygen binding than are usual in teleost fishes. He argues that this property is of adaptive value since it reduces the possibility of premature oxygen dissociation and diffusional loss from the blood as the blood crosses the thermal gradients in the fishes' muscles. It also helps to ensure that the warmest, most active muscles receive the oxygen supplies that they need to maintain their activity. [Graham (1973) does not state it, but an implication of this theory could be that the aerobic red muscles are the most important in this connection; thus, that regional endothermy in these fishes relates largely to cruising speed maintenance, not to burst speed amplitude, as is postulated by Graham (1975)].

Stevens *et al.* (1974) studied the physical dimensions of the blood vessels in the central heat exchangers of the skipjack tuna (*Katsuwonus pelamis*). They compared these dimensions with those of the blood vessels in the gas-secreting glands of eel swim bladders. The vessels of the heat exchanger are much larger in diameter and much longer than those of the gas glands. They infer that it is unlikely that the heat exchanger also functions significantly as a gas exchanger.

Roberts and Graham (1974) have made a preliminary report of a study of body temperatures, heart rates, and electrical activity patterns of swimming muscles in the evolutionarily primitive mackerel, *Scomber japonicus,* swimming at different controlled speeds. This species does not possess vascular heat exchangers. It generated small, but significant rises in muscle temperatures at higher swimming speeds and much higher heart rates. Electromyograms indicated that mainly the red muscles functioned at slow speeds, but both red and white muscles functioned at higher speeds. Studies like this may contribute to an understanding of the mechanisms involved in the functioning of vascular heat exchangers in more advanced scombrids and also of the evolutionary mechanisms which led to the development of the exchangers.

IV. Addendum

In the time which has passed since the completion of this chapter significant additions have been made to the literature in many of the topic areas covered. Integration of this new material into the text of the chapter would have caused additional delay in the production of the book. As a way of avoiding this undesirable outcome, while simultaneously providing access to the newer material, bibliographic information on the most relevant more recent publications follows. These references are to the vertebrate literature only. B. W. Belman was not available to produce a similar list for the invertebrates.

Two books have been published (Hoar and Randall, 1978; Sharp and Dizon, 1978) which include important reviews containing new material on pelagic fishes relating to all three major topic areas included in this chapter. The individual chapters in these books are not listed separately here.

Acknowledgment

This chapter was written while M. S. Gordon was holder of a senior Queen's Fellowship in Marine Science from the Department of Science, Government of the Commonwealth of Australia. We wish to thank Prof. G. A. Horridge, Department of Neurobiology, Research School of Biological Sciences, Australian National University, Canberra, Australia for making available space and secretarial assistance invaluable in the preparation of the chapter. Financial and facilities support for B. W. Belman were provided by the Zoology-Fisheries program, UCLA. We thank T. Falconer, F. Jackson, and J. Belman for their expert secretarial work.

Suggested Reading

Belman, B. W., and Gordon, M. S. (1979). Comparative studies on the metabolism of shallow-water and deep-sea marine fishes. V. The effects of temperature and hydrostatic pressure on oxygen consumption in the mesopelagic zoarcid *Melanostigma pammelas*. *Mar. Biol.* **50,** 275–281.

Brill, R. W. (1979). The effect of body size on the standard metabolic rate of skipjack tuna, *Katsuwonus pelamis*. *Fish. Bull.* **77,** 494–498.

Childress, J. J., and Somero, G. N. (1979). Depth related decreases in enzymic activity in tissues of deep living pelagic marine teleost fishes. *Mar. Biol.* **52,** 273–283.

Childress, J. J., Taylor, S., Cailliet, G., and Price, M. H. (1980). Patterns of growth, reproduction and energy usage in some meso- and bathypelagic fishes from off California. *Mar. Biol.* (in press).

Dizon, A. E., and Brill, R. W. (1979). Thermoregulation in tunas. *Am. Zool.* **19,** 249–265.

Freadman, M. A. (1979). Swimming energetics of striped bass (*Morone saxatilis*) and bluefish (*Pomatomus saltatrix*): Gill ventilation and swimming metabolism. *J. Exp. Biol.* **83,** 217–230.

George, J. C., and Stevens, E. D. (1978). Fine structure and metabolic adaptation of red and white muscles in tuna. *Environ. Biol. Fishes* **3,** 185–191.

Gordon, M. S., Putnam, R. W., Tarifeno, E., and Vojkovich, M. (1977). Anaerobic metabolism during high speed swimming and in fatigue in the eastern Pacific jack mackerel (*Trachurus symmetricus*). *Proc. Int. Union Physiol. Sci.* **13,** 275 (abstr.).

Gordon, M. S., Loretz, C., Chow, P., and Vojkovich, M. (1979). Patterns of metabolism in marine fishes using different modes of locomotion. *Am. Zool.* **19,** 897 (abstr.).

Graham, J. B. (1979). Effect of swimming speed on the excess temperatures and activities of heart and red and white muscles in the mackerel, *Scomber japonicus*. *Fish. Bull.* **76,** 861–867.

Guppy, M., Hulbert, W. C., and Hochachka, P. W. (1979). Metabolic sources of heat and power in tuna muscles. II. Enzyme and metabolite profiles. *J. Exp. Biol.* **82,** 303–320.

Hoar, W. S., and Randall, D. J., eds. (1978). "Fish Physiology." Vol. 7. Academic Press, New York.

Hulbert, W. C., Guppy, M., Murphy, B., and Hochachka, P. W. (1979). Metabolic sources of heat and power in tuna muscles. I. Muscle fine structure. *J. Exp. Biol.* **82,** 289–301.

Low, P. S., and Somero, G. N. (1976). Adaptation of muscle pyruvate kinases to environmental temperatures and pressures. *J. Exp. Zool.* **198,** 1–12.

Neill, W. H., Chang, R., and Dizon, A. (1976). Magnitude and ecological implications of thermal inertia in skipjack tuna, *Katsuwonus pelamis* (Linnaeus). *Environ. Biol. Fishes* **1,** 61–80.

Noble, R. W., Pennelly, R. R., and Riggs, A. (1975). Studies of the functional properties of the hemoglobin from the benthic fish, *Antimora rostrata*. *Comp. Biochem. Physiol. B* **52,** 75–81.

Sharp, G. D., and Dizon, A. E., eds. (1978). "The Physiological Ecology of Tunas." Academic Press, New York.

Siebenaller, J. F., and Somero, G. N. (1978). Pressure-adaptive differences in lactate dehydrogenases of congeneric fishes living at different depths. *Science* **201,** 255–257.

Siebenaller, J. F., and Somero, G. N. (1979). Pressure-adaptive differences in the binding and catalytic properties of muscle-type (M_4) lactate dehydrogenases of shallow- and deep-living marine fishes. *J. Comp. Physiol.* 129(4): 295–300.

Somero, G. N. (1979). Interacting effects of temperature and pressure on enzyme function and evolution in marine organisms. *Biochem. Biophys. Perspect. Mar. Biol.* **4,** 1–27.

Somero, G. N., and Childress, J. J. (1980). A violation of the metabolism-size scaling paradigm: Activities of glycolytic enzymes in muscle increase in larger size fishes. *Physiol. Zool.* (in press).

Somero, G. N., and Hochachka, P. W. (1976). Biochemical adaptations to pressure. *In* "Adaptation to Environment: Essays on the Physiology of Marine Animals'' (R. C. Newell, ed.), pp. 481-510. Butterworth, London.

Somero, G. N., Siebenaller, J. F., and Hochachka, P. W. (1980). Physiological and biochemical adaptations to the deep sea. *In* ''The Sea'' (G. T. Rowe, ed.), Vol. 8 (in press).

Torres, J., Belman, B. W., and Childress, J. J. (1979). Oxygen consumption rates of mid-water fishes as a function of depth of occurrence. *Deep-Sea Res., Part A* **26,** 185-197.

Webb, P. W. (1977). Effects of size on performance and energetics of fish. *In* ''Scale Effects in Animal Locomotion'' (T. J. Pedley, ed.), pp. 315-331. Academic Press, New York.

References

Ache, B. (1972). Amino acid reception in the antennules of *Homarus americanus*. *Comp. Biochem. Physiol.* **42,** 807-811.

Alexander, R. M. (1970). ''Functional Design in Fishes,'' 2nd ed. Hutchinson, London.

Alexander, R. M. (1972). The energetics of vertical migration by fishes. *In* ''The Effects of Pressure on Organisms'' (M. A. Sleigh and A. G. Macdonald, eds.), pp. 273-294. Academic Press, New York.

Ameyaw-Akumfi, C. (1975). The functional morphology of the body and tail muscles of the tuna *Katsuwonus pelamis* Linnaeus. *Zool. Anz.* **194,** 367-375.

Anctil, M. (1975). Prospects in the study of interrelationships between vision and bioluminescence. *In* ''Vision in Fishes: New Approaches in Research'' (M. A. Ali, ed.), pp. 657-671. Plenum, New York.

Andersen, H. T. (1966). Physiological adaptations in diving vertebrates. *Physiol. Rev.* **46,** 212-243.

Andersen, H. T., ed. (1969). ''The Biology of Marine Mammals.'' Academic Press, New York.

Anonymous (1969). Underwater tuna school tracked by sonar. *Commer. Fish. Rev.* **31** (11), 9-10.

Arnold, G. P. (1974). Rheotropism in fishes. *Biol. Rev. Cambridge Philos. Soc.* **49,** 515-576.

Ashmole, N. P. (1971). Sea bird ecology and the marine environment. *In* ''Avian Biology'' (D. P. Farner and J. R. King, eds.), Vol. 1, pp. 223-286. Academic Press, New York.

Baldridge, H. D., Jr. (1972). Accumulation and function of liver oil in Florida sharks. *Copeia* No. 2, 306-325.

Banse, K. (1964). On the vertical distribution of zooplankton in the sea. *Prog. Oceanogr.* **2,** 53-125.

Barber, S. B. (1961). Chemoreception and thermoreception. *In* ''Physiology of Crustacea'' (T. H. Waterman, ed.), Vol. 2, pp. 109-131. Academic Press, New York.

Barham, E. G. (1971). Deep sea fishes: Lethargy and vertical orientation. *In* ''Biological Sound Scattering in the Ocean'' (G. B. Farquhar, ed.), pp. 100-118. Maury Center for Ocean Sciences, U.S. Department of the Navy, Washington, D.C.

Baudinette, R. V., and Schmidt-Nielsen, K. (1974). Energy cost of gliding flight in herring gulls. *Nature (London)* **248,** 83-84.

Belman, B. W. (1978). Respiration and the effects of pressure on the mesopelagic vertically migrating squid *Histioteuthis heteropsis*. *Limnol. Oceanogr.* **23,** 735-739.

Belman, B. W., and Childress, J. J. (1976). Circulatory adaptations to the oxygen minimum layer in the bathypelagic mysid *Gnathophausia ingens*. *Biol. Bull.* (*Woods Hole, Mass.*) **150,** 15-37.

Berger, M., and Hart, J. S. (1974). Physiology and energetics of flight. *In* ''Avian Biology'' (D. S. Farner and J. R. King, eds.), Vol. 4, pp. 415-477. Academic Press, New York.

Berkson, H. (1966). Physiological adjustments to prolonged diving in the Pacific green turtle (*Chelonia mydas Agassizi*). *Comp. Biochem. Physiol.* **18,** 101-119.

Berkson, H. (1967). Physiological adjustments to deep diving in the Pacific green turtle (*Chelonia mydas*). *Comp. Biochem. Physiol.* **21,** 507-524.

Bernstein, M. H., Thomas, S. P., and Schmidt-Nielsen, K. (1973). Power input during flight of the fish crow, *Corvus ossifragus*. *J. Exp. Biol.* **58,** 401–410.
Bilinski, E. (1974). Biochemical aspects of fish swimming. *Biochem. Biophys. Perspect. Mar. Biol.* **1,** 239–288.
Blaxter, J. H. S. (1969). Swimming speeds of fish. *United Nations FAO Fish. Rep.* **62,** 69–100.
Blaxter, J. H. S., Wardle, C. S., and Roberts, B. L. (1971). Aspects of the circulatory physiology and muscle systems of deep-sea fish. *J. Mar. Biol. Assoc. U.K.* **51,** 991–1006.
Boddeke, R., Slijper, E. J., and Van der Stelt, A. (1959). Histological characteristics of the body musculature of fishes in connection with their mode of life. *Proc. K. Ned. Akad. Wet., Ser. C* **62,** 576–588.
Boersma, D. (1975). Adaptation of Galapagos penguins for life in two different environments. *In* "The Biology of Penguins" (B. Stonehouse, ed.), pp. 101–114. Macmillan, New York.
Bone, Q. (1966). On the function of the two types of myotomal muscle fibre in elasmobranch fish. *J. Mar. Biol. Assoc. U.K.* **46,** 321–350.
Brauer, R. W., ed. (1972). "Barobiology and the Experimental Biology of the Deep Sea." North Carolina Sea Grant Program, Chapel Hill.
Brett, J. R. (1964). The respiratory metabolism and swimming performance of young sockeye salmon. *J. Fish. Res. Board Can.* **21,** 1183–1226.
Brett, J. R. (1971). Temperature: Fishes. *In* "Marine Ecology" (O. Kinne, ed.), Vol. 1, Part 1, pp. 515–568. Wiley, New York.
Brett, J. R. (1972). The metabolic demand for oxygen in fish, particularly salmonids, and a comparison with other vertebrates. *Respir. Physiol.* **14,** 151–170.
Brett, J. R. (1973). Energy expenditure of sockeye salmon, *Oncorhynchus nerka,* during sustained performance. *J. Fish Res. Board Can.* **30,** 1799–1809.
Brett, J. R., and Glass, N. R. (1973). Metabolic rates and critical swimming speeds of sockeye salmon (*Oncorhynchus nerka*) in relation to size and temperature. *J. Fish. Res. Board Can.* **30,** 379–387.
Brown, C. E., and Muir, B. S. (1970). Analysis of ram ventilation of fish gills with application to skipjack tuna (*Katsuwonus pelamis*). *J. Fish Res. Board Can.* **27,** 1637–1652.
Brown, F. A. (1973). Bioluminescence. *In* "Comparative Animal Physiology" (C. L. Prosser, ed.), pp. 951–966. Saunders, Philadelphia, Pennsylvania.
Bullock, T. H. (1955). Compensation for temperature in the metabolism and activity of poikilotherms. *Biol. Rev. Cambridge Philos. Soc.* **30,** 311–342.
Bullock, T. H., and Horridge, G. A. (1965). "Structure and Function in the Nervous System of Invertebrates," 2 vols. Freeman, San Francisco, California.
Calder, W. A., and King, J. R. (1974). Thermal and caloric relations of birds. *In* "Avian Biology" (D. S. Farner and J. R. King, eds.), Vol. 4, pp. 259–413. Academic Press, New York.
Capen, R. L. (1967). "Swimbladder Morphology of Some Mesopelagic Fishes in Relation to Sound Scattering," Res. Rep. 1447. Naval Electronics Laboratory, San Diego, California.
Carey, F. G. (1973). Fishes with warm bodies. *Sci. Am.* **228,** 36–44.
Carey, F. G., Teal, J. M., Kanwisher, J. W., Lawson, K. D., and Beckett, J. S. (1971). Warm-bodied fish. *Am. Zool.* **11,** 135–145.
Charniaux-Cotton, H. (1960). Sex determination. *In* "Physiology of Crustacea" (T. H. Waterman, ed.), Vol. 1, pp. 411–447. Academic Press, New York.
Cheng, L. (1975). Pleuston-animals of the sea-air interface. *Oceanogr. Mar. Biol.* **13,** 45–76.
Childress, J. J. (1968). Oxygen minimum layer, vertical distribution and respiration of the mysid *Gnathophausia ingens*. *Science* **160,** 1242–1243.
Childress, J. J. (1969). The respiratory physiology of the oxygen minimum layer mysid *Gnathophausia ingens*. Ph.D. Dissertation, Stanford University, Stanford, California.
Childress, J. J. (1971a). Respiratory adaptations to the oxygen minimum layer in the bathypelagic mysid *Gnathophausia ingens*. *Biol. Bull.* (*Woods Hole, Mass.*) **141,** 109–121.

Childress, J. J. (1971b). Respiratory rate and depth of occurrence of midwater animals. *Limnol. Oceanog.* **16,** 104–106.

Childress, J. J. (1975). The respiratory rates of midwater crustaceans as a function of depth of occurrence and in relation to the oxygen minimum layer off southern California. *Comp. Biochem. Physiol.* **50,** 787–799.

Childress, J. J. (1976). Physiological approaches to the biology of midwater crustaceans. *In* "Prediction of Sonic Scattering Layers" (N. Anderson, ed.), pp. 37–80. Plenum, New York.

Childress, J. J. (1977). The effects of pressure, temperature and oxygen on the oxygen consumption rate of the midwater copepod *Gaussia princeps*. *Mar. Biol.* **39,** 19–24.

Childress, J. J., and Nygaard, M. H. (1973). The chemical composition of midwater fishes as a function of depth of occurrence off southern California. *Deep-Sea Res.* **20,** 1093–1109.

Childress, J. J., and Nygaard, M. H. (1974). Chemical composition and buoyancy of midwater crustaceans as a function of depth of occurrence off southern California. *Mar. Biol.* **27,** 225–238.

Clarke, M. R. (1969). A new mid-water trawl for sampling discrete depth horizons. *J. Mar. Biol. Assoc. U.K.* **49,** 945–960.

Clarke, W. D. (1963). Function of bioluminescence in mesopelagic organisms. *Nature (London)* **198,** 1244–1246.

Clutter, R. I. (1969). The microdistribution and social behavior of some pelagic mysid shrimps. *J. Exp. Mar. Biol. Ecol.* **3,** 125–155.

Conover, R. J. (1968). Zooplankton-life in a nutritionally dilute environment. *Am. Zool.* **8,** 107–118.

Dando, P. R. (1969). Lactate metabolism in fish. *J. Mar. Biol. Assoc. U.K.* **49,** 209–223.

Dennell, R. (1955). Observations on the luminescence of bathypelagic Crustacea, Decapoda, of the Bermuda area. *J. Linn. Soc. London, Zool.* **42,** 393–406.

Denton, E. J., and Gilpin-Brown, J. B. (1959). On the buoyancy of the cuttlefish. *Nature (London)* **184,** 1330–1332.

Denton, E. J., and Gilpin-Brown, J. B. (1961). The distribution of gas and liquid within the cuttlebone. *J. Mar. Biol. Assoc. U.K.* **41,** 365–381.

Denton, E. J., and Gilpin-Brown, J. B. (1966). On the buoyancy of the pearly Nautilus. *J. Mar. Biol. Assoc. U.K.* **46,** 723–759.

Denton, E. J., and Gilpin-Brown, J. B. (1971). Further observations on the buoyancy of *Spirula*. *J. Mar. Biol. Assoc. U.K.* **51,** 363–373.

Denton, E. J., and Nicol, J. A. C. (1964). The choroidal tapeta of some cartilaginous fishes (Chondrichthys). *J. Mar. Biol. Assoc. U.K.* **44,** 219–258.

Denton, E. J., and Shaw, T. I. (1962). The buoyancy of gelatinous marine animals. *J. Physiol. (London)* **161,** 14P–15P.

Denton, E. J., and Warren, F. J. (1957). The photosensitive pigments in the retinas of deep sea fish. *J. Mar. Biol. Assoc. U.K.* **36,** 651–662.

Denton, E. J., Gilpin-Brown, J. B., and Shaw, T. I. (1969). A buoyancy mechanism found in cranchid squid. *Proc. R. Soc. London, Ser. B* **174,** 271–279.

Denton, E. J., Gilpin-Brown, J. B., and Wright, P. G. (1970). On the "filters" in the photophores of mesopelagic fish and on a fish emitting red light and especially sensitive to red light. *J. Physiol. (London)* **208,** 72–75.

Douglas, E. L., Friedl, W. A., and Pickwell, G. V. (1976). Fishes in oxygen-minimum zones: Blood oxygenation characteristics. *Science* **191,** 957–959.

Dunson, W. A., ed. (1975). "The Biology of Sea Snakes." University Park Press, Baltimore, Maryland.

Eakin, R. M. (1972). Structure of invertebrate photoreceptors. *In* "Handbook of Sensory Physiology" (H. J. A. Dartnall, ed.), Vol. 7, Part 1, pp. 623–684. Springer-Verlag, Berlin and New York.

Elsner, R., Pirie, J., Kenney, D. D., and Schemmer, S. (1974). Functional circulatory anatomy of cetacean appendages. *In* "Functional Anatomy of Marine Mammals" (R. J. Harrison, ed.), Vol. 2, pp. 143–159. Academic Press, New York.

Emlen, S. T. (1975). Migration: Orientation and navigation. *In* "Avian Biology" (D. S. Farner and J. R. King, eds.), Vol. 5, pp. 129–219. Academic Press, New York.

Fahrenbach, W. H. (1964). The fine structure of a nauplius eye. *Z. Zellforsch. Mikrosk. Anat.* **62,** 182–197.

Fierstine, H. L., and Walters, V. (1968). Studies in locomotion and anatomy of scombroid fishes. *Mem. South. Calif. Acad. Sci.* **6,** 1–29.

Fretter, V., and Graham, A. (1964). Reproduction. *In* "Physiology of Mollusca" (K. M. Wilbur and C. M. Yonge, eds.), Vol. 1, pp. 127–164. Academic Press, New York.

Fry, F. E. J. (1971). The effect of environmental factors on the physiology of fish. *In* "Fish Physiology" (W. S. Hoar and D. J. Randall, eds.), Vol. 6, pp. 1–98. Academic Press, New York.

Fuzzessery, Z. M., and Childress, J. J. (1976). Comparative chemosensitivity to amino acids and their role in the feeding activity of bathypelagic and littoral crustaceans. *Biol. Bull.* (*Woods Hole, Mass.*) **149,** 522–538.

Ghiretti, F. (1966). Respiration. *In* "Physiology of Mollusca" (K. M. Wilbur and C. M. Yonge, eds.), Vol. 2, pp. 175–208. Academic Press, New York.

Gibbs, R. H., Jr., and Collette, B. B. (1967). Comparative anatomy and systematics of the tunas, genus *Thunnus. Fish. Bull.* **66,** 65–130.

Giese, A. C., and Pearse, J. S., eds. (1974a). "Reproduction of Marine Invertebrates," Vols. 1 and 2. Academic Press, New York.

Giese, A. C., and Pearse, J. S. (1974b). Introduction and General Principles. *In* "Reproduction of Marine Invertebrates" (A. C. Giese and J. S. Pearse, eds.), Vol. 1, pp. 1–49. Academic Press, New York.

Goldsmith, T. H. (1972). The natural history of invertebrate visual pigments. *In* "Handbook of Sensory Physiology" (H. J. A. Dartnall, ed.), Vol. 7, Part 1, pp. 685–719. Springer-Verlag, Berlin and New York.

Gordon, M. S. (1968). Oxygen consumption of red and white muscles from tuna fishes. *Science* **159,** 87–90.

Gordon, M. S. (1972a). Comparative studies on the metabolism of shallow water and deep sea marine fishes. I. White muscle metabolism in shallow water fishes. *Mar. Biol.* **13,** 222–237.

Gordon, M. S. (1972b). Comparative studies on the metabolism of shallow water and deep sea marine fishes. II. Red muscle metabolism in shallow water fishes. *Mar. Biol.* **15,** 246–250.

Gordon, M. S. (1975). Effects of temperature and pressure on the oxidative metabolism of fish muscle. *In* "Comparative Physiology—Functional Aspects of Structural Materials" (L. Bolis, H. P. Maddrell, and K. Schmidt-Nielsen, eds.), pp. 211–222. North-Holland Publ., Amsterdam.

Gordon, M. S., and Chow, P. H. (1974). Unusual patterns of aerobic activity metabolism in fishes. *Am. Zool.* **14,** 1258 (abstr.).

Gordon, M. S., Belman, B. W., and Chow, P. H. (1976). Comparative studies on the metabolism of shallow-water and deep-sea marine fishes. IV. Patterns of aerobic metabolism in the mesopelagic deep-sea fangtooth fish, *Anoplogaster cornuta. Mar. Biol.* **35,** 287–293.

Graham, J. B. (1973). Heat exchange in the black skipjack, and the blood-gas relationship of warm-bodied fishes. *Proc. Natl. Acad. Sci. U.S.A.* **70,** 1964–1967.

Graham, J. B. (1975). Heat exchange in the yellowfin tuna, *Thunnus albacares,* and skipjack tuna, *Katsuwonus pelamis,* and the adaptive significance of elevated body temperatures in scombrid fishes. *Fish. Bull.* **73,** 219–229.

Graham, J. B., Gee, J. H., and Robison, F. S. (1975). Hydrostatic and gas exchange functions of the lung of the sea snake *Pelamis platurus. Comp. Biochem. Physiol. A* **50,** 477–482.

Gray, I. E. (1954). Comparative study of the gill area of marine fishes. *Biol. Bull. (Woods Hole, Mass.)* **107,** 219-225.

Greer-Walker, M., and Pull, G. A. (1975). A survey of red and white muscle in marine fish. *J. Fish Biol.* **7,** 295-300.

Halcrow, K. (1963). Acclimation to temperature in the marine copepod *Calanus finmarchicus. Limnol. Oceanogr.* **8,** 1-8.

Hall, F. G. (1930). The ability of the common mackerel and certain other marine fishes to remove dissolved oxygen from sea water. *Am. J. Physiol.* **93,** 417-421.

Hara, T., and Hara, R. (1972). Cephalopod retinochrome. *In* "Handbook of Sensory Physiology" (H. J. A. Dartnall ed.), Vol. 7, Part 1, pp. 720-746. Springer-Verlag, Berlin and New York.

Hardy, A. (1956). "The Open Sea: Its Natural History," Part I. Collins, London.

Hardy, M. G. (1962). Photophore and eye movement in the euphausiid *Meganyctiphanes norvegica* (G. O. Sars). *Nature (London)* **196,** 790-791.

Harrison, R. J., ed. (1972). "Functional Anatomy of Marine Mammals," Vol. 1. Academic Press, New York.

Harrison, R. J., ed. (1974). "Functional Anatomy of Marine Mammals," Vol. 2. Academic Press, New York.

Harvey, E. N. (1961). Light production. *In* "Physiology of Crustacea" (T. H. Waterman, ed.), Vol. 2, pp. 171-187. Academic Press, New York.

Hasler, A. D. (1971). Orientation and fish migration. *In* "Fish Physiology" (W. S. Hoar and D. J. Randall, eds.), Vol. 6, pp. 429-510. Academic Press, New York.

Heatwole, H., and Seymour, R. (1975). Diving physiology. *In* "The Biology of Sea Snakes" (W. A. Dunson, ed.), pp. 289-327. University Park Press, Baltimore, Maryland.

Hochachka, P. W., ed. (1971). Pressure effects on biochemical systems of abyssal fishes: The 1970 ALPHA HELIX expedition to the Galápagos Archipelago. *Am. Zool.* **11,** 399-576.

Hochachka, P. W. (1974). Enzymatic adaptation to deep sea life. *In* "Biology of the Oceanic Pacific" (C. B. Miller, ed.), pp. 107-136. Oregon State University, Corvallis.

Hochachka, P. W., ed. (1975). Pressure effects on biochemical systems of abyssal and midwater organisms: The 1973 Kona expedition of the ALPHA HELIX. *Comp. Biochem. Physiol. B* **52,** 1-200.

Hochachka, P. W., and Somero, G. N. (1971). Biochemical adaptation to the environment. *In* "Fish Physiology" (W. S. Hoar and D. J. Randall, eds.), Vol. 6, pp. 99-156. Academic Press, New York.

Hochachka, P. W., and Somero, G. N. (1973). "Strategies of Biochemical Adaptation." Saunders, Philadelphia, Pennsylvania.

Hochachka, P. W., Owen, T. G., Allen, J. F., and Whittow, G. C. (1975). Multiple end products of anaerobiosis in diving vertebrates. *Comp. Biochem. Physiol. B* **50,** 17-22.

Hopkins, T. L., Baird, R. C., and Milliken, D. M. (1973). A messenger-operated closing trawl. *Limnol. Oceanogr.* **18,** 488-490.

Hudson, R. C. L. (1973). On the function of the white muscles in teleosts at intermediate swimming speeds. *J. Exp. Biol.* **58,** 509-522.

Hughes, G. M. and Morgan, M. (1973). The structure of fish gills in relation to their respiratory function. *Biol. Rev. Cambridge Philos. Soc.* **48,** 419-475.

Hunter, J. R. (1971). Sustained speed of jack mackerel, *Trachurus symmetricus. Fish. Bull.* **69,** 267-271.

Hunter, J. R., and Zweifel, J. R. (1971). Swimming speed, tail beat frequency, tail beat amplitude, and size in jack mackerel, *Trachurus symmetricus,* and other fishes. *Fish. Bull.* **69,** 253-266.

Irving, L. (1969). Temperature regulation in marine mammals. *In* "The Biology of Marine Mammals" (H. T. Andersen, ed.), pp. 147-174. Academic Press, New York.

Jørgensen, C. B. (1966). "Biology of Suspension Feeding." Pergamon, Oxford.

Kampa, E. M. (1955). Euphausiopsin, a new photosensitive pigment from the eye of euphausiid crustaceans. *Nature (London)* **175,** 966–998.

Kampa, E. M., and Boden, B. D. (1957). Light generation in a sonic scattering layer. *Deep-Sea Res.* **4,** 73–92.

Kinne, O. (1970). Temperature-animals-invertebrates. *In* "Marine Ecology" (O. Kinne, ed.), Vol. 1, Part 1, pp. 407–514. Wiley, New York.

Kishinouye, K. (1923). Contributions to the comparative study of the so-called scombroid fishes. *J. Coll. Agric., Tokyo Imp. Univ.* **8,** 293–475.

Kooyman, G. L. (1972). Deep diving behavior and effects of pressure in reptiles, birds, and mammals. *In* "The Effects of Pressure on Organisms" (M. A. Sleigh and A. G. Macdonald, eds.), pp. 295–311. Academic Press, New York.

Kooyman, G. L. (1975). Behavior and physiology of diving. *In* "The Biology of Penguins" (B. Stonehouse, ed.), pp. 115–137. Macmillan, New York.

Lang, T. G., and Norris, K. S. (1965). Swimming speed of a Pacific bottlenose porpoise. *Science* **151,** 588–590.

Lang, T. G., and Pryor, D. (1966). Hydrodynamic performance of porpoises, *Stenella attenuata. Science* **152,** 531–533.

Lee, R. F., and Hirota, J. (1973). Wax esters in tropical zooplankton and nekton and the geographical distributions of wax esters in marine copepods. *Limnol. Oceanogr.* **18,** 227–239.

Lee, R. F., Hirota, J., and Barnett, A. M. (1971). Distribution and importance of wax esters in marine copepods and other zooplankton. *Deep-Sea Res.* **18,** 1147–1165.

Levandowsky, M., and Hodgson, E. S. (1965). Amino acid and amine receptors of lobsters. *Comp. Biochem. Physiol.* **16,** 159–161.

Lewis, R. W. (1970). The densities of three classes of marine lipids in relation to their possible role as hydrostatic agents. *Lipids* **5,** 151–153.

Lipskaya, N. Y. (1974). Metabolic rates of young fish belonging to tropical species. [Intensivnost' obmena u molodi nekotorykh vidov tropicheskikh ryb.] *Vopr. Ikhtiol.* **14,** 1076–1086.

Locket, N. A. (1975). Some problems of deep-sea fish eyes. *In* "Vision in Fishes: New Approaches in Research" (M. A. Ali, ed.), pp. 645–655. Plenum, New York.

Lythgoe, J. A. (1972). The adaptations of visual pigments to the photic environment. *In* "Handbook of Sensory Physiology" (H. J. A. Dartnall, ed.), Vol. 7, Part 1, pp. 564–603. Springer-Verlag, Berlin and New York.

Macdonald, A. G. (1975). "Physiological Aspects of Deep Sea Biology." Cambridge Univ. Press, London and New York.

McFarland, W. N. (1971). Cetacean visual pigments. *Vision Res.* **11,** 1065–1076.

McFarland, W. N., and Munz, F. W. (1975). The evolution of photopic visual pigments in fishes. *Vision Res.* **15,** 1071–1080.

McLeese, D. W. (1970). Detection of dissolved substances by the American lobster *Homarus americanus* and olfactory attraction between lobsters. *J. Fish. Res. Board Can.* **27,** 1371–1378.

Magnuson, J. J. (1970). Hydrostatic equilibrium of *Euthynnus affinis,* a pelagic teleost without a gas bladder. *Copeia No. 1,* pp. 56–85.

Magnuson, J. J. (1973). Comparative study of adaptations for continuous swimming and hydrostatic equilibrium of scombroid and xiphoid fishes. *Fish. Bull.* **71,** 337–356.

Magnuson, J. J., and Prescott, J. H. (1966). Courtship, locomotion, feeding and miscellaneous behavior of Pacific bonito (*Sarda chiliensis*). *Anim. Behav.* **14,** 54–67.

Malins, D. C. (1975). Cetacean biosonar. Part 2. The biochemistry of lipids in acoustic tissues. *Biochem. Biophys. Perspect. Mar. Biol.* **2,** 237–290.

Marshall, N. B. (1972). Swimbladder organization and depth ranges of deep-sea teleosts. *In* "The Effects of Pressure on Organisms" (M. A. Sleigh and A. G. Macdonald, eds.), pp. 261–272. Academic Press, New York.

Marshall, S. M. (1973). Respiration and feeding in copepods. *Adv. Mar. Biol.* **11,** 57–120.
Mauchline, J. (1971). Seasonal occurrence of mysids (Crustacea) and evidence of social behavior. *J. Mar. Biol. Assoc. U.K.* **51,** 809–825.
Mauchline, J. (1972). The biology of bathypelagic organisms, especially Crustacea. *Deep-Sea Res.* **19,** 753–780.
Mauchline, J., and Fisher, L. R. (1969). The biology of euphausiids. *Adv. Mar. Biol.* **7,** 1–454.
Mednikov, B. M. (1961). On the sex ratio in deep water Calanoida. *Crustaceana (Leiden)* **3,** 105–109.
Meek, R. P., and Childress, J. J. (1973). Respiration and the effect of pressure in the mesopelagic fish *Anoplogaster cornuta* (Beryciformes). *Deep-Sea Res.* **20,** 1111–1118.
Menzies, R. J., George, R. Y., and Rowe, G. T. (1973). "Abyssal Environment and Ecology of the World Oceans." Wiley, New York.
Modigh, M., and Tota, B. (1975). Mitochondrial respiration in the ventricular myocardium and in the white and deep red myotomal muscles of juvenile tuna fish (*Thunnus thynnus* L.). *Acta Physiol. Scand.* **93,** 289–294.
Morris, R. W. (1974). Respiratory metabolism and ecology of fishes. *Chem. Zool.* **8,** 447–469.
Muir, B. S., and Hughes, G. M. (1969). Gill dimensions for three species of tunny. *J. Exp. Biol.* **51,** 271–285.
Muir, B. S., and Kendall, J. I. (1968). Structural modifications in the gills of tunas and some other oceanic fishes. *Copeia* No. 2, 388–398.
Munk, O. (1966). The ocular anatomy of some deep sea teleosts. *Dana Rep.* **13,** Pap. 70, 1–15.
Munz, F. W. (1958). Photosensitive pigments from the retina of certain deep sea fishes. *J. Physiol. (London)* **140,** 220–225.
Munz, F. W. (1964). The visual pigments of epipelagic and rocky shore fishes. *Vision Res.* **4,** 441–454.
Munz, F. W., and McFarland, W. N. (1973). The significance of spectral position in the rhodopsins of tropical marine fishes. *Vision Res.* **13,** 1829–1874.
Munz, F. W., and McFarland, W. N. (1975). Presumptive cone pigments extracted from tropical marine fishes. *Vision Res.* **15,** 1045–1062.
Muravskaya, Z. A. (1972). [Rate of nitrogen excretion and oxygen consumption in certain Black Sea fish with different ecologies.] Intensivnost' ekskretsii azota i potrebleniya kisloroda u nekotorykh Chernomorskikh ryb s razlichnoi ekologiei. *Biol. Nauki (Moscow)* **15,** 39–42.
Nicol, J. A. C. (1962). Animal luminescence. *Adv. Comp. Physiol. Biochem.* **1,** 217–273.
Nichol, J. A. C. (1968). Observations on luminescence in pelagic animals. *J. Mar. Biol. Assoc. U.K.* **37,** 25–36.
Norris, K. S. (1968). The evolution of acoustic mechanisms in odontocete cetaceans. *In* "Evolution and Environment" (E. T. Drake, ed.), pp. 275–289. Yale Univ. Press, New Haven, Connecticut.
Norris, K. S. (1975). Cetacean biosonar. Part 1. Anatomical and behavioral studies. *Biochem. Biophys. Perspect. Mar. Biol.* **2,** 215–236.
Omori, M. (1971). Preliminary rearing experiments on the larvae of *Sergestes lucens* (Penaeidae, Natantia, Decapoda). *Mar. Biol.* **9,** 228–234.
Omori, M. (1974). The biology of pelagic shrimps in the ocean. *Adv. Mar. Biol.* **12,** 233–324.
Packard, A. (1972). Cephalopods and fish: The limits of convergence. *Biol. Rev. Cambridge Philos. Soc.* **47,** 241–307.
Peaker, M., and Linzell, J. L. (1975). "Salt Glands in Birds and Reptiles." Cambridge Univ. Press, London and New York.
Pearcy, W. G., and Small, L. F. (1968). Effects of pressure on the respiration of vertically migrating crustaceans. *J. Fish. Res. Board Can.* **25,** 1311–1316.
Pennycuick, C. J. (1975). Mechanics of flight. *In* "Avian Biology" (D. S. Farner and J. R. King, eds.), Vol. 5, pp. 1–75. Academic Press, New York.

Pora, E. A., Wittenberg, C., Stoicovici, F., and Rusdea, D. (1955). The biology of *Trachurus trachurus mediterraneus* of the Black Sea. *XI*. Contribution to understanding of the role of the central nervous system in oxygen consumption. *Acad. Repub. Bul. Stiint., Pop. Rom., Sect. Biol. 7:* 633–656.

Pora, E. A., Stoicovici, F., and Wittenberg, C. (1959). La biologie du chinchard de la Mer Noire. Note XII. Contributions à l'étude de la fécondation et de l'eclosion des oeufs en fonction des facteurs du milieu. *Trav. Stn. biol. Mar. Agigea* **4,** 107–110.

Pora, E. A., Abraham, A. D., and Sildan-Rusu, N. (1967). La biologie du chinchard de la Mer Noire (*Trachurus mediterraneus ponticus*). XIV. Action des hormones stéroidigues sur quelques processus du métabolisme intermédiaire. *Mar. Biol.* **1,** 33–35.

Prange, H. D. (1976). Energetics of swimming of a sea turtle. *J. Exp. Biol.* **64,** 1–12.

Pritchard, A. W., Hunter, J. R., and Lasker, R. (1971). The relation between exercise and biochemical changes in red and white muscle and liver in the jack mackerel, *Trachurus symmetricus*. *Fish. Bull.* **69,** 379–386.

Prosser, C. L., ed. (1967). "Molecular Mechanisms of Temperature Adaptation," Publ. No. 84, pp. 351–376. Am. Assoc. Adv. Sci., Washington, D.C.

Quetin, L. B., and Childress, J. J. (1976). Respiratory adaptations of *Pleuroncodes planipes* Stimpson to its environment off Baja, California. *Mar. Biol.* **38,** 327–334.

Randall, D. J. (1970). Gas exchange in fish. *In* "Fish Physiology" (W. S. Hoar and D. J. Randall, eds.), Vol. 4, pp. 253–292. Academic Press, New York.

Renaud, D. L., and Popper A. N. (1975). Sound localization by the bottlenose porpoise, *Tursiops truncatus*. *J. Exp. Biol.* **63,** 569–585.

Renfro, W. C., and Pearcy, W. G. (1966). Food and feeding apparatus of two pelagic shrimps. *J. Fish. Res. Board Can.* **23,** 1971–1975.

Roberts, J. L. (1974). Control of ram gill ventilation: Thermal and hypoxic stresses. *Am. Zool.* **14,** 1258 (abstr.).

Roberts, J. L. (1975). Active branchial and ram gill ventilation in fishes. *Biol. Bull.* (*Woods Hole, Mass.*) **148,** 85–105.

Roberts, J. L., and Graham, J. B. (1974). Swimming and body temperature of mackerel. *Am. Zool.* **14,** 1258 (abstr.).

Rokop, F. J. (1974). Reproductive patterns in the deep sea benthos. *Science* **186,** 743–745.

Sather, B. T., and Rogers, T. A. (1967). Some inorganic constituents of the muscles and blood of the oceanic skipjack, *Katsuwonus pelamis*. *Pac. Sci.* **21,** 404–413.

Sewell, R. S., and Fage, L. (1948). Minimum oxygen layer in the ocean. *Nature (London)* **162,** 949–951.

Shchepkin, V. Y., Shul'man, G. E., and Goncharova, L. I. (1974). Muscle lipid composition in scad, *Trachurus mediterraneus ponticus,* and scorpionfish, *Scorpaena porcus*. *In* "Physiology and Biochemistry of Lower Vertebrates" (E. M. Kreps, ed.), pp. 62–67. Science Press, Moscow (in Russian).

Shul'man, G. E., Sigaeva, T. G., and Shchepkin, V. Y. (1973). [Fat expenditures as an index of the level of active metabolism of the horse mackerel during swimming.] Zatraty zhira kak pokazatel' urovnya aktivnogo obmena stavridy vo vremya plavaniya. *Dokl. Akad. Nauk SSSR, Ser. Biol.* **211,** 1482–1484; *Dokl. Biol. Sci.* (*Engl. Transl.*) pp. 356–358 (1973).

Sleigh, M. A., and Macdonald, A. G., eds. (1972). "The Effects of Pressure on Organisms." Academic Press, New York.

Smith, K. L., Jr. (1973). Energy transformations by the sargassum fish, *Histrio histrio* (L.). *J. Exp. Mar. Biol. Ecol.* **12,** 219–227.

Smith, K. L., Jr., and Hessler, R. R. (1974). Respiration of benthopelagic fishes: *In situ* measurements at 1230 meters. *Science* **184,** 72–73.

Smith, K. L., Jr., and Teal, J. M. (1973). Temperature and pressure effects on respiration of thecostomatous pteropods. *Deep-Sea Res.* **20,** 853–858.

Steen, J. B. (1970). The swim bladder as a hydrostatic organ. *In* "Fish Physiology" (W. S. Hoar and D. J. Randall, eds.), Vol. 4, pp. 414-444. Academic Press, New York.

Steen, J. B. (1971). "Comparative Physiology of Respiratory Mechanisms." Academic Press, New York.

Stevens, E. D. (1972). Some aspects of gas exchange in tuna. *J. Exp. Biol.* **56,** 809-823.

Stevens, E. D., Lam, H. M., and Kendall, J. (1974). Vascular anatomy of the counter-current heat exchanger of skipjack tuna. *J. Exp. Biol.* **61,** 145-153.

Sverdrup, H. U., Johnson, M. W., and Fleming, R. H. (1942). "The Oceans: Their Physics, Chemistry and General Biology." Prentice-Hall, Englewood Cliffs, New Jersey.

Tchnidonova, J. G. (1959). (Feeding of some groups of macroplankton in the northwestern Pacific). *Tr. Inst. Okeanol., Akad. Nauk SSSR* **30,** 166-189.

Teal, J. M. (1971). Pressure effects on the respiration of vertically migrating decapod crustacea. *Am. Zool.* **11,** 571-576.

Teal, J. M., and Carey, F. G. (1967a). Respiration of a euphausiid from the oxygen minimum layer. *Limnol. Oceanogr.* **12,** 548-550.

Teal, J. M., and Carey, F. G. (1967b). Effects of pressure and temperature on the respiration of euphausiids. *Deep-Sea Res.* **14,** 725-733.

Trusevich, V. V. (1974). Utilization of energy-rich phosphates by white skeletal muscles of scad, *Trachurus mediterraneus ponticus,* under conditions of measured muscular loads and fatigue. *In* "Physiology and biochemistry of lower vertebrates" (E. M. Kreps, ed.), pp. 55-61. Science Press, Moscow (in Russian).

Tucker, V. A. (1972). Metabolism during flight in the laughing gull, *Larus atricilla. Am. J. Physiol.* **222,** 237-245.

Vlymen, W. J. (1970). Energy expediture of swimming copepods. *Limnol. Oceanogr.* **15,** 348-356.

Wahrenbrock, E. A., Maruschak, G. F., Elsner, R., and Kenney, D. W. (1974). Respiration and metabolism in two baleen whale calves. *Mar. Fish. Rev.* **36,** 3-9.

Walters, V., and Fierstine, H. L. (1964). Measurements of swimming speeds of yellowfin tuna and wahoo. *Nature (London)* **202,** 208-209.

Waterman, T. H. (1961). Light sensitivity and vision. *In* "Physiology of Crustacea" (T. H. Waterman, ed.), Vol. 2, pp. 1-64. Academic Press, New York.

Wells, M. J. (1966). Cephalopod sense organs. *In* "Physiology of Mollusca" (K. M. Wilbur and C. M. Yonge, eds.), Vol. 2, pp. 523-545. Academic Press, New York.

Wittenberg, J. B. (1960). The source of carbon monoxide in the float of the Portuguese man-of-war, *Physalia physalis* L. *J. Exp. Biol.* **37,** 698-705.

Wittenberger, C. (1968). Biologie du chinchard de la Mer Noire (*Trachurus mediterraneus ponticus*). XV. Recherches sur le métabolisme d'effort chez *Trachurus* et *Gobius. Mar. Biol.* **2,** 1-4.

Wittenberger, C. (1972). The glycogen turnover rate in mackerel muscles. *Mar. Biol.* **16,** 279-280.

Wolvekamp, H. P., and Waterman, T. H. (1960). Respiration. *In* "Physiology of Crustacea" (T. H. Waterman, ed.), Vol. 1, pp. 35-100. Academic Press, New York.

Yuen, H. S. H. (1966). Swimming speeds of yellowfin and skipjack tuna. *Trans. Am. Fish. Soc.* **95,** 203-209.

Functional Adaptations of Deep-Sea Organisms

8

Robert Y. George

I. Introduction to Deep-Sea Biosphere and Biota

The deep-sea biosphere is an extensive and voluminous environment that embraces the ocean floor and the water column from the Continental Slope to the abyssal plain. Despite its enormity, the abyss of the ocean is the least understood segment of Earth's biosphere, particularly in relation to life strategies and functional adaptations of the biota. In fact, there is no clear cut definition for the deep-sea environment since the parameters to delineate its very boundaries have not become universally acceptable to those biologists who are primarily concerned with the ecology and physiology of deep-sea animals. In the deep sea the most striking physical features that exert apparent impact on functional adaptations are high hydrostatic pressure, low temperature, absence of light, diminished food supply, and the constant character of the environment.

A. HYDROSTATIC PRESSURE

Obviously hydrostatic pressure is the most striking factor in the deep-sea environment. An increase of 100 m in depth results in an increase of ambient pressure from 1 to 10 atm. One atmosphere is equivalent to 14.696 pounds per square inch (psi) or 1.033 kg/cm^2. The enormous hydrostatic pressure at trench depths beyond 6000 m is recognized as the principal cause of demarcating an uniquely hyperbaric hadal or ultra-abyssal zone (Belyaev, 1972).

For the purpose of convenience an isobath of 2000 m is often assumed as the upper limit of the deep sea according to the several classification schemes for vertical zones of the marine environment (Hedgpeth, 1957). This arbitrary definition implies that any benthic or benthopelagic regime with ambient hydrostatic pressure in excess of 200 atm is characterized as deep sea. At these depths, abyssal animals adapted to the prevailing conditions are invariably present. Special physiological and biochemical adaptations have evidently occurred in the evolution of these deep-sea organisms, for shallow water organisms exhibit excitory and convulsive reactions of demonstrable neurophysiological implications upon exposure to deep-sea simulation, i.e., to hydrostatic pressure exceeding 100 atm. Euplanktonic organisms are also known to undergo stressful reactions of hyperactivity, convulsions, and paralysis as cogent evidence of physiological trauma induced by increased hydrostatic pressure (George and Marum, 1974). The earlier barobiological studies are summarized by Regnard (1891), Fontaine (1930) and Ebbecke (1935). The recent studies are discussed in a symposium volume on experimental biology of the deep sea (Brauer, 1972).

B. LOW TEMPERATURE

In general, the deep-sea environment is cold, with temperature relatively constant at levels usually less than 5°C. Nevertheless, low temperature per se is not unique to the deep sea since the polar and subpolar shallow marine provinces experience a thermal regime ranging from 5 to −1.8°C. Because of the uniformly low temperature in the deep sea, this biotope is devoid of any seasonal thermal stimuli for triggering any phenomena, such as growth or reproduction, that has been demonstrated in temperate shallow seas.

The deep-sea biota is cold stenothermal and obviously adapted to function within narrow thermal ranges. A definition of deep sea was once put forth by

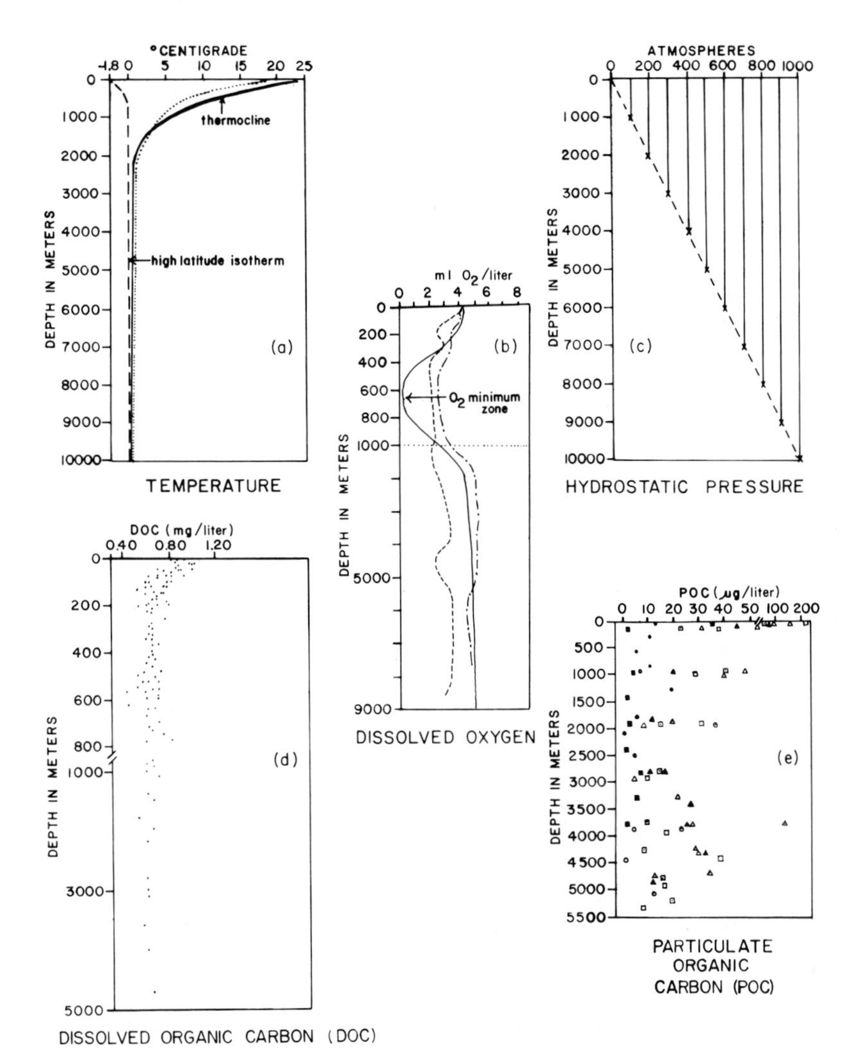

Fig. 1. Vertical profiles of (a) temperature, (b) dissolved oxygen, (c) hydrostatic pressure, (d) dissolved organic carbon (DOC), and (e) particulate organic carbon (POC) in relation to depth. Data derived from various sources to present a generalized trend in the Northwestern Atlantic Ocean. (DOC data from Barber, 1968; POC data from Gordon, 1970).

Bruun (1957) on the basis of 10° and 4°C isotherms. Bruun proposed that the bathyal or bathypelagic zone was at the 10° isotherm, whereas the abyssal or the abysso-pelagic zone was at the 4° isotherm. The vertical distribution of temperature in the ocean is shown in Fig. 1, which also illustrates other factors that may or may not directly influence activities and life strategies of deep sea organisms. Subtle changes in temperature within the various vertical zones of deep-sea environment essentially reflect the nature or origin of the water masses and currents that gently flow over the ocean floor.

There are certain exceptions to the generally conceived notion that the deep sea is always cold. Enclosed ocean basins such as the Mediterranean Sea are indeed known to contain warm water at temperatures exceeding 13°C in depths between 600 and 4000 m. The warm stenothermal conditions of the Mediterranean deep-sea biosphere have also been known to accommodate a number of warm-water bathypelagic and benthic organisms that generally occupy the upper warm layers of the eastern Atlantic Ocean (George and Menzies, 1968b). In the hydrothermal vent areas of the abyss close to Galapagos Islands, warm deep-sea conditions are now known to support opportunistic species that include giant clams (vesico mysid bivalve, *Calyptogena*), vestimentiferan pogonophorans and chemosynthetic bacteria (Karl *et al.*, 1980). Surprisingly high temperature exceeding 40° to 50°C is encountered at depths of 200 m in the Red Sea, where no life is known to exist. These depressions in the Red Sea contain hot brines with salinity as high as 340‰ (Degens and Ross, 1970). Hypersaline condition, coupled with high pressure and high temperature, is certainly not conducive for the presence and proliferation of marine life.

C. LACK OF LIGHT

Solar light is totally absent in the deep sea, and therefore primary production is entirely lacking. Nevertheless, a diverse life is found to exist without depending on light as a source of energy. In this perpetually dark environment a vast number of abyssal inhabitants have only rudimentary visual organs. Others, however, possess normal or enlarged eyes with retinas which have an exceedingly large number of rods of relatively long length. This adaptation permits sensitive perception of any dim transient light originating from bioluminescence. In sightless forms, other tactile organs or structures have attained a high degree of specialization. The prompt response of blind amphipods to baited traps in the abyss is indicative of their efficient chemoreceptive capacities (Dahl, 1979).

The penetration of visible light from the sun-lit upper layer of the ocean varies from low- to high-latitude marine environment. Furthermore, the quality of light tends to vary as the depth increases since water absorbs light rays in varying degrees. Some vague correlation between the light entering the depths and the coloration of the animals appears to exist, but the adaptive value is obscure. As early as 1888, Agassiz, on the subject of "physiology of deep-sea life," pointed out that animal coloration appears not to have been influenced by the presence or absence of light since a diverse array of colors is encountered by animals from euphotic and aphotic zones. However, in the upper limits of the lightless deep sea (depth ranging from 400 to 1000 m), brilliant red, scarlet, or orange tinge tend to dominate the colors of bathypelagic and archibenthal crustaceans and echinoderms, as markedly evident from the numerous observations on deep-sea samples from the northwestern Atlantic Ocean.

Light of biological origin, termed "bioluminescence," is not uncommon among abyssal animals. Although light production among terrestrial or fresh water animals is extremely rare, several marine groups of various phyla are known to emit light, presumably for reasons of adaptive significance. Originally symbiotic bacteria were considered the principal agents for producing light. The anatomical details of the light producing organs in deep-sea fishes (Brauer, 1908) illustrate the special adaptive development of the structures that enable these animals to capture their prey or to recognize their mates in the dark. The light emitted by the bioluminescent deep-sea organisms is devoid of heat and without any rays of infrared or ultraviolet wave length. The production of the light is caused by the oxidation of a proteinaceous substance, called luciferin, which is formed by living cells. Another substance, luciferase, is also present as a catalyst to promote or accelerate the oxidation. The detailed biochemical mechanism of the light generated by a variety of abyssal fishes, crustaceans, and coelenterates is still not fully understood, primarily because of the complexities involved in studying living individuals from the deep sea.

D. OXYGEN

Dissolved oxygen in the cold abyssal depths appears to be near saturation levels (6.3–6.4 ml/liter) in deep-sea zones that are bathed by currents of varying velocities. However, some ocean depths are known to be in zones of tranquillity or stagnation where oxygen levels reach low concentrations or even anoxic situations, such as in the extreme depths of the Black Sea or Cariaco Trench in the Caribbean. These anoxic zones are also azoic. In the Black Sea, for example, the waters at depths of 2000 m are precluded from any mixing through convection currents due to the thermal stratification and freshwater flow. The deep-sea environment in the Black Sea is inhabited only by anaerobic bacteria. Free oxygen in this abyss is completely used up by the decomposition of the abundant organic material that accumulates on the sea floor from the productive surface zone. As a result hydrogen sulfide is formed over the sea floor from 200 to 2000 m. Only the sulfide-reducing anaerobic bacteria are adapted to dwell in these toxic abyssal conditions.

Along the eastern edge of the Pacific Ocean, off California and Peru, oxygen minimum zones occupy significant areas of the ocean down to 800 m. A special circulatory adaptation of bathypelagic mysids inhabiting the oxygen minimum layer has been the subject of a recent study by Belman and Childress (1976). The dissolved oxygen level in this O_2 minimum zone is as low as 0.5 ml/liter, but these crustaceans have been demonstrated to be efficient in utilizing oxygen at this low tension. The studies on the respiratory physiology of the deep-dwelling mysid, *Gnathophausia ingens,* adequately document the ability of these crustaceans to generate rapid water flow over the gills at rates as high as 8 ml/g wet wt/min, while removing up to 80% of the oxygen during this process (Childress, 1971b). Anatomically these mysids have proportionately enlarged heart and arteries in contrast with similar crustaceans from oxygen-enriched regions. In terms of blood velocity, the cardiac output and the turnover time seem to be extremely effective. This physiological manifestation is evidently a functional adaptation to derive maximum oxygen in the zone of low oxygen tension.

Although irregularities in distribution of dissolved oxygen are often encountered because of a variety of factors, the deep sea is sufficiently supplied with the oxygen necessary for sustaining normal metabolic activities. Experimental data on respiration of deep-sea animals, based on studies of abyssal animals under simulated deep-sea conditions (Teal, 1971; Teal and Carey, 1967; Childress, 1971a,b; George, 1979a) and on a few *in situ* measurements (Smith and Hessler, 1974), provide information about the low metabolism of deep-sea organisms. Therefore, the depletion of oxygen by the low biomass of the deep-sea biosphere should be relatively insignificant in comparison with the shallow water biosphere of greater biomass and higher metabolic activity. The low metabolic activity of deep-sea organisms is interpreted as a physiological adaptation to conserve energy in a nutritionally impoverished abyssal environment.

E. FOOD

The diminished food supply and a drastic reduction in total biomass is the deep-sea biosphere are evidently consequences of a food-chain that does not incorporate the primary producers. The fundamental source of input of organic matter into the abyss is located outside the boundaries of the deep sea, either in the shallow euphotic zones or in the upper sun-lit layers of the open ocean where the total organic matter oxidized in respiration is far less than what is synthesized. The organic matter in the deep sea can occur in many forms, such as dissolved organic matter, particulate substances in suspension, bacterial populations, planktonic or bathypelagic animals, and benthic and benthopelagic animals. Organic matter enters the abyss as a continuous rain of the remains of corpses, excreta, and metabolites through the water column or from the adjacent slope through turbidity flows. A careful analysis of the energy budget of the deep-sea biosphere may reveal that dissolved organic material constitutes a sizeable portion of metabolic needs of abyssal animals, which are particularly well-adapted to utilize this source of food by means of specialized heterotrophic processes. A perusal of the dominant feeding types among deep-sea benthos (Sokolova, 1959, 1972) demonstrated that the detritophagous mode of feeding is the chief method of gathering food among the mud-dwelling animals. Heterotrophism is also reported for the deep-sea pogonophorans (Southward and Southward, 1968, 1970). However, filter-feeders as well as carnivorous animals are also represented in the spectrum of deep-sea life. Although the standing stock of the deep-sea biosphere is low (0.01–1.00 g/m^2) the relative diversity of marine life in the abyss is found to be remarkably high (Sanders, 1968; Dayton and Hessler, 1972).

F. CONSTANCY

Since deep sea hydrographic conditions are probably seasonless, annual rhythms in behavior, growth, and reproduction commonly observed in temperate environments should be absent in the abyss. On the contrary, calendar events such as peaks of reproductive activity or seasonal breeding in deep-sea biosphere have come to light in the course of studies during the past decade (George and Menzies, 1967, 1968a; Schoener, 1968; Rokop, 1974; George and Nesbitt, 1980). Moreover, the question of biotime in deep-sea life, with special reference to longevity or age of deep-sea organisms, is an extremely interesting problem in view of the contention that a deep sea clam has been found to have a longer life and slower growth rate than shallow water bivalves, as determined by ^{228}Ra chronology (Turekian *et al.*, 1975). From a physiological standpoint, the problem becomes singularly important because of the prevailing hypothesis that the low metabolism of deep-sea organisms in an environment of constant low temperature and high hydrostatic pressure has resulted in reduced growth rate, delayed maturity, and prolonged life. It should be noted that there are few quantitative measurements from experimental data to support this hypothesis.

G. DEEP-SEA ZONATION

As an introduction to the discussion of different physiological processes, a prelude is presented pointing out generalities in respect to the types of biota that are now known to live in the different zones of the deep-sea biosphere. The very fact that vertical zones are distinguishable within the deep-sea biosphere provides indirect evidence of adaptations of organisms within narrow depth limits or in restricted ranges of hydrostatic pressure. In an elaborate investigation of abyssal faunal zones in different geographic regions (Menzies *et al.*, 1973), discrete bathymetric limits were established for deep-sea zones on the basis of species composition. The recognized deep-sea zones are the upper abyssal zone over the continental slope, the mesoabyssal zone at the junction of slope and rise, the lower abyssal zone over the continental rise, and the lower abyssal red clay zone over the abyssal plain. Similarly Belyaev (1966) established three distinct "hadal" or ultra abyssal zones in trenches between depths of 6000 and 10,915 m. At present, data are not available on the pressure tolerance of these deep sea inhabitants to levels higher or lower than those of the natural environment. In contrast, eurybathial species with wide vertical depth ranges are known in the deep sea. Recent studies on the behavioral responses of the deep-sea eurybathial decapod *Parapagurus pilosimanus* demonstrate that individuals from 800 to 1000 m show convulsive reactions when maintained at 220 atm for 5 days in the high pressure aquarium. Since this deep-sea crab is distributed down to 5000 m, more than twice the convulsive threshold pressure for populations from 1000 m, the data provide clues as to physiological variations between populations inhabiting different depth ranges (R. Y. George, unpublished data). Further studies are necessary to test this concept of vertical physiological races in the deep-sea environment. Latitudinal physiological races are recognized in many other marine environments (Vernberg and Vernberg, 1972).

Two distinct types of animals dwell in the deep-sea biosphere, one type occupying the various depths of the water column and the second type inhabiting the soft sediments over the sea floor. The midwater animals are generally termed as pelagos and the bottom-living organisms are termed as benthos.

H. DEEP-SEA PELAGOS

The deep-sea pelagic life can be further subdivided into mesopelagic, bathypelagic, and hadopelagic, depending on the depth of occurrence as illustrated in the accompanying scheme of deep-sea zones (Fig. 2). Pelagic organisms are usually good swimmers, with somewhat stream-lined body configurations and paired appendages or fins for buoyancy or swift propulsion in the water. Fishes, squids, and a variety of crustaceans are typical examples of pelagic animals that live in the deep-sea water column. Apart from macroscopic animals, microscopic life prevails throughout the water column down to the trench floor. Depending on the size, these pelagic microorganisms are termed as microplankton (cells of 250–500 μm), nannoplankton (10–15 μm), and ultraplankton (0.5–10 μm). Apparently these microorganisms in the mid-water environment play a significant role in the overall energy flux since they have appreciable metabolic activity. Bathypelagic deep-water bacteria have been isolated from the abyssal waters of the Indian Ocean (Johnson *et al.*, 1968). As a matter of fact, these authors considered certain strains of *Bacillus* and *Spirillum* endemic to the deep-sea environment. Kriss (1960) reported the abundance of bacteria in boundary layers between water masses of different density as a consequence of increased organic matter. However Sieburth (1971) presented different views re-

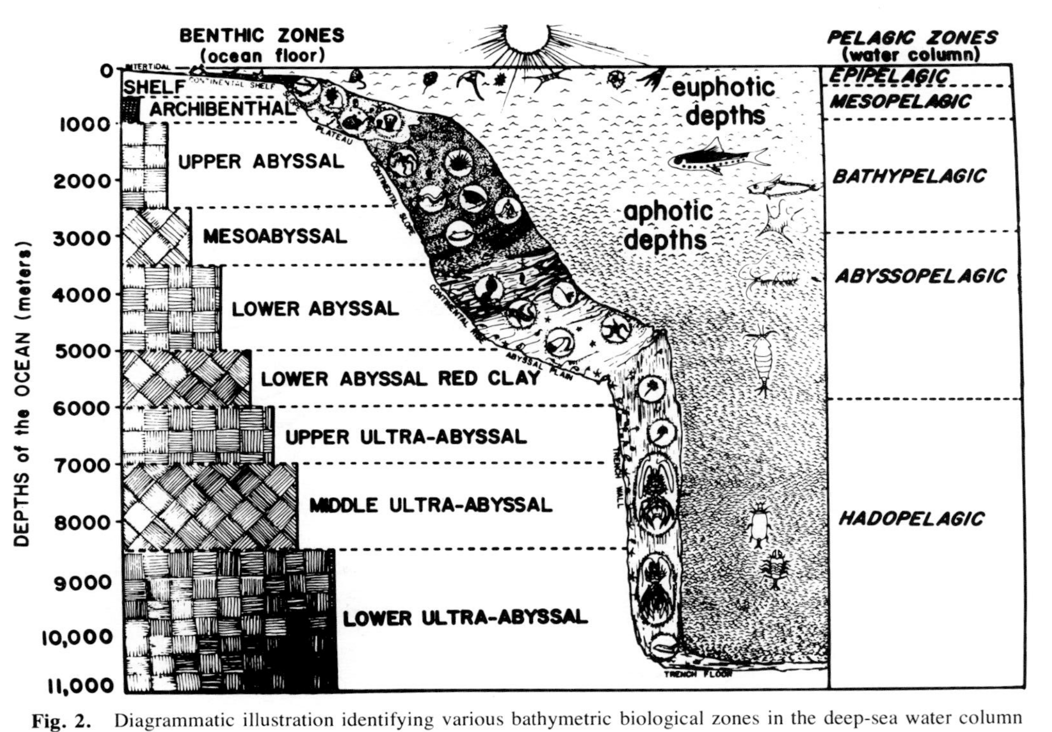

Fig. 2. Diagrammatic illustration identifying various bathymetric biological zones in the deep-sea water column and over deep-ocean floor.

garding the distribution and activity of oceanic bacteria. In addition to bacteria, unicellular flagellates have been reported throughout the aphotic zone below 1000 m down to 5000 m in the North Atlantic (Fournier, 1966). These flagellates in the deep-sea water column exist as a viable and actively metabolizing community.

In addition to the finding of flagellates in deep water, Hamilton *et al.*, (1968) discovered very high concentrations of spheric cells at depths between 1000 and 3000 m, with peak densities as high as 1–60 × 10^3 cells. This density was based on direct microscopic counts of filtered cells. Analysis of deep-sea water samples revealed ATP levels indicative of concentrations of living carbon three times as high as the values of direct counts. These findings clearly show the presence and abundance of pelagic microorganisms. The deep-sea biosphere is also known to support marine fungi (Kohlmeyer, 1968). The marine yeast *Rhodosporidium malvinellum* has been found commonly at depths of 4000 m (Fell, 1970).

The deep-sea water column supports a variety of zooplankton down to the hadal depths of the trenches (Vinogradov, 1962). There is, however, a direct relationship between decline of plankton biomass and increase in depth (Fig. 3).

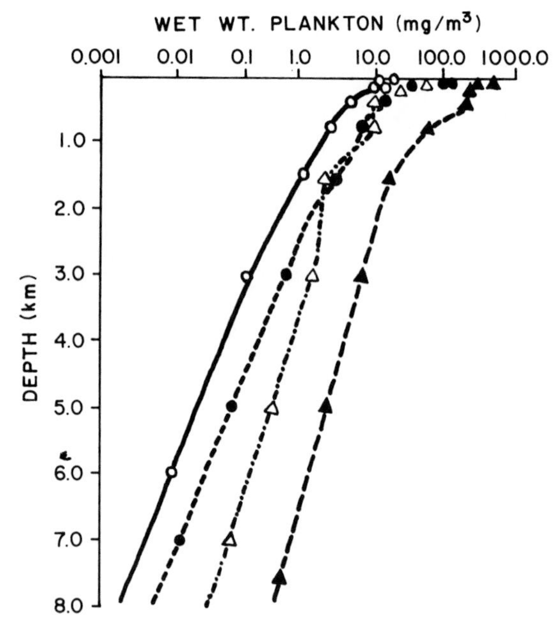

Fig. 3. Vertical distribution of plankton biomass in deep-sea trenches. Kurile-Kamchatka Trench (○——○); Kermadec Trench (●- - - -●); Bougainville Trench (△-·-·-△); Mariana Trench (▲- - - -▲). (From Vinogradov, 1962).

Furthermore, there appears to be profound seasonal variations in biomass and plankton even at a depth of 2000 m (Menzel and Ryther, 1961). The patchiness seen in the distribution of plankton in the deep sea presents problems pertaining to quantification of their relative density.

I. DEEP-SEA BENTHOS

The benthic animals inhabiting the deep-sea floor can also be subdivided into epifauna living on the sediment and infauna living within the sediment. The epifauna includes three different types of organisms; the swimming-predatory megabenthos (e.g., fishes, amphipods, decapods), the attached sessile megabenthos (e.g., crinoid, sea pen), and the sedentary megabenthos (asteroid, ophiuroid, holothurians). The infauna includes the macrobenthos (polychaetes, isopods), the meiobenthos (0.04–1.0 mm in length, e.g., nematodes, kinorhynchs, copepods), and the microbenthos (bacteria). The size structure of the deep-sea benthos was the subject of a recent report by Thiel (1975).

The predatory megabenthos such as the abyssal amphipods, *Eurythenes gryllus* (George, 1979a) and *Orchomene* sp. (Shulenberger and Hessler, 1974), are abundant in the deep-sea biosphere. These amphipods evidently play an important role in the trophic structure of deep-sea community. Such predators must possess very sensitive chemoreceptors, which aid them to detect bait or dead bodies of large animals such as whales and giant squids on which these scavengers feed. The epifauna, sedentary or sessile organisms, are sparsely distributed over the abyssal plain or in the slope environment of low surface productivity (Fig. 4). On the other hand, the distribution of sedentary epifauna attains peak density and biomass in deep-sea floor in canyons (Fig. 5) as well as on trench floor where favorable feeding conditions prevail because of the accumulation of enriched organic matter from turbidity currents. For example, the floor of the eutrophic Banda Trench at depths of 6580–7270 m supports a biomass as great as 10 g/m^2. Rich macrofauna in association with wood and plant material are also known from hadal depths of the Gilliss Deep within the Puerto Rican Trench (George and Higgins, 1979). The floor of the oligotrophic Cayman Trench, however, supports a biomass as low as 0.01 g/m^2 (George and Wright, 1977). Despite the diminished biomass, the deep-sea benthic environment exhibits a high species diversity. It is not uncommon to find as many as eight species of a given genus even in a single sample from the deep-sea biosphere. This situation is encountered often in the case of infaunal animals in the deep-sea bottom sediments of *Globigerina* ooze or the red clay of the abyssal plain. This is perhaps indicative of higher rate of speciation in the deep sea biosphere as opposed to that of shallow seas.

Thus far we have dealt primarily with the environmental background and the

Fig. 4. Deep-sea bottom photograph of the Continental Slope Environment (2600 m) off New Jersey, northwestern Atlantic Ocean. Note the occurrence of ophiuroids, echinoids, the deep-sea chimaerid fish *Hydrolagus affinus,* and some signs of bioturbation.

various types of biota inhabiting the deep-sea biosphere. In the following pages special focus will be placed on the functional adaptations of deep-sea organisms on the basis of our limited knowledge of this subject. Whenever possible an effort is made to emphasize areas of study that will enhance our knowledge toward a better understanding of the functional adaptations of the deep-sea organisms. The different aspects of deep-sea biology discussed in this section include nutritional modes, respiratory adaptations, circulatory patterns, excretion, reproductive strategies, molecular adaptations, perception of the environment and an overview on the state of knowledge to emphasize gaps and significant areas that deserve special studies.

Fig. 5. Bottom photograph showing peak density of epifauna in the Hudson Canyon at 2100 m in the northwestern Atlantic Ocean. Note the abundance of the abyssal holothurian *Peniagone willemoesia* and the few individuals of *Echinus affinus*. Total area of photo approximately 2.85 m^2. (Photo courtesy of Naval Research Laboratory, W. L. Brundage.)

II. Digestive Dynamics and Energetics

The energy required for life processes largely depends on food input to support two fundamental needs, the first directed toward synthetic processes of production and the second involving supply of sufficient energy for a whirlpool of activities that constantly go on within the organism. If food availability is limited, both the synthetic production and level of activity should be reduced. The general scarcity of good in the deep-sea environment is reflected in the extremely low biomass or standing stock which at depths greater than 3000 m ranges between 0.01 and 1.0 g/m^2 (Filatova, 1960). Various adaptations have evolved which allow abyssal animals to cope with the nutritional limitations of their environment. One such adaptation appears to be the low rate of oxygen consumption of deep-sea animals. Oxygen uptake during respiration provides the fuel necessary for growth, repair work, reproductive process, excretion, secretion of digestive glands, or production of hormones for regulating vital physiological activities. The work performed seems to be at a very low tempo and, if this metabolic depression is the ''norm'' for deep-sea organisms, the food limitation is perhaps a casual factor, despite the impact and interaction of other factors such as high hydrostatic pressure and low temperature. Two distinct aspects of the digestive dynamics of deep-sea animals, i.e., the source of food supply and the types of digestive adaptations are considered below on the basis of the limited information available.

A. SOURCES OF FOOD IN THE DEEP SEA

Although there is no primary production in the abyssal biotope, some material that is synthesized as organic material in sun-lit environments does reach the deep sea. In regions isolated from the continental margin, such as the mid-oceanic abyssal plains at depths exceeding 5000 m, the deposition of organic material from land or shallow water is almost negligible because of their location beyond the reach of any turbidity flow. Furthermore, the rain of dead plankton or nekton is also dissipated well before reaching these abyssal plains. Such poor organic regimes are characterized as oligotrophic deep-sea environments (Sokolova, 1972). In the oligotrophic regions, the organic matter available as food on the sea floor consists primarily of insoluble residual substances. The amount of particulate and dissolved organic carbon in the water column is low, and the sediment contains meager concentrations of organic carbon, i.e., less than 0.3% of dry weight (Sokolova, 1972). High positive values of redox potential (+600 Mv) have been reported, and there is a thick oxidized upper layer (Bezrukov, 1955). The sedimentary bed is composed of eupelagic and ceolithic clay. In oligotrophic situations, bottom-dwelling organisms are adapted to derive their food largely from the small quantities of suspended food particles as filter-feeders. The organic matter in the sediment is decomposed and has little digestive value; the feeding conditions consequently are not favorable for deposit-feeding animals (Carey, 1972). Along the continental slope or beneath the productive equatorial oceanic regions, food availability for deep-sea animals is relatively greater. Organic matter accumulates rapidly and remains buried in the sediments in a nutritionally utilizable state. These deep-sea environments are characterized by relatively high sediment concentrations of organic carbon (0.5–1.5% dry weight), neutral or reducing sediments (E_h ranging between +70 and −300 mV), a thin or negligible upper oxidized layer, and biogenic sediments composed of diatomaceous, radiolarian, or foraminiferal oozes. In these "eutrophic" areas of the deep sea, a dietary layer of organic matter accumulates in depressions or on flat areas of the ocean floor such as may be found in the trenches. The favorable feeding conditions induced by organic enrichment promote the proliferation of deposit-feeding organisms. The source of food for deep-sea animals includes one or more of the following categories: (1) benthic and benthopelagic populations; (2) deep-sea zooplankton; (3) midwater animals; (4) abyssal bacterial populations; (5) particulate organic carbon (POC); and (6) dissolved organic carbon (DOC).

B. BENTHIC AND BENTHOPELAGIC POPULATIONS

In terms of their mode of feeding, deep-sea animals can be classified as carnivores or omnivores, filter feeders, and deposit feeders. Benthopelagic crustaceans are also cannibalistic, with adults often feeding on juveniles. These carnivores derive their carbohydrates either from the protein or fat that constitutes a significant composition of the prey. The energy stored in animals is distributed in inert skeletal material that is not utilizable as food (e.g., calcium carbonate shells of mollusks, cirripeds, and corals), and it is also distributed in the soft parts of the body in the form of proteins, carbohydrates, lipids, or minerals of nutritive value. When an organism dies in the cold and hyperbaric deep-sea environment, the soft body components constitute a source of food for scavengers, such as the abyssal fishes, amphipods, and shrimps. The abundance of these organisms in the abyss is well documented by the recent time-lapse photography of these scavengers, which are attracted toward baits because of their sensitive chemoreceptive adaptations. In the central part of the Arctic Ocean, over 800 individuals of the abyssal amphipod, *Eurythenes gryllus,* were attracted to a baited trap in less than 1 h at 1800 m (George, 1979a). Scavenging amphipod *Hirundellea gigas* tends to crowd around the bait in a very short time and secure a substantial food supply (Fig. 6). The absence of sexually mature adult amphipods in the baited traps suggested the hypothesis that "in scavengers such as these one large meal might be sufficient to carry a juvenile into reproductive maturity" (Shulengerger and Hessler, 1974). In the food-limited deep-sea

Fig. 6. Photograph showing the abundance of scavenger amphipod *Hirundellea gigas* around a baited trap at a depth of 9604 m in the Philippine Trench, Lat. 10° 36.3′ N, Long. 126° 38.1′ W, about 12 hr after lowering of bait. (Photo courtesy of Dr. Robert R. Hessler, Scripps Institution of Oceanography.)

environment, these omnivorous animals may utilize a single meal for prolonged periods in their life history. The efficient utilization of intermittently available food over an extensive time period is perhaps a special digestive adaptation of deep-sea animals. Similarly, some deep-sea fishes are equipped with highly extensible jaws and therefore can swallow a sizable amount of food during a single meal, which may be utilized over an extensive period. Most often the stomach of captured deep-sea fishes is either empty or packed fully with food.

The most dominant trophic mode among deep-sea animals is deposit-feeding or detritophagy. This mode of food gathering from the mud or sediment is encountered in a variety of groups, including bivalves, isopods, tanaids, polychaetes, holothurians, ophiuroids, asteroids, and even benthic fishes, as revealed in analysis of gut contents. In the abyssal bivalve, *Abra profundorum,* a considerably enlarged hindgut storage area is found to accumulate organic matter in the digestive tract and possibly extract nutrients from this reservoir with the aid of bacterial commensals (Allen and Sanders, 1966). In a recent study on the deep-sea anemone, *Actinauge rugosa,* it was revealed that special anatomical modifications have occurred in the pedal disc with the development of an "anal" opening (R. Y. George, unpublished). Anemones in shallow environments are invariably carnivorous and always occur on a hard substrate as a sessile organism. This deep-sea anemone inhabits a soft bottom of the pteropod or globigerina ooze. With the aid of the "annal" opening in the pedal disc, the anemone accumulates a large mud ball in the proximal part that becomes bulbous (Fig. 7). Under pressures of 1 atm in the laboratory, the deep-sea anemone expels the mud ball and tends to accummulate a mud ball on exposure to an organically rich soft mud bottom. Does the anemone derive any nutrients from the mud? Is the mud ball used as a physical ballast? These are questions that still remain to be answered. Nonetheless, the anatomical shift for accommodation of a large size mud ball in the deep-sea anemone is an adaptation that may possibly be linked to an altered mode of digestion.

The sestanophagous or filter feeders in the deep-sea environment are relatively rare in comparison with detritophages or carnivores. This mode of feeding is seen in the abyssal cirriped, *Scalpellum,* which has a well-developed basket-shaped crown of cirri and a movable peduncle for trapping suspended food particles that are primarily borne in bottom currents. The sessile deep-sea brachiopods, the stalked crinoids, and perhaps the stalked sea pen, *Umbellula,* are other examples of filter feeders in the deep sea. One exception to the rule is the bryozoan, *Kinetoskias,* which is associated with a hyaline stalk that is partially buried in the soft sediment. This bryozoan bends the stalk to enable zoaria to reach the bottom sediments for possible feeding (Menzies *et al.*, 1973). Here again we encounter a shift from the normal filter-feeding mode of bryozoans to the sediment-feeding adaptations of the deep sea.

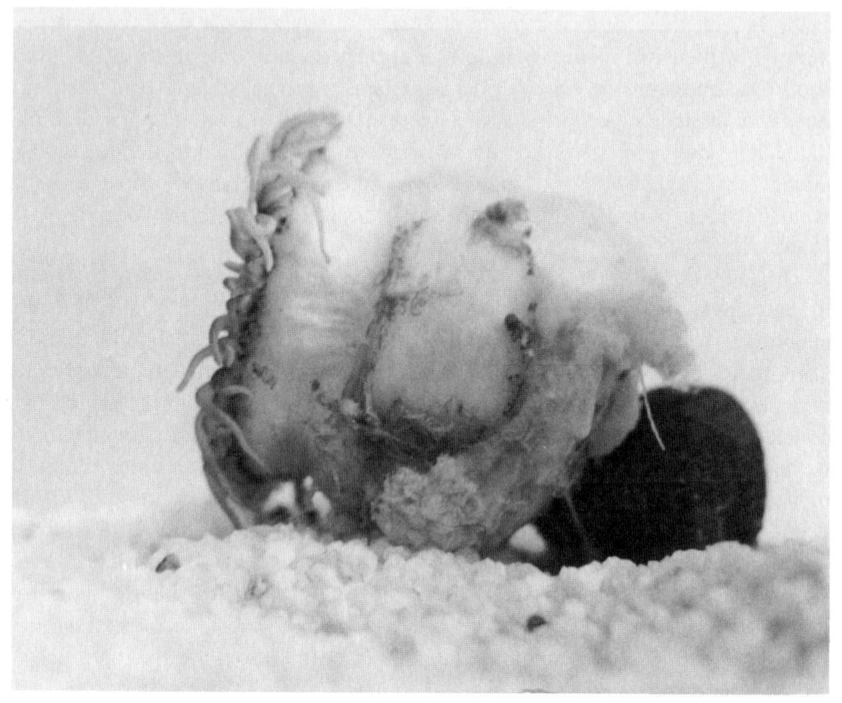

Fig. 7. Deep-sea anemone, *Actinauge rugosa*, from the Northern Blake Plateau off North Carolina. Note the large "mud ball" expelled through the "pedal opening" of the anemone.

C. DEEP-SEA ZOOPLANKTON

It has been suggested that deep-sea zooplankton obtain food through an unique phenomenon of "ladder of migration." This theory was originally postulated by Vinogradov (1962) to imply that an overlapping chain or ladder of vertical migration ranges of planktonic animals offers an ideal route of transfer of energy from the euphotic upper layer of the oceans to the depths of the abyss. Presumably this mechanism of vertical food transport is far more efficient than the rain of plankton and its products from the surface to the sea floor.

It is known that some of the planktonic species of copepods exhibit a very wide bathymetric range. For example, *Spinocalanus abyssalis* and *S. magnus* occupy a depth range of 200–5000 m in the north Atlantic Ocean (Grice and Hulseman, 1965). There are also endemic deep-sea copepods with a restricted depth distribution within the abyss. The abyssal copepod, *Lucicutia curta,* exhibits a vertical distribution between 4000 and 5000 m. Beyond the abyssal depths, the ultra-abyssal copepod *Parascaphocalanus zenkevitchi,* occupies the deep trenches of the ocean to depths as great as 8500 m. Along the steep vertical ladder of the water column there exist zones where populations from minimum depths of the deeper zones mingle with populations from maximum depths of upper zone, thus permitting confluence in terms of energy exchange. Therefore, the presence of a vertical plankton ladder, however meager in biomass, constitutes not only a source of food for predators, but also a mode of transfer of energy to the abyssal and ultra abyssal zones.

D. MIDWATER MACRO-ORGANISMS

Midwater animals, such as the shrimp, the squid, and bathypelagic fishes, seem to occur at various depths, but quantitative data are lacking on the bulk of their biomass and their relative role in the deep-sea food chain. Very often the bottom sediment contains fish otoliths and squid beaks, suggesting that the corpses of these animals contribute to the organic enrichment of the deep sea biotope. There is some interesting information on the chemical composition of midwater animals which suggests that body water content varies from minimum values of 63% to as high as 95%. This variation is generally the reflection of the buoyancy characteristics of the animals. The low water content is usually encountered in the vertical migratory fishes, which also have a high lipid content. The high water content is found in the nonmigratory deep-living fishes (Blaxter *et al.*, 1971; Childress and Nygaard, 1973, 1974). Deep-sea midwater crustaceans are also known to have a body water content as great as 95%. In terms of food value, high water content of midwater animals does not provide sufficient caloric input in trophic transfer. However, there is an obvious adaptive significance since the body fluids are considerably less dense than sea water, which results in positive buoyancy in a variety of midwater animals, including squids (Denton *et al.*, 1969) and fishes (Denton and Marshall, 1958).

The chemical composition of midwater fishes and crustaceans seems to show depth-related trends. The vertically migrating myctophid fishes possess high caloric contents and hence are presumably important in the trophic organization of the deep sea as transporters of energy to deep depths. Lipid is apparently the chief energy store in these midwater animals. Apparently highest levels of lipid seem to occur in vertical migrators, with peak values as high as 50% of the ash-free dry weight of the body (Culkin and Morris, 1970). Low levels of lipid have been reported in deeper living species; these animals generally contain a gas-filled swim bladder and have lipid levels lower than 50% of the ash-free dry body weight. A definite trend of decline of lipid and also caloric value with increasing depth of occurrence was established by Childress and Nygaard (1973) in midwater fishes, as illustrated in Fig. 8. These authors also pointed out a decline of protein with increasing depth of occurrence in both midwater fishes and crustaceans. This decrease with depth was attributed to a general reduction of muscle tissue (Blaxter *et al.*, 1971). The deep-dwelling species are almost neutrally buoyant, and have reduced skeletons and high water content.

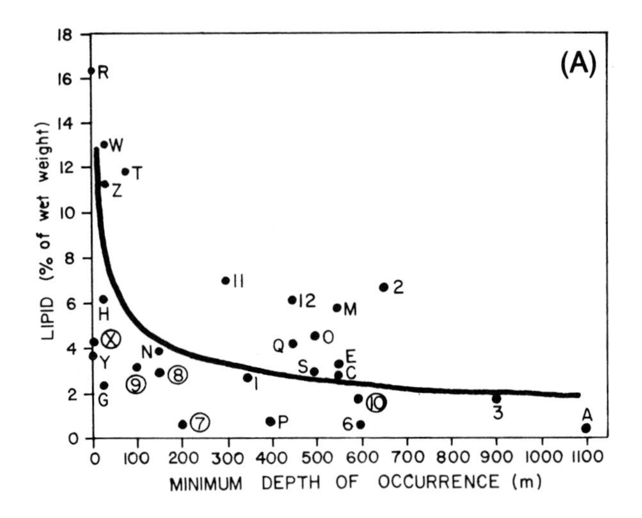

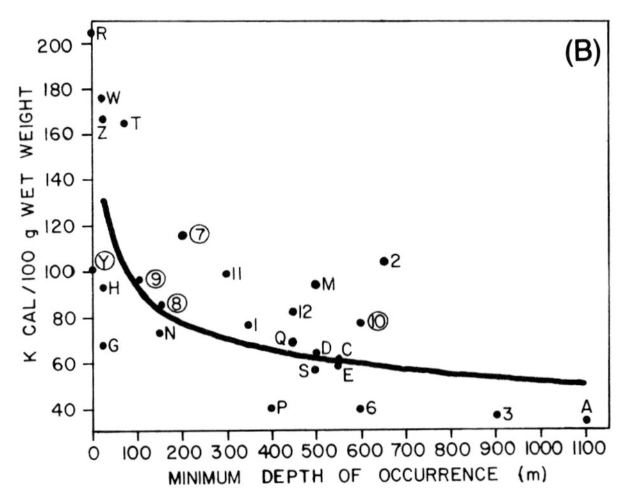

Fig. 8. (A) Lipid level and (B) caloric content as a function of depth of occurrence in midwater fishes: A, *Talismania bifurcata;* C, *Anoplogaster cornuta;* D, *Borostomias panamensis;* E, *Bathylagus milleri;* G, *Bathylagus wesethi;* H, *Leuroglossus stilbius;* M, *Cyclothone acclinidens;* N, *Idiacanthus antrostomus;* P, *Poromitra crassiceps;* Q, *Scopelogadus mizolepis;* R, *Diaphus theta;* S, *Lampanyctus regalis;* T, *L. ritteri;* W, *Stenobrachius leucopsaurus;* Ⓧ, *Symbolophorus californiensis;* Ⓨ, *Tarletonbeania crenularis;* Z, *Triphoturus mexicanus;* 1, *Avocettina* sp; 2, *Scopelengys tristis;* 3, *Oneirodes acanthias;* 6, *Sagamichthys abei;* ⑦, *Argyropelecus affinis;* ⑧, *A. lychnus;* ⑨, *A. sladeni;* ⑩, *sternoptyx obscura;* 11, *Stomias atriventer;* 12, *Melanostigma pammelas.* (From Childress and Nygaard, 1973).

Midwater animals are mainly omnivorous or carnivorous. In view of the fact that they spend their life history within the water column, a rather simple food chain operates for midwater organisms, which is distinct from that of benthopelagic animals. Our knowledge on the adaptive strategies of deep-sea midwater biota is presently very limited, and further studies on nutritional adaptations are necessary to increase our understanding of this special mode of deep-sea life.

E. ABYSSAL BACTERIAL POPULATIONS

It is a well-documented fact that bacteria prevail in the deep-sea water column without any depth-related trend in biomass (Johnson *et al.*, 1968; Jannasch and Wirsen, 1973). However, there appear to be higher concentrations of bacteria in boundary zones where water masses of two different densities come into confluence, and this increase in bacterial biomass is attributed to accumulation of organic material at the interface of two different water masses (Kriss, 1960). In the deep ocean floor, from surface sediment down to 100 ft core depth, bacterial density with values as high as 38 million bacteria per gram of deep-sea sediment has been reported from the Pacific Ocean. Barophilic bacteria have also been known to occur in the extreme depths of the deep-sea trenches (ZoBell and Morita, 1957). More recently, cultures of barophilic bacteria (*Spirillum*) were isolated from amphipods trapped and retrieved at *in situ* pressure conditions from 5700 m (Yayanos, 1978; Yayanos *et al.*, 1979). These organisms exhibited increased growth rate and generation time of 4–13 hr at *in situ* pressure (570 atm) at 2°–4°C, and they showed reduced growth rate and a generation time of 3–4 days at 1 atm pressure at 2°C. Bacterial numbers in the sediments of the oligotrophic regions of deep-sea red clay and globigerina ooze were found to range from 10 to 100,000 bacteria per gram of the sediment (Morita and ZoBell, 1955). Evidently the wide variation in bacterial count is a reflection of organic content of the deep-sea sediment.

Do the detritophagous or deposit feeding macroorganisms utilize bacteria from the deep-sea sediments as a dietary source? Recent studies indicate that the bacteria in the intestine of deep-sea fish and invertebrates are viable and metabolically far more active than microbes inhabiting the water and sedimentary environment of these abyssal animals (J. Oliver personal communication; Schwarz *et al.*, 1975). The precise role of deep-sea bacteria in the abyssal food chain requires detailed elucidation to place the interactions between microbial populations and macroorganisms in proper perspective.

F. PARTICULATE ORGANIC CARBON (POC)

An important dietary source for deep-sea organisms is comprised of the nonliving minute flakes and fragments of organic matter that largely result from breakdown of material originating from slumping and mixing processes. The metabolites and excreta of animals contribute to this category of organic matter. The endproducts of carcass disintegration also constitute considerable quantities of organic particles. According to Williams (1971), particulate organic carbon in the deep sea is rather insignificant, since POC constitutes only about 2% of the total organic material. In studying the vertical distribution of POC in the temperate northern Atlantic Ocean, Gordon (1970) reported 5–35 mg/liter between 4000 and 5500 m with some values exceeding 100 mg/liter at approximately 4000 m. Analysis of the chemical composition of POC illustrates that there is a total lack of lipid in this source of food. However, protein and carbohydrates are found in the various types of particulate matter. Deep-sea benthic animals that are equipped with a filter-feeding mode of eating, such as the cirriped, *Scalpellum,* probably utilize the suspended organic particles as a dietary source.

G. DISSOLVED ORGANIC CARBON (DOC)

The bulk of organic carbon in the marine environment occurs in the form of dissolved organic carbon (Sutcliffe, *et al.*, 1963), and apparently the deep sea is no exception. Why do we find rich quantities of organic matter in dissolved form? From what does the DOC originate? What is the role of DOC in the digestive dynamics of deep-sea animals? These are pertinent questions related to the nutritional adaptations and bioenergetics of abyssal animals.

The immense volume of deep-sea water, in more or less constant motion induced by current flows, promotes dissolution of substances released by living animals in excretory process or emittance of clouds of material during bioluminescence. In addition, substances such as amino acids and polypeptides are also contributed from decaying organic tissue. In a study of the ATP concentration in the deep-sea water column of the north Atlantic Ocean, it was found that there is a six-fold increase in ATP levels at the sediment water interface at depths of 4165 and 6339 m (Karl *et al.*, 1976). Dissolved organic carbon also showed a two-fold increase at the deep-sea sediment water interface. Clearly the DOC shows peak levels in the upper photic zone, with concentrations ranging from 0.5 to 1.2 mg/liter. Nevertheless, the DOC is apparently ubiquitous throughout the water column down to depths of about 5000 m, where it is approximately 0.5 mg/liter. Barber (1968) hypothesized that the dissolved organic carbon from the abyssal depths resists microbial oxidation. This hypothesis hinges on the determination of the precise role of barophilic bacteria and their power of degradation in the hyperbaric conditions of the deep-sea environment. Despite the implications of microbial action, the available data do suggest the significant occurrence of DOC in the abyss. New evidence is also emerging on the nutritional role of DOC and its utilization by deep-sea animals that are adapted to derive food by heterotrophic processes.

In elucidating the original theory of August Pütter, Johnston (1972) focused on the importance of DOC as an unique food source for a variety of organisms including gut parasites and bacteria. An extraordinary example of heterotrophism among deep-sea animals can be seen in the case of the pogonophoran, *Siboglinum*. Anatomically these wormlike organisms are completely devoid of a gut, nor do they possess a mouth or an anus. In the absence of a normal alimentary tract, how do they accomplish the intake of food? The body wall of pogonophorans is adapted to derive nutrients as dissolved organic matter from the ambient environment. The body wall is composed of an epidermis surrounded by a thin cuticle, and the epidermis consists of secretory cells. The uptake of nutrients through the body wall of pogonophorans was recently investigated by Little and Gupta (1970). They monitored the uptake of labelled amino acids, [^{14}C] phenylalanine and [^{14}C] glycine, in the continental slope species of pogonophoran, *Siboglinum ekmani,* and they demonstrated the uptake of these labelled amino acids through the body wall. Similar studies on the permeability of the body wall of a deeper species, *Siboglinum atlanticum,* recovered from a depth of 1800 m, was carried out with labelled amino acids and glucose at 1 atm and 6°C (Southward and Southward, 1970). These authors pointed out a high rate of uptake during the first hour of exposure and a considerable reduction in uptake as the concentration of the medium decreased. Pogonophorans are generally found to live in deep-sea benthic environments of relatively rich organic conditions (Southward and Southward, 1967). Furthermore, this group of animals is characteristic of the deep sea, particularly over the eutrophic trench floor (Ivanov, 1963) where nutritional conditions are favorable with high concentrations of solutes.

Perhaps other soft-bodied deep-sea organisms, including coelenterates and polychaetes, possess similar adaptations involving a heterotrophic mode of nutritional uptake. Recent observations on aquarium maintenance of deep-sea animals, such as the deep-sea solitary coral *Thecopsammia,* and the polychaete, *Hyalonoecia,* recovered from 1000 to 1800 m, suggest that these upper abyssal animals live for prolonged periods without solid food when they are maintained in waters of high dissolved organic carbon. It will be exceedingly interesting to test whether heterotrophic uptake of dissolved organic carbon has evolved as a special feeding adaptation in organisms inhabiting the deep-sea environment where decanting of solutes takes place. Evidently the pogonophorans are genuine examples of this feeding adaptation among abyssal organisms.

H. DEEP-SEA ENERGETICS

From an ecophysiological standpoint, the deep-sea community seems to operate metabolically at a considerably lowered tempo in comparison with shallow water communities. The low overall metabolic activity appears to be the outcome of long evolutionary adaptation to low food availability, low temperature, and high hydrostatic pressure. Despite the absence of photosynthesis in the abyss, other components of the food chain are fully represented in the deep-sea community. The omnivores and carnivores are largely seen among the predatory megafaunal animals which occur as bathypelagic or benthopelagic animals. The detritus-feeding organisms are seen among the large epifaunal and small infaunal benthic animals. Although a hypothetical deep-sea food chain can be developed on the basis of our knowledge on the composition of the deep-sea community, we do not yet have quantitative data on several important questions bearing on the digestive dynamics and energetics of deep-sea biota. Several key questions still remain unanswered. What is the normal nutritional need for abyssal animals? Is there any temporal pattern in their food uptake? How much food is required as the minimum need for sustaining metabolic activities? What is the fate of food in terms of accretion and partition for growth and reproductive activities? Is the economy of food supply responsible for general low metabolism in deep sea? What role do bacteria play in the flux of energy in the deep-sea food chain? All these questions await added studies. The low oxygen consumption rate of deep-sea animals provides only a clue as to the reduced food requirements. The very question of low metabolic rate is discussed in the following section with emphasis on respiratory adaptations.

III. Respiratory and Metabolic Adaptations

Respiration in an organism is incontravertably a sign of life. The amount of oxygen consumed provides a clue as to the proportion of the chemical energy released for vital processes involving food conversion, accretion, attrition, and heat production due to muscular or other activities (Krogh, 1941). The questions related to respiratory adaptations of abyssal animals are discussed below in comparison to shallow water animals to delineate any discernible types of metabolic adaptation that is characteristic of life in deep-sea conditions.

Our knowledge of the respiration of deep-sea animals stems largely from measurements of oxygen uptake, which is not necessarily the most reliable index of metabolic activity or energy production in an animal. Without any supply of oxygen, dehydrogenation of a stored compound, such as glucose, may result in the release of energy. It may be that glycolysis occurs more frequently in organisms than is generally suspected, perhaps more so in conditions of stress that inhibit performance of ventilative activities, such as gill movements and pleopod beats, or in conditions of temporary anoxia. However, it is known that organisms can live aerobically at very low oxygen concentrations (as low as 0.26 ml of oxygen per liter) in the oxygen minimum layers at great depths by regulating the rate of oxygen consumption (Childress, 1968).

The metabolic adaptations of deep-sea animals should be discussed from two

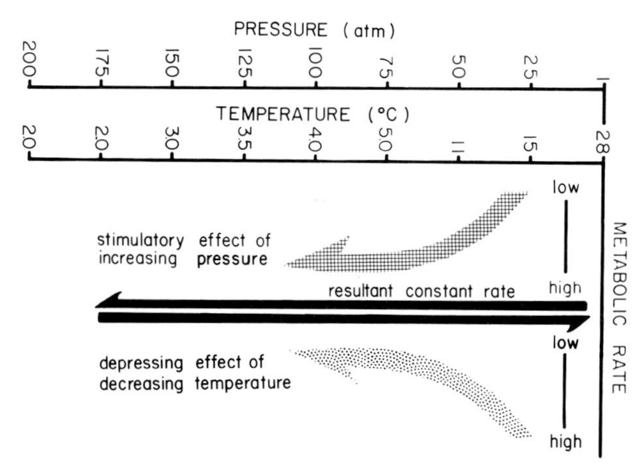

Fig. 9. A hypothetical model of respiratory adaptation of vertical migrators with constant metabolic rate as a result of increasing pressure and decreasing temperature in their normal vertical range of occurrence in the water column. (After George, 1979b.)

distinct views to place the problem in proper perspective; i.e., the types of respiratory adaptations that are necessary to cope with the low temperature and high pressure and the "low" metabolic performance that is possibly induced by conditions of food scarcity in the abyss.

There are at least two distinct models of respiratory adaptations. In the case of vertically migrating animals, respiration is sustained at a more or less constant level with the accelerating influence of increasing hydrostatic pressure counteracted by the depressing influence of the decreasing temperature. This trend is illustrated in a simple diagram which distinguishes between the balanced metabolic performance due to interaction of decreasing temperature and that due to increasing pressure (Fig. 9). The second model implies pressure-insensitive respiration, which is primarily governed by temperature (Fig. 10). This second pattern of metabolic adaptation is observed in stenothermal deep-sea animals with a wide vertical distribution in the abyssal environment.

The stimulatory effect of hydrostatic pressure on the respiration of shallow water animals was elucidated clearly in a series of experiments by Naroska (1968). The influence of pressure and temperature on the respiration of the eurythermal arctic amphipod, *Anonyx nugax*, revealed the obvious effect of pressure on metabolic activity (George, 1979a). The physiological responses, as indicated in the pattern of oxygen consumption, point out that pressure effect is less pronounced at the lowest acclimation temperature (Fig. 11). The maximum rate of O_2 uptake in all three experimental temperature occurs at about 140 atm when amphipods are found to be in the hyperactive phase. The pressure effect on respiration is far more pronounced at 6°C and at 14°C than at −1.8°C, the

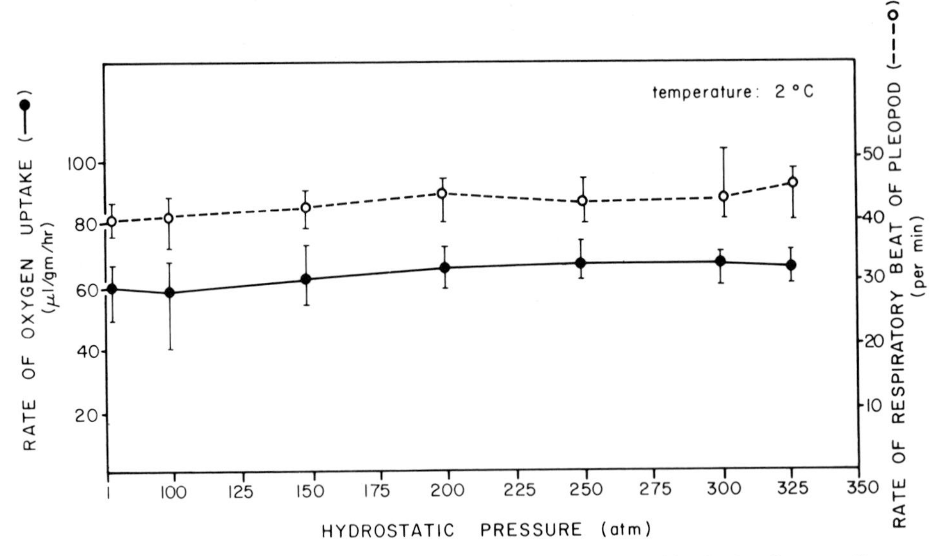

Fig. 10. Pressure-sensitive pattern of response of the deep-sea amphipod, *Eurythenes gryllus*. Note the absence of any pronounced influence of pressure on the pleopod activity and respiratory rate. (Data from George, 1979a.)

temperature at which the experimental populations were captured in the field and acclimated in the laboratory. *Anonyx nugax* is readily acclimated to higher thermal levels up to 14°C, but sharp sensitivity to pressure is seen at elevated temperature. At all experimental temperatures, the biphasic mode of the curve is apparent, with the onset of an inhibitory influence of hydrostatic pressure on respiration of the amphipods beyond 150 atm. Prolonged exposure to elevated pressure leads to suffocation and respiratory failure.

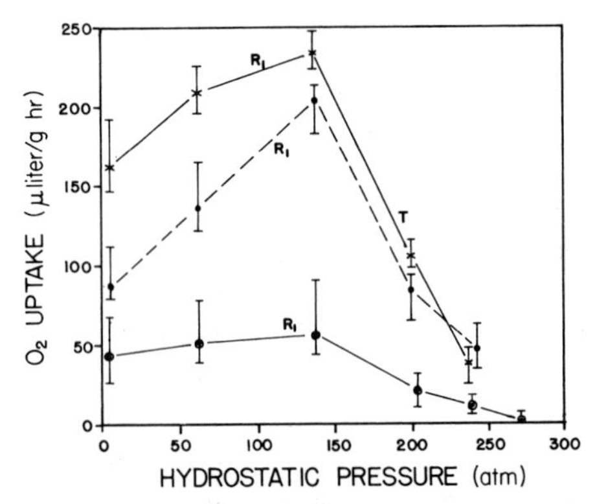

Fig. 11. Synergistic effect of hydrostatic pressure and temperature on the respiratory rate of the arctic shelf amphipod, *Anonyx nugax*. ×, 14°C; ●, 6°C; ◉, −18°C. (After George, 1979b.)

There have been few observations and experimental studies on living abyssal animals that inhabit depths beyond 3000 m. In a recent study it was found that the rate of oxygen consumption of the arctic abyssal amphipod, *Eurythenes gryllus,* presents a picture which is in striking contrast to that of shallow-water amphipods. *Eurythenes gryllus* is a cosmopolitan deep-sea amphipod known to dwell at depths between 1000 and 6500 m. The experimental populations were captured at a depth of 1800 m in the central part of the Arctic Ocean near the Drifting Ice Island T-3. Living individuals were maintained at various simulated hydrostatic pressure in hyperbaric metabolism chambers that monitored their respiration. Rate of oxygen uptake was studied at five different pressure levels between 80 and 340 atm (George, 1979a). In this deep-sea amphipod, respiration appears to be maintained at a fairly constant level within this range of more than 250 atm. Apparently this species is adapted to regulate respiratory rate to one constant level at the various depths of its occurrence. Furthermore, the pleopod activity is also sustained within a constant level throughout this pressure range. However, *E. gryllus* is extremely sensitive to subtle changes in ambient temperature. Respiration, as well as pleopod activity, is drastically reduced beyond 3° C (Fig. 12a,b), whereas the inhibition of respiration is evident at 4° and 6° C. This information confirms that there is a very narrow thermal range within which normal activity persists. Such a stenothermal feature is possibly a ''norm'' for some deep-sea species that inhabit the cold depths below the thermocline in the temperate–tropical marine environment.

In the case of vertically migrating animals that spend a period of time both in the warm surface waters and beneath the thermocline in the deep cold waters, special respiratory adaptations are recognized. Teal and Carey (1967) showed the significance of shifts in the metabolic rate of euphausids; i.e., they exhibit an elevated rate during the night while feeding in the warm surface waters and a suppressed rate of oxygen uptake during the day while resting in the deeper cold waters. Here again we see clearly a temperature-dependent respiration that seems to typify diurnal migrators in general. This reduction in metabolic activity is a transient phenomenon of considerable adaptive advantage for the purpose of conserving energy. Although temperature appears to exert a strong influence on respiration, pressure variation does not impose any significant impact on metabolism. It should be noted that the bathymetric limits of the daily excursion of these diurnal migrators fall within a pressure level that approximates less than 100 atm, a level that apparently does not exert irreversible damage to the organisms.

Another trend is seen in epipelagic animals which show a relatively constant metabolic rate over increasing hydrostatic pressure (Napora, 1964; Teal, 1971). In species such as the North Atlantic pteropods, *Diacrid trispinosa, Cuvierina columnella,* and *Clio pyramidata,* respiration is largely governed by temperature over their normal depths of occurrence. However, a stimulatory effect on the respiration of these epipelagic organisms is seen at pressures beyond the maxi-

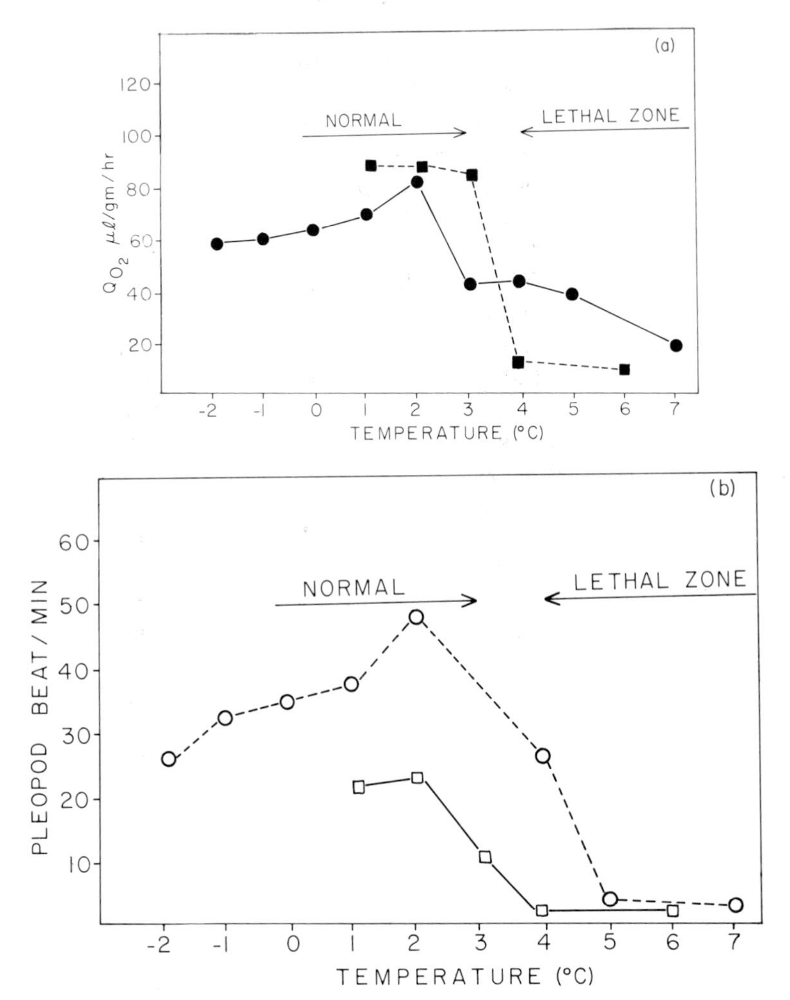

Fig. 12. Respiratory-activity response of selected polar and deep-sea poikilotherm to thermal stress beyond upper lethel level. (a) Thermal sensitivity in oxygen consumption rate in the polar isopod, *Glyptonotus acutus* (●) and the abyssal amphipod *Eurythenes gryllus* (■). Note the abrupt decrease in respiratory rate in deep-sea amphipod at 3°C. (b) Temperature effect on pleopod activity in the same polar isopod (o) and abyssal amphipod (□). (After George, 1979b.)

mum depth of their occurrence, which is normally about 800 m (Smith and Teal, 1973). In Fig. 13, the oxygen uptake rate of these three species as a function of depth pressure is illustrated to make a comparison with a deeper mesopelagic pteropod, *Limacina helicoides*. This deep species is bathypelagic, occupying depths below 1000 m. Respiration in the deeper species is relatively uniform within the vertical depth range and is primarily temperature-dependent (Fig. 13).

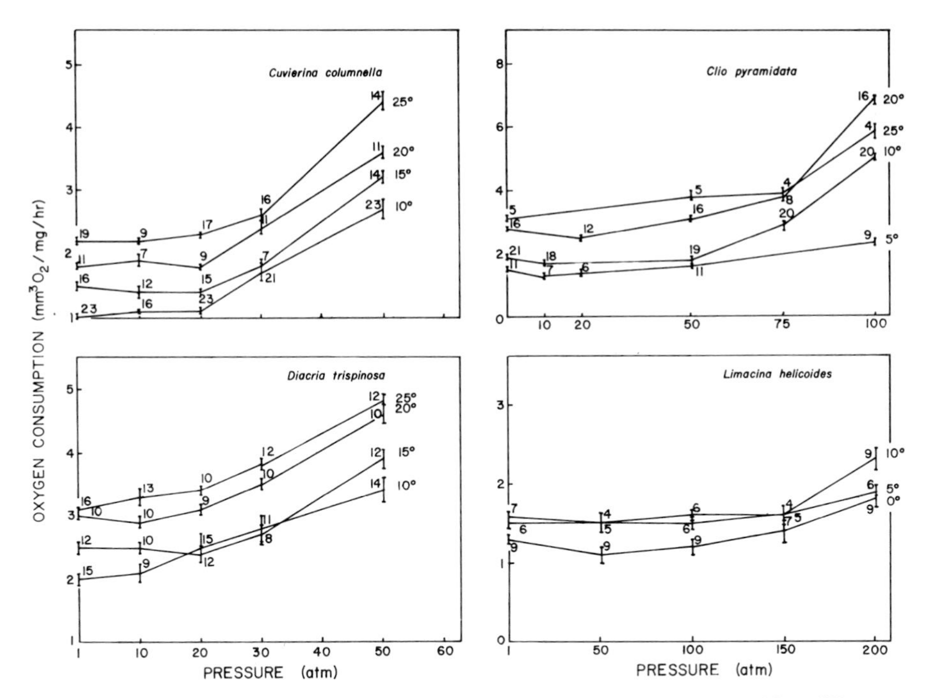

Fig. 13. Influence of hydrostatic pressure on the rate of oxygen consumption of four different species of pteropods at different temperatures. Number of experimental determinations indicated for each selected combination of pressure and temperature. (After Smith and Teal, 1973.)

The respiratory rate of the giant ostracod, *Gigantocypris,* also exhibited variation at different temperatures but suggested constancy between 0 and 500 atm (MacDonald *et al.,* 1972). Such patterns of respiratory adaptations characterize the vertical migratory animals.

The respiratory adaptations of the bathypelagic mysid, *Gnathophausia ingens,* enables this organism to live aerobically at very low oxygen concentrations in the O_2 minimum layer (Childress, 1971b). This large mysid crustacean has physiological mechanisms that regulate respiratory rate, such as large gill surface and an enhanced ventilation rate. It would be of interest to study the types of blood pigments in these deep animals and their oxygen carrying capacity in relation to temperature, pressure, and oxygen tension.

The respiratory rates of several midwater crustaceans and two species of midwater fish were determined as a function of depth of occurrence in a recent study (Childress, 1975). The oxygen uptake rate of these species was found to exhibit a rapid decline with increasing depth of occurrence. The regressions of maximum and minimum oxygen consumption in relation to depth are depicted in Fig. 14. The decrease in the respiratory rate was attributed to the concomitant

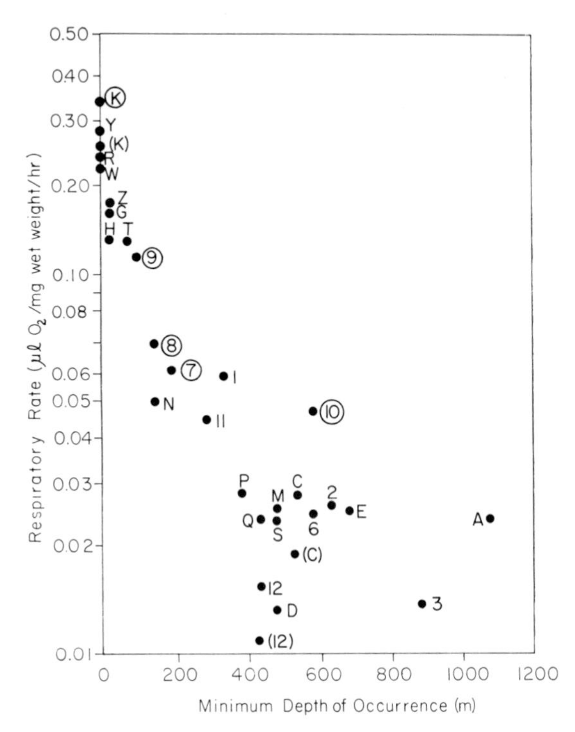

Fig. 14. Correlation between rate of oxygen consumption and bathymetric distribution of several midwater deep-living animals. Species identified with letters and numbers are same as in Fig. 8. (After Childress, 1975.)

lowering of the maintenance metabolic needs of the animals at greater depths.

In elucidating the respiratory responses of the fish, *Anoplogaster cornuta,* to the effects of pressure, it was pointed out that at habitat temperature there is no appreciable influence of pressure on oxygen consumption rate (Meek and Childress, 1973). These authors reported a significant difference between shallow and deep-sea fish. The shallow-water fish is sensitive to pressure and the deep-water fish is insensitive to various pressures within the habitat range. This deep-dwelling fish lacks a swimbladder and therefore comes up alive without any physical damage. This fish also is able to regulate its oxygen consumption down to 0.5 ml/liter. In an earlier study, a hypothesis was postulated to correlate the protein content of this midwater species with oxygen consumption (Childress and Nygaard, 1974). Temperature and protein content accounted for about 20.5% of the difference in Q_{O_2} between pressure levels of 1 and 120 atm.

In a series of recent experimental studies on abyssal benthic organisms from the slope depths, it was found that the deep-sea decapod, *Parapagurus pilosimanus,* which occurs between 800 and 3600 m, normally exhibits very little locomotory activity at habitat pressure (R. Y. George, unpublished data). Perhaps this reduced level of general activity or work per unit time in deep-sea animals is responsible for the low metabolic rate.

The second view of metabolic adaptation concerns the overall low metabolic tempo of deep-sea animals. The reduced metabolic activity of deep-sea bacteria was brought into focus by Jannasch *et al.* (1971), whose work showed microbial degradation to be 100 times slower at the deep-sea conditions than in controls at the same low temperature at 1 atm. In an *in situ* experiment at a depth of 1850 m over the continental slope environment, oxygen uptake of deep-sea sediment was found to be about 0.5 ml/m²/hr in comparison with 22.9–67.8 ml/m²/hr in shallow sedimentary regimes (Smith and Teal, 1973). This study led to the conclusion that biological community respiration in the deep sea is two orders of magnitude less than that in a shallow environment. The difference here is not necessarily a reflection of pressure effect, but rather the results of significantly different magnitudes of metabolizing biomass between the deep sea and shallow sediment. On the basis of *in situ* benthic community respiration measurements from 40 to 5200 meters depth, Smith (1978) postulated the hypothesis that only 15–29% of the organic carbon in the deep sea, which represents 2–7% of organic carbon fixed at the surface, is utilized by the benthic community at 2000, 3000, and 3650 meters. Another deep sea *in situ* measurement, made at a depth of 1230 m in the San Diego Trough, involved the monitoring of the respiration of two benthopelagic fish, the rat-tail fish, *Coryphenoides acrolepis,* and the hag fish, *Eptatretus deani* (Smith and Hessler, 1974). The respiratory rate of the deep-sea rat tail was found to be two orders of magnitude lower than the rates of the shallow-water fishes. Similarly, the respiratory rate of the deep-sea hag fish was found to be much lower (2.2 ml/g/hr) than that of the shallow-water hag fish, *Petromyzon marinus* (75.5 ml/g/hr), at similar temperature.

The few measurements on the metabolic rate of deep-sea animals do reinforce the emerging view that life activities in some abyssal organisms go on at much slower tempo in the energy-limited abyssal environment. An attempt is made in Table I to tabulate the metabolic rates of seven animals that represent different phylogenetic positions each of which demonstrate an apparently reduced respiratory rate. However, it should be realized that deep-living crustaceans have a respiratory rate (Table II) one order of magnitude higher than that of the animals listed in Table I. All these animals are either mesopelagic, from depths between 400 and 1800 m in the water column, or archibenthal, from the upper continental slope. The only exception is the abyssal amphipod, which is known to occur at more than 5200 m. In interpreting the data on low metabolic rates of deep-sea animals, we should bear in mind that these animals were decompressed during capture and repressurized to habitat pressure for monitoring oxygen uptake. It is not clear whether this methodology has imposed any deleterious influence on the experimental animal. Nevertheless, recent data on the archibenthal decapod,

TABLE I

Respiratory Rate of Deep-Living Animals with Low Oxygen Uptake Measured at Habitat Temperature[a]

Number	Group	Species	Minimum and maximum depth of occurrence (m)	Experimental temperature (°C)	Average rate of respiration (μl/gm/hr)
1	Copepod	*Bathycalanus bradyi*	900	3	3.90
2	Isopod	*Anuropus bathypelagicus*	200–800	3	8.00
3	Ostracod	*Gigantocypris agassizii*	900–1300	—	1.50
4	Hagfish	*Eptratretus deani*	1230	3.5	2.20
5	Macrurid	*Coryphaenoides acrolepis*	1230	3.5	2.40
6	Polychaete	*Hyalinoecia artifex*	800–1600	4	2.00
7	Solitary coral	*Thecopssammia socialis*	700–1200	4	4.62

[a] Data from different sources: 1–3 from Childress, 1975; 4–5 from Smith and Hessler, 1974; 6 from Magnum, 1972; 7 from George, 1979a,b.

TABLE II

Respiratory Rate of Deep-Living Crustaceans with High Oxygen Uptake Measured at Habitat Temperature[a]

Number	Group	Species	Minimum and maximum depth of occurrence (m)	Experimental temperature (°C)	Average rate of respiration (μl/gm/hr)
1	Mysid	*Gnathophausia ingens*	400–4000	3	54.00
2	Decapod	*Notostomus* sp. (*patentissimus* ?)	1200	3	54.00
3	Amphipod	*Paracallisoma coecus*	500–1100	—	45.00
4	Brachyuran	*Geryon quinquedens*	400–1400	5	22.40
5	Amphipod	*Eurythenes gryllus*	1800–6200	2	62.00

[a] Data from different sources: 1–3 from Childress, 1975; 4–5 from George, 1979a,b

Parapagurus pilosimanus, point out that decompression and repressurization below 100 atm do not result in any adverse effects on the organisms. The absence of any irreversible effect on exposure of these animals to pressure variation of about 100 atm is also reflected in the high survival of these upper abyssal decapods during retrieval from their habitat at 1000 m (R. Y. George, unpublished data). However, it must be realized that pressure conservation becomes essential for recovery of lower abyssal populations that occur at depths beyond 2500 m.

In making comparisons of O_2 uptake rates, several factors should be considered with reference to reproductive stage, size, age, sex, nutritional status, and several experimental constraints. Furthermore, efforts to draw comparisons between deep and shallow-water species of distant taxonomic affinity, including species of different genera within the same family, bring in certain inherent artifacts. Another problem in comparative studies is the diversity of expression of metabolic rates of animals in the literature, such as oxygen uptake rate in relation to wet weight, dry weight, ash-free dry weight, protein, and nitrogen. In addition, care must be taken to understand the metabolic rate of animals in terms of maximum and minimum levels as well as the physiological implications of resting and active metabolism. Significant results can only be obtained in controlled simulated conditions lasting for prolonged periods of several days, so that any periodism or definite pattern in the uptake of oxygen can be distinguished. The question still remains whether a low metabolic level reflects the nutritional limitations of the deep sea or whether it is the norm for deep-sea animals.

IV. Circulatory Adaptations

Circulatory adaptations of deep-sea animals are very poorly known. Circulation of blood within the organisms serves several vital physiological functions, including the transport of nutrients from digestive tract to other parts of the body, the removal of excretory products to renal or extrarenal organs, the passage of hormones to specific sites, and the performance of the two-way traffic of respiratory gases between respiratory surfaces and the rest of the body. The flow of blood in the blood vessels is maintained by peristaltic movement and there appear to be regularities in the velocity of flow in different animals. For example, the velocity of blood flow ranges from 10 cm/sec in large arteries to less than 1.0 cm/sec in the sinuses of the spiny lobster, *Panulirus interruptus* (Belman, 1975). However, the circulatory physiology of deep-sea organisms is virtually unexplored in terms of blood pressure, blood volume, and velocity. It is not known whether the diminished food supply affects circulatory dynamics.

A recent study on the circulatory adaptation of a deep-dwelling bathypelagic mysid, *Gnathophausia ingens,* sheds light on some anatomical modification of functional significance (Belman and Childress, 1976). It was pointed out that the circulatory system of this deep water mysid has relatively high rates of blood flow, although anatomically the system is typical of the basic crustacean pattern. There appears to be a certain enlargement in the diameter of the arteries in comparison with those of shallow-water decapods. Furthermore the heart in this deep-water mysid is very large for a crustacean of its size. The data on the blood pressure of this mysid also suggest apparent circulatory modifications of possible adaptive value. The circulation is mainly heart driven. The blood pressure pattern in this mysid differs from shallow-water crustaceans in that about 40% of the preferred resistance occurs across the gills. This species tends to put a significant portion of its cardiac work into movement of the blood through the gills.

Other interesting features are the blood volume pumped as a function of time and the turnover time which reflects on the flow rate in relation to the total blood volume. It was estimated that in *Gnathophausia,* in which blood volume approximates 20–30% of the body weight and cardiac output ranges between 55 and 225 ml/kg/min, the turnover time is 0.9–5 min in an animal weighing about 10 gm. This information confirms a relatively rapid blood flow which is comparable to that of the lobster, *Homarus* (Burger and Symthe, 1953), or the decapod, *Cancer maenas* (Blatchford, 1971). Belman and Childress (1976) also made an attempt to compute the metabolic cost of such a rapid cardiac output in the mysid, *Gnathophausia.* On the basis of the circulatory physiological and biophysical considerations, the heart tissue is assumed to be about 15% efficient and, therefore, the cardiac work for a mysid of 10 g is estimated to be approximately $3.5/10^{-4}$ Cal/min. The cardiac work in the lobster, *Panulirus,* weighing 600 g is calculated to be about $2.4/10^{-3}$ Cal/min. The relative efficiency in the mysid was attributed to its proportionately larger diameter and shorter blood vessels. It was also pointed out that the higher efficiency in this deep-water species is of adaptive significance as a special energy conserving mechanism in a food-limited environment. Adaptations are not necessarily related to depth, per se, since the circulatory adaptation of these mysids is definitely a response to their occurrence in the deeper oxygen minimum layers. In addition, the gill surface areas are larger in comparison to other crustaceans, with an approximate range of 5–15 cm^2/g wet weight. For example, the gill surface area of the shallow-water mud shrimp, *Callinassa californiensis,* was found to be a mean value of 4.13 ± 0.72 cm^2/g wet body weight (Torres *et al.,* 1977). Perhaps this deep water mysid has developed an efficient circulatory system for effectively utilizing the available oxygen while passing large volumes of water over its gills. Hydrostatic pressure is also known to influence the active and passive sodium transport, as well as the potassium transport (Moon, 1975). Pressure appears to exert significant effect on $(Na^+ + K^+)$ −ATPase and Mg^{2+}−ATPase in the gills of marine teleost fish (Pfeiler, 1978). In a recent study on pressure effect on sodium transport in deep-living gammarid amphipods from Lake Baikal, Brauer *et al.*

TABLE III

Concentrations of Potassium and Sodium in Muscles of Deep-Water and Shallow-Water Fishes[a]

Experimental Species	m*M* K/100gww	m*M* Na/100gww
Deep-sea fishes		
Synaphobranchus (deep eel)	16.6	147
Antimora sp.	65.1	34.1
Antimora heart	40.2	64.1
Coryphaenoides (from 2000 m)	65.8	43.2
Shallow-water fishes		
Anisarchus sp. (from 200 m, 8°–9° C)	122	42.5
Lepidopsetta (subsurface, 12° C)	84	34
Carassius sp. (5° acclimated)	97.8	44
(25° acclimated)	102.7	47.6

[a] After Prosser *et al.* (1975).

(1980) concluded that abyssal species, when exposed to 1 atm, go into negative sodium balance, even though shallow-water species remain in neutral or positive balance under similar conditions.

The physiological state and the contractile property of the heart of shallow and deep-sea fishes became the subject of a recent comparative study (Prosser *et al.*, 1975). Hearts of the codling, *Antimora,* and the rat tail, *Coryphaenoides,* were compressed to about 200 atm, which is close to the depth of their occurrence. These deep-sea fishes, which possess swimbladders, were killed in the process of decompression during retrieval, and no heart beat was observed. Repressurization to habitat pressure failed to cause any recovery of heart beat. Unusually high concentrations of sodium and low potassium contents were found in the muscles. It was suggested that permeability to ions drastically increased during retrieval without any pressure protection (Table III). Evidence of leaky membranes was seen while examining the blood, which was 80–90% hemolyzed. Obviously decompression inactivated the contraction and dilation of the heart. Appropriate methods of capture involving thermal and pressure protection are necessary to yield healthy deep-sea fishes and invertebrates for well-defined experimental studies concerning the circulatory dynamics and other pertinent physiological adaptations of abyssal animals.

V. Processes of Excretion

There is a definite dearth of knowledge about the structure and function of excretory organs in deep-sea animals. We do not know whether the nephridia of polychaetes in the deep sea perform nitrogenous excretion in a manner comparable to their shallow-water counterparts. In a predatory carnivorous animal such as the scavenging amphipods, the food which is largely protein, is absorbed at some unknown rate into body tissue as amino acids and used as a source of energy. In this process, amino acids undergo conversion into non-amino organic acids. This deamination leads to the production of excretory ammonia. Since the accumulation of ammonia in the body is toxic because of the high alkalinity, the animal eliminates ammonia into the ambient medium. Very little is known about the types of organs or sites of excretion of ammonia in deep-sea animals.

In shallow-water decapods, the excretory organ is the antennal gland. The urine of the crab is isotonic with blood; however, it contains much more magnesium than potassium. In the deep-sea decapods, the process of excretion is not at all understood. Because of the slight variation in salinity within the abyss, the deep-sea poikilotherms are generally considered to be stenohaline and poikilosmotic. The question of ionic regulation in abyssal animals has not been raised thus far, although ionic concentrations of deep-sea waters vary from one water mass to another, particularly on either side of the calcium carbonate dissolution depth. Future studies are needed to shed light on the types of excretory organs, mode, and rate of excretion in deep-sea animals. The waste products of metabolism, including feces as the nondigested portion of food consumed, nitrogenous waste in the form of urine, and exudates, such as mucus or milk, contribute to the energy flux in the deep sea as discussed in the earlier section.

VI. Adaptative Strategies in Reproduction

In the coastal marine environment of the temperate latitudes, most animals exhibit highly synchronous annual reproductive cycles in tune with various seasonal cues. In invertebrates, pelagic larval production is the normal mode of development. It is generally believed that in the seasonless deep sea, larval production is rare and direct development is dominant. Although seasonal breeders are uncommon, evidence of breeding cycles in abyssal animals has been known in deep-sea isopods (George and Menzies, 1967, 1968a), in ophiuroids (Schoener, 1968), and in scaphopods and brachiopods (Rokop, 1974). Recent studies on the deep-sea brachiopod, *Frieleia halli,* illustrated that the population is genetically highly variable (Valentine and Ayala, 1975). High genetic variability is also reported in deep-sea echinoderms (Murphy *et al.*, 1976), and its relation to environmental variability or stability is still not clear (Gooch and Schopf, 1972). Genetic strategies of adaptations may explain the existance of seasonal breeders in a nonseasonal deep-sea environment. High genetic variability is also correlated with adaptive flexibility and ecological opportunism (Dobzhansky, 1970).

The deep-sea asellote isopods in the northwestern Atlantic Ocean exhibit a reproductive peak during the late fall–winter period. These asellotes are only found in the deep-sea environment. In the same deep-sea habitats the isopods, *Cirolana* and *Astacilla,* which have a greater representation in the shelf environment, are summer breeders (George and Menzies, 1968a).

The abyssal ophiuroids, *Ophiura ljungani* and *Ophimusium lymani,* were also found to show definite periodism in reproduction (Schoener, 1968). Winter samples contained specimens with well-developed gonads and eggs. The breeding activities of eleven invertebrate species from the San Diego Trough at a depth of 1240 m were carefully examined by Rokop (1974). He pointed out that nine species showed year-around reproduction. However, the brachiopod, *Frieleia halli,* and the scaphopod, *Cadulus californicus,* showed evidence of seasonal spawning.

A detailed study was conducted of the reproductive pattern of the deep-sea decapod, *Parapagurus pilosimanus* at a depth of 900–1000 m in the northern part of the Blake Plateau (R. Y. George and D. Nesbitt, unpublished). Large population samples were procured from the same site during the four surface seasons of the same year. These deep-sea decapods exhibit an asynchronous year-round breeding activity, but they also showed a highly synchronous seasonal spawning in the winter. Ovigerous females containing undifferentiated eggs and eggs in embryonic stages were commonly seen in all four seasons, but only in the winter months did the female carry eggs which had developed to advanced larvae. The populations maintained at 1 atm in laboratory aquaria also showed spawning during the winter period between December and February. Evidently these deep-sea crabs are adapted to initiate spawning in the cold surface season.

Several views are offered to explain the rhythmic reproductive activity in supposedly constant physical conditions. In the shallow habitats breeding cycles are controlled by exogenous physical cues. However, in the deep sea, external cues are lacking if one assumes uniform hydrographic conditions of the abyss. Although the temperature shows little variation year-round at the bottom, it is possible that nutrient flow may vary seasonally as a consequence of changes in the productivity of the upper layers of the ocean. Also, variation in the near bottom current in relation to tidal cycles has been attributed as the necessary physical cue in the San Diego Trough. In the Blake Plateau, seasonal shift in the thermocline as a result of the undulating movement of the eastern boundary of the Gulf Stream is perhaps the physical cue to induce cyclic reproductive activity.

Deep-sea fishes are known to have special adaptations for mate location in the dark abyss. For example, eels have well developed olfactory organs, macrurids have organs of sound production, and clupeoid fishes have photophores producing luminescences. Ceratoids exhibit extreme sexual dimorphism with dwarfed parasitic males on large females (Mead *et al.,* 1964).

The reproductive trends discussed above emphasize the broad adaptive feature in relation to the coordination between the breeding mode or season of the deep-sea animals and the demands of the external conditions. Reproduction in these deep-sea animals is undoubtedly a prolonged and intricate process involving hormonal control for initiation of gametogenesis, oogenesis, and metamorphosis. We need to understand the mode and mechanisms underlying the transfer of food reserves, such as glycogen and lipid, from sites of storage to developing oocytes. The reproductive physiology of deep-sea animals is certainly an area of research that warrants far more detailed studies.

VII. Perception of the Environment

Deep-sea animals are evidently sensitive to external stimuli which may include sonic vibrations or subtle concentrations of dissolved substances. In the absence of light, sensory physiological adaptations with particular specialization in the faculties of chemoreception and mechanoreception would be expected.

A. ACOUSTIC SENSES

Underwater sound pulses are perceived in deep-sea fishes with the aid of well-developed lateral line organs and auditory labyrinths (Marshall, 1971). Knowledge of the auditory sensitivity of otolith organs in deep-sea fishes is very limited. There is indication that the noise level in the deep-sea environment is reduced to very low frequency noises at depths of 3000 m, as observed in the sound monitoring experiments from the bathyscaph *Trieste*.

Deep-sea fishes use swimbladders for sound reception and transformation. The accoustic properties of the swimbladder are bound to vary in relation to increase in depth with concurrent increase in density of the compressed gas (usually oxygen). We do not know the sound threshold of deep-sea fishes as to the maximum and minimum range of perception. In the abyssal fish from the Puerto Rico Trench. *Bassogigas profundissimus,* the swimbladder is apparently ''gas-free'' and is known to contain a hydrate formation. This is probably the reason for the intact arrival of these fishes after recovery from 7000 m without any rupture of the swimbladder (Hemmingsen, 1975). It is not quite clear what functional role such a swimbladder plays in these ultra abyssal fishes. However, it is known that in abyssal fishes not only does it function in the perception of sound, but the swimbladder also aids in intraspecific communication.

B. CHEMORECEPTION

Another significant sensory adaptation in deep-sea animals is chemoreception. Detection of food and recognition of mating partners or predators within the community is usually accomplished by chemoreception. The attraction of numerous scavenger-type predatory organisms to baited traps in the oligotrophic abyssal plain suggests the well developed faculty of chemoreception in these animals.

There is very little experimental data on chemoreception in deep-sea animals. In an effort to establish threshold responses, sensitivity of the midwater mysid, *Gnathophausia ingens,* from 600 to 900 m was examined to a range of amino acid concentrations (Fuzessery and Childress, 1976). The response threshold for dactyl receptors was $6 \times 10^{-8} M$ and the threshold for antennal receptors was $5 \times 10^{-7} M$. It was also found that the response intensity and duration increased with concentration. The low amino acid concentration, even in the range of 10^{-7} *M*, induces spontaneous feeding responses in deep-sea animals. The fact that feeding responses occur as a reflex at such concentrations of amino acids suggests that chemical cues do play a prominent role in the life strategy of deep-sea animals.

C. SENSE OF SIGHT

The vision of deep-sea animals is an intriguing subject since we do encounter organisms with prominent eyes in the aphotic abyssal zone. Whether the eye in the abyssal animal is functionless or highly developed for perception of bioluminescent light poses an interesting question. Most deep-sea animals, primarily the infaunal benthic organisms, are as blind as animals living in a cave environment. The proportion of the eyeless to eye-bearing isopods increases with depth and the curve of ocular index OI_{50} (50% of the species blind) shows a trend related to latitude and depth (Menzies *et al.*, 1973). In the deep-sea fishes, eyes are usually large and contain a large number of light-sensitive rods in the retina. The visual pigments seem to have maximum light absorption wave length at approximately 480 nm. The retinal pigment in deep-sea fishes was characterized as a chrysopsin as opposed to a rhodopsin in shallow-water fish (Denton and Warren, 1956).

In the deep ocean bioluminescence is predominently seen at depths between 500 and 1500 m, both in the mesopelagic and in the archibenthal upper abyssal benthic regimes. The luminescent secretion of the bathypelagic fish, *Searsia,* was investigated by Nicol (1958). The intensity of the luminescence was estimated as a radiant flux of 430×10^{-9} mW/m^2 receptor surface at a distance of 1 m. An adaptive value of bioluminescence in this bathypelagic fish is the ability to blind a predator with the release of the luminescent cloud. In a recent study on bioluminescence of deep-sea echinoderms, Herring (1974) pointed out that in response to osmotic stress the crinoid, *Annacrinus wyvillethomsoni,* increased the intensity of luminescence. KC1 stimulation of deep-sea ophiuroid, *Amphiura grandisquama,* produced a brilliant synchronous flashing, followed by luminous waves of low intensity. Luminescence is seen in bathypelagic elasipod holothurians and in deep-sea nekton as well as in zooplankton. Luminescence is obviously a widespread phenomenon among abyssal metazoans (Herring, 1974). Intestinal bacteria isolated from abyssal fishes are also found to show luminescence (K. Ohwada, personal communication). To date there is no report of a detailed experimental study on the phenomenon of bioluminescence in deep-sea animals. The luminescent property of the photophores appears to be pressure-dependent and, therefore, simulation of both high pressure and low temperature are necessary components of any experimental design in studies on luminescence.

D. PRESSURE PERCEPTION

Although the effects of hydrostatic pressure on the activity and respiration of shallow-water organisms have recently been studied rather extensively, very few studies have been done on pressure susceptibility (Giller, 1971) and responses of deep-sea animals. Results from vertical zonation studies of various groups of animals and from data on survival of deep-sea animals during retrieval from their habitat depths provide only indirect evidence of the pressure tolerance limits of abyssal animals. It is well documented that shallow and surface dwelling marine animals exhibit a pressure-induced hyperactivity on exposure to increasing pressure (50–80 atm). At elevated pressure levels exceeding 100 atm, the shallow-water crustaceans go into convulsive reactions that may reflect possible stress on the nervous system. The convulsion threshold for littoral or pelagic crustaceans occurs between 60 and 150 atm (Flugel and Schlieper, 1970; Schlieper, 1972; Menzies and George, 1972; George and Marum, 1974; McDonald, 1973, 1975 and McDonald and Teal, 1975). In mammals, such hyperbaric conditions have also elicited convulsions that characterize the high pressure neurological syndrome (HPNS) (Brauer *et al.*, 1974, 1975). The HPNS is basically a reflection of changes in the central nervous system and involves early effects as manifested in tremors and motor disturbances and later effects with outbursts of convulsive reactions consisting of significant electroencephalogram changes. This syndrome is seen in various mammals between 30 and 150 atm.

The hyperexcitability behavior at low pressure and convulsive reactions at moderate pressure are apparently observed in all shallow-water marine animals. However, observations on deep-sea animals indicate that the hyperexcitability phenomenon appears to be absent in mesopelagic and archibenthal animals. The experimental study on the pressure effect on the abyssal amphipod, *Eurythenes gryllus,* from 1800 m demonstrated that there is no obvious hyperexcitability behavior in response to increasing pressure; however, convulsive reactions occur at 580 atm, which is about 400 atm higher than the pressure of the experimental population (George, 1979a,b). The absence of a hyperexcitability behavior was also revealed in the investigations of pressure effects on the deep-sea ostracod, *Gigantocypris* (MacDonald, 1972). These observations indicate that deep-sea animals show special pressure adaptations that are distinctly different from those of shallow-water animals.

Deep-sea bony fishes use the swimbladder for sensing the depth as they move up and down the water column. These mesopelagic fishes commonly possess a fairly large swimbladder for maintaining buoyancy in the water column (Marshall, 1960). If the floating fish at a depth of 1000 m descends to 2000 m, the equilibrium pressure of gas in the swimbladder is bound to increase exponentially with the increase of ambient pressure from 100 to 200 atm. This process enables the fish to maintain higher gas pressure in the swimbladder at greater depths. During vertical movements up and down the water column, gas pressure in the swimbladder is apparently lost through diffusion or regulated by secretion (Scholander and VanDam, 1953). High concentrations of oxygen are found in the swimbladder of deep-sea bathypelagic fishes due to the efficient mechanisms of gas secretion against a gradient. The P_{O_2} of the gas can reach levels several orders of magnitude higher than the ambient P_{O_2}. Hadal fish from extreme depths of the ocean possess functional swimbladder (Nielson and Munk, 1964).

The deep-sea crustaceans, such as the isopod, *Macrostylis,* possess statocysts on the telson. However, we do not know whether these are pressure sensory organs. The high mortality of abyssal animals retrieved from beyond 2000 m suggests that these organisms are adapted to a high pressure environment so that decompression proves lethal. In the absence of experimental data, it is generally believed that temperature is the critical factor and not hydrostatic pressure. There is evidence that the stenothermal deep-sea organisms succumb to transient thermal shock during passage through the warm layers of the thermocline. Recent retrieval studies also point out that decapod crustaceans from upper slope depths of less than 1500 m become moribund on exposure to high temperature, but recover on immediate transfer to low temperature. The deep sea decapod, *Parapagurus pilosimanus,* withstands rapid decompression during retrieval and can survive for prolonged periods (over a year) at 1 atm (George, 1979a). However, animals from depths beyond 2000 m are invariably killed, possibly because of the effect of decompression and thermal change. Obviously we need to develop advanced methods to capture deep-sea animals without thermal or pressure variations in order to conduct more definite studies on pressure–temperature perception.

VIII. Some Molecular Aspects of High Pressure Adaptation

It is becoming increasingly clear that high hydrostatic pressure exerts a profound influence on biochemical reactions at the cellular level and consequently affects the physiological processes that occur in organisms under normal conditions. A general understanding of the pressure effects at the molecular level comes from a series of recent biochemical studies (Hochachka, 1975a). It is well documented that temperature directly affects the kinetics of biochemical reactions and the cellular chemistry of the organism (Wieser, 1973). Similarly, hydrostatic pressure impinges on the structure and function of the cells. The effects of pressure fall into two categories: (1) the rate effect of pressure on metabolic reaction rates; and (2) the weak-bond structural effects involving the biochemical processes that are dependent on "weak" chemical bonds. As salient examples, different structures or processes are considered pressure-susceptible and these include the higher orders of protein structure, membrane structure, lipid–lipid interactions, nucleic acid interactions, nucleic acid–protein interaction, and hormone–receptor protein binding (Hochachka and Somero, 1973). These authors pointed out that pressure influences the cellular balance by introducing a change in volume in a biochemical reaction. The basic change is essentially the difference between ΔV, the overall volume change in a reaction, and $\Delta V \neq$, the volume change in the formation of the activated complex. On the basis of the volume changes, biochemical reactions were classified into three categories: (1) pressure-activated reactions with a negative $\Delta V \neq$; (2) pressure-inhibited reactions with a positive $\Delta V \neq$; and (3) pressure-independent reactions with no net volume change.

In the past decade some experimental studies have been conducted to understand the complex mechanism of high pressure adaptation in deep-sea animals. Earlier studies (Marsland, 1970) indicated increased pressure resistance of fertilized sea urchin eggs with increased addition of ATP. However, it has not been determined whether deep-sea animals must expend metabolic energy to sustain normal activities and even to maintain protoplasmic gel structures. Hochachka (1975a,b) has proposed that the ability of deep-sea organisms to withstand the enormous hydrostatic pressures of the abyssal region may be linked to biochemical adaptations. It was pointed out by Penniston (1971) that quarternary protein structure of "normal" enzymes of "normal" shallow-water animals could not be sustained under the high pressure and low temperature conditions of the abyss. Does this mean that in the course of evolution special types of enzymes originated in the deep-sea animals permitting them to function in their unique physical environment? Hochachka (1975a) contends that the enzyme M_4 lactate dehydrogenase from the abyssal organisms greatly reduces hydrophobic contributions to binding coenzymes. He also found that the enzyme citrate synthase from the gill extracts of the deep-sea fish, *Antimora rostrata,* performed its activity at 1 atm of pressure as an enzyme polymer with multiple subunits and a molecular weight of about 270,000. However, the same enzyme, when exposed to modest pressure, appears to depolymerize into a more active two-subunit enzyme of about 100,000 molecular weight. A model depicting enzyme activity of gill citrate synthase is illustrated in the following simple diagram (Fig. 15).

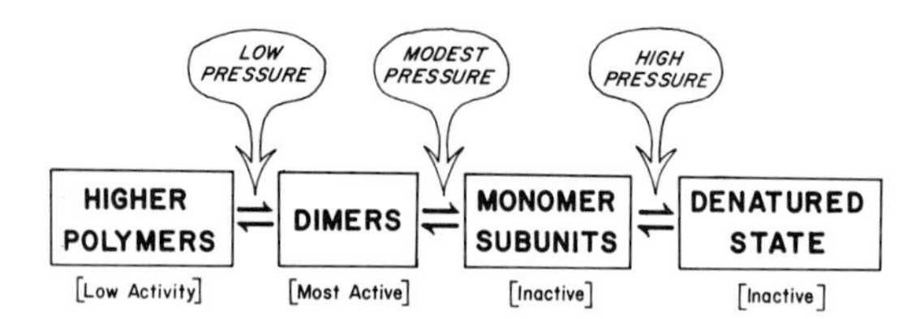

Fig. 15. A diagrammatic representation of various polymerization states of *Antimora* gill citrate synthase enzyme in relation to pressure effects. (After Hochachka, 1975c.)

Hochachka (1975a) postulated that when invading a high pressure environment, abyssal animals use a number of biochemical strategies, such as monomeric enzymes, multimeric enzymes with subunit–subunit interactions requiring modest pressure for stabilization of the functional optimal quaternary structure, or enzyme designs in which the formation of the active multi-subunit enzyme proceeds with a zero or net negative volume change. Another apparent enzymatic adaptation to pressure has been reported in the deep-sea fish, *Antimora rostrata,* at the citrate synthase locus. This deep-sea fish does not occur at extreme depths where the pressure is 500 atm or more. It is not known whether the mechanism of enzyme adaptation will be homologous with the *Antimora* gill enzyme in abyssal animals from depths exceeding 4000 m. In an earlier study, it was found that the enzyme activity of diphosphatase (FDPase) in the abyssal macrurid fish, *Coryphaenoides acrolepis,* was inhibited at hydrostatic pressures that were higher than it normally encountered. However, the homologous enzyme from a surface-dwelling fish, *Pimelometopon pulchrum,* exhibited a much greater sensitivity to high pressure (Hochachka *et al.,* 1970). This information on the structure and function of enzymes points out that the process of evolution of animals at the various depth zones of the ocean has resulted in distinct differences of enzyme activity between deep-sea and shallow-water organisms (Fig. 16). There is some evidence of changes in fatty acid composition of animals from different depths (Lewis, 1962, 1967).

The underlying mechanisms for these effects remain to be elucidated. However, an interesting area for determining differences in cellular mechanisms of adaptation in shallow and deep-sea animals may be accessible in studies of the neuromuscular transmission process. It is clearly demonstrated that high pressure exposure induces hyperactivity and other manifestations of the high pressure neurological syndrome (HPNS) in shallow and surface-dwelling marine animals (George and Marum, 1974). In a recent investigation (Campenot, 1975), data was obtained to evaluate the comparative pressure effects on the transmission at the neuromuscular junction of the shallow-water lobster, *Homarus americanus,* and the deep-living red crab, *Geryon quinquedens*. This study indicated that high hydrostatic pressure (50–200 atm) depresses the amplitude of excitatory junctional potentials (ejp's) at the neuromuscular junction in lobsters. The

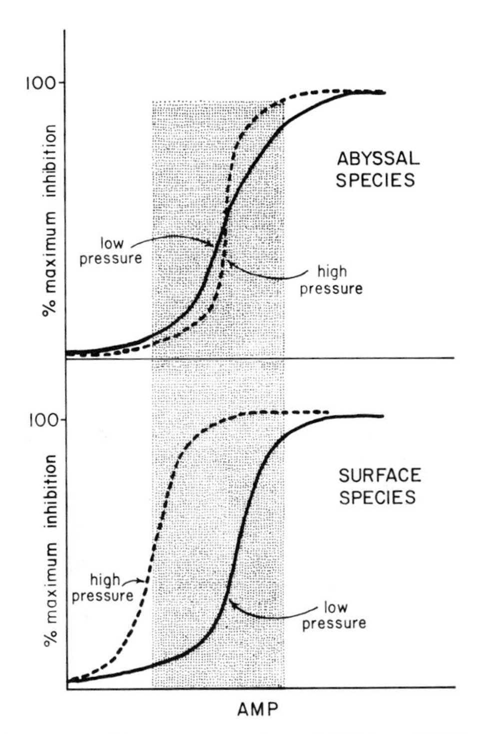

Fig. 16. Effect of pressure on the percentage maximum inhibition of EDPases by AMP. Approximate physiological range of AMP is shown by shading. In the abyssal species, the enzyme-modulator affinity is pressure-independent and AMP is an equally good inhibitor at all pressures encountered in nature. In the surface species, AMP inhibition is accentuated at high pressure owing to an increase in enxyme-modulator affinity. (After Hochachka and Somero, 1973.)

mechanism of the pressure-induced depression was interrupted as a result of decrease in the amount of transmitter substance released by the nerve endings. Nevertheless, the results of the deep-sea red crab study provided clues to the possible existence of a compensatory mechanism for adaptation to high pressure. Pressure-induced depression was seen at the neuromuscular junction of the red crab at low frequencies of nerve stimulation, but pressure at the high frequencies significantly increased the amount of facilitation at this synapse, resulting in little or no overall deficit in the net transmitter release. This enhancement of a process normally seen in shallow-water animals to compensate for a pressure-induced reduction in function provides an interesting clue to conservative adaptive mechanisms. Further studies are necessary to elucidate the effects of hydrostatic pressure on synaptic transmission and the function of the nervous systems of animals adapted to perform normal activities in the deep-ocean environment at pressures of 500–1000 atm.

It has been well established in earlier studies that not only low temperature, but also high hydrostatic pressure at normal physiological temperatures rapidly and reversibly depolymerize spindle microtubules (Zimmerman and Marsland, 1964; Zimmerman, 1970; Salmon, 1975). The striking effect of pressure on microtubules is illustrated in Fig. 17. Microtubules are minute, measuring 0.00002 mm in diameter but variable in length. These rigid and hollow cylindrical structures are involved in determining cell shape, and they participate in the process of movement of material within the cells, sperm mobility, ciliary movement, and cell division. Hydrostatic pressure exceeding 200 atm is known to perturb the equilibrium of the mitotic spindle and to abruptly dissolve the mitotic apparatus.

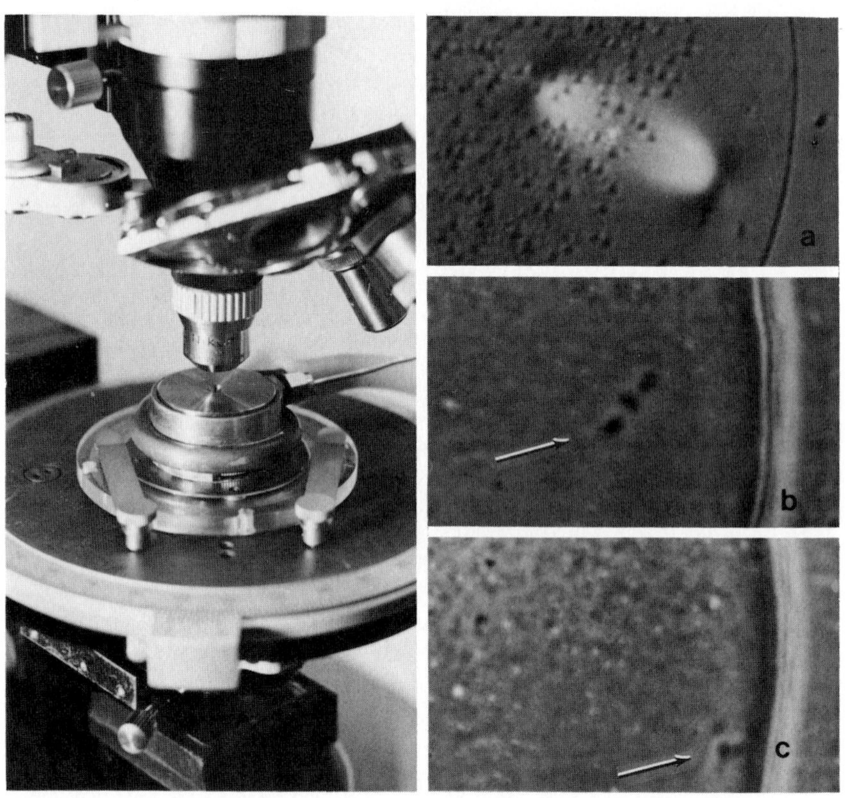

Fig. 17. Photograph showing a living mitotic spindle of the shallow water polychaete, *Chaetopterus,* oocyte at metaphase at 1 atm as viewed in polarized-light (a) and in phase contrast microcopy (b). Note the complete depolymerization of the birefrigent spindle fiber microtubules within 5 min after application of a pressure of 200 atm. Also note the shift in the position of the chromosomes (c) to the cell surface at the site of the tightly anchored pole (arrow). On the left side is the miniature hydrostatic pressure chamber used in this study by Dr. E. D. Salmon. (Photo courtesy of Dr. Salmon.)

Evidently, high pressure inhibits cell division in shallow-water animals at metaphase because of the profound effect on the normal functioning of the microtubules (Fig. 17). What is not known is how abyssal animals are so adapted that cell division occurs at extreme hyperbaric conditions and also how microtubule spindles in deep-sea organisms are stable under high hydrostatic pressure. Studies on the ultrastructure of deep-sea animals in controlled high pressure experimental conditions can offer explanation and pertinent answers to the questions raised above.

For the purpose of understanding any ultrastructural changes in deep-sea animals during the process of recovery from the abyss, some preliminary studies have been conducted to examine the ultrastructure of gill, liver, and skeletal muscle in two deep-sea teleosts; *Antimora rostrata,* and the deep-sea eel, *Synophobranchus* (Hulbert and Moon, 1975). They found that major ultrastructural changes occurred only during rapid, *in vitro* decompression. Such changes are particularly observable in gill and muscle tissues with indications of disrupted membranes and loss of structural protein integrity. The authors also pointed out the shortcoming of their experimental results in that the deep-sea fishes were killed in the process of retrieval because of a multitude of factors, including decompression effect and thermal shock. Here again, there is a great need to perform retrieval of deep-sea animals with controlled temperature protection to discern decompression effects. There is also the need to recover deep-sea animals in *in situ* pressure and temperature for obtaining suitable material for laboratory studies. We can only achieve a realistic insight into problems of molecular adaptations of abyssal animals when appropriate methods are developed to perform prolonged studies on living animals from the deep-sea environment.

IX. Overview on Physiological Adaptations of Abyssal Animals

A perusal of the various aspects discussed above on the functional strategies of deep-sea animals reveals the emergence of an array of physiological and biochemical adjustments that have evolved as a price of adaptation to abyssal life. We are now beginning to understand the role of hydrostatic pressure in the physiology of deep-sea animals despite the widely accepted earlier view that it is not to the enormous range of pressure but to the subtle differences in temperature one should look for explanation of deep-sea animal adaptation. The synergism between high hydrostatic pressure and low temperature is often neglected. The synergistic effect of various combinations of these two environmental parameters on biochemical reactions and physiological processes in deep-sea animals must be studied. Another major functional modification in the deep-sea life may not necessarily be the consequence of low temperature or high pressure, but rather an impact of the limited food availability or nutritional deficiency in the deep-sea biotope. Low metabolism is one such adaptation for conserving energy and as such, has obvious genetic implications. The following aspects of deep-sea animal adaptations pose intriguing questions that call for specific experimental studies in the future.

(1) The nutritional strategies in the food-limited abyssal environment with focus on the energy flux in benthic and bathypelagic food chain.

(2) The metabolic adjustments of deep-sea animals with particular emphasis on the adaptive mechanism responsible for the reduced physiological performance of activities.

(3) The precise influence of high hydrostatic pressure on normal functional activities related to survival, growth, and reproduction in deep-sea organisms.

(4) The sensory physiology of abyssal animals and the perception of sensory stimuli that are attenuated in the enormous deep-sea environment. The problems of chemoreception, responses to acoustic pulses, and pressure–temperature perception in deep-sea animals are important areas of study in adaptive physiology.

(5) Abyssal animals inhabiting the red clay and radiolarian ooze are and have been constantly exposed to chronic low level radiation from radium for generations. This unique natural condition calls for experimental studies on genetic aspects and physiological adaptations to low level natural radiation.

(6) The problem of physiological races in widely bathymetric or eurybathial deep-sea animals and the events of speciation in the abyss are challenging areas of research.

(7) The question of high genetic variability and polymorphism in deep-sea populations as a consequence of the environmental stability of the deep-sea biotope requires further investigation.

(8) Finally, there is need for controlled, long-term laboratory experiments on living populations of deep-sea animals in high pressure aquaria to study various biochemical and physiological adaptive strategies.

To understand the spectrum of functional adaptations of deep-sea animals it is essential that live, healthy populations from the abyss are captured and retrieved in a pressure–temperature controlled deep-retrieval system and transferred to a deep-ocean simulation facility for prolonged studies at *in situ* habitat conditions. Such a development is bound to pave the way for physiological studies of the functional adaptations of abyssal organisms.

Acknowledgment

This chapter on the functional adaptations of deep-sea animals contains partly the results of present and past research projects sponsored by the Office of Naval Research and the National Science Foundation. I am grateful to the Oceanic Biology Program of ONR and Division of Polar Program of NSF for support to conduct studies on the physiology of deep-sea animals in the Arctic,

Antarctic and Atlantic Oceans. I wish to graciously thank Dr. Ralph W. Brauer for the several stimulating discussions on the subject of adaptive strategies of abyssal animals. My wife, Chandra George, helped me in the preparation of the manuscript. I also express my gratitude to Gloria Crowell for preparing the figures and typing the manuscript.

References

Agassiz, A. (1888). Three cruises of the United States Coast and Geodetic Survey Steamer. Blake, Vol. 1, Chapter 13. The physiology of deep sea life. *Bull. Mus. Comp. Zool.* **14,** 294–314.

Allen, J. A., and Sanders, H. L. (1966). Adaptations to abyssal life as shown by the bivalve *Abra profundorum* (Smith). *Deep-Sea Res.* **13,** 1175–1184.

Barber, R. T. (1968). Dissolved organic carbon from deep waters resists microbial oxidation. *Nature (London),* **220,** 274–275.

Belman, B. W. (1975). Some aspects of the circulatory physiology of the spiny lobster *Panulirus interruptus. Mar. Biol.* **29,** 295–305.

Belman, B. W., and Childress, J. J. (1976). Circulatory adaptations to the oxygen minimum layer in the Bathypelagic Mysid *Gnathophausia ingens. Biol. Bull. (Woods Hole, Mass.)* **150,** 15–37.

Belyaev, G. M. (1966). Bottom fauna of the ultra-abyssal depths of the world ocean (in Russian). *Tr. Inst. Okeanol., Akad. Nauk SSSR* **591,** 1–248.

Belyaev, G. M. (1972). "Bottom Fauna of the Ultra-Abyssal Depths of the World Ocean." Acad. Nauk Press, Moscow. (English transl. by Israel Program for Scientific Translations.)

Bezrukov, P. L. (1955). Bottom deposits on the Kurile-Kamchatka trench. *Tr. Inst. Okeanol., Akad. Nauk SSSR* **12,** 97–129.

Blatchford, J. G. (1971). Haemodynamics of *Carcinus maenus* (L). *Comp. Biochem. Physiol.* **39,** 193–202.

Blaxter, J. H. S., Wardle, C. S., and Roberts, B. L. (1971). Aspects of circulatory physiology and muscle systems of deep sea fish. *J. Mar. Biol. Assoc. U. K.* **51,** 991–1006.

Brauer, A. (1908). Die Tiefsee-Fische II. Anatomischer Teil. *Wiss. Ergeb. 'Valdivia'* **15,** 1–266.

Brauer, R. W., ed. (1972). "Barobiology and the Experimental Biology of the Deep Sea," N.C. Sea Grant Program, Spec. Publ. Univ. North Carolina Press.

Brauer, R. W., Beaver, R. W., Hogue, C. D., Ford, B., Goldman, S. M., and Venters, R. T. (1974). Intra-and interspecies variability of vertebrate high-pressure neurological syndrome. *J. Appl. Physiol.* **37,** 239–250.

Brauer, R. W., Beaver, R. W., Mansfield, W. M., O'Connor, F., and White, L. W. (1975). Rate factors in development of the high-pressure neurological syndrome. *J. Appl. Physiol.* **38,** 220–227.

Brauer, R. W., Bekman, M. Y., Keyser, J. B., Nesbitt, D. L., Shvetzov, S. G., Sidelev, G. N., and Wright, S. L. (1980). Comparative studies of sodium transport and its relation to hydrostatic pressure in deep and shallow water gammarid crustaceans from Lake Baikal. *Comp. Biochem. Physiol.* **65A,** 119–127.

Bruun, A. F. (1957). Deep sea and abyssal depths. Chapter 22. *Mem., Geol. Soc. Am.* **67,** 641–672.

Burger, J. W., and Smythe, C. M. (1953). The general form of circulation in the lobster *Homarus. J. Cell. Comp. Physiol.* **42,** 369–383.

Burnett, B. R. (1973). Observation of the microfauna of the deep sea benthos using light and scanning electron microscopy. *Deep-Sea Res.* **20,** 413–427.

Campenot, R. B. (1975). The effects of high hydrostatic pressure on transmission at the crustacean neuromuscular junction. *Comp. Biochem. Physiol. B***52,** 133–140.

Carey, A. G. (1972). Food sources of sublittoral, bathyal and abyssal asteroids in the Northeast Pacific Ocean. *Ophelia* **10,** 35–47.

Childress, J. J. (1968). Oxygen minimum layer: Vertical distribution and respiration of the Mysid *Gnathophausia ingens*. *Science* **160,** 1242–1243.
Childress, J. J. (1971a). Respiratory rate and depth of occurrence of midwater animals. *Limnol. Oceanog.* **16,** 104–106.
Childress, J. J. (1971b). Respiratory adaptations to the oxygen minimum layer in the Bathypelagic Mysid *Gnathophausia ingens. Biol. Bull. (Woods Hole, Mass.)* **141** (1), 109–121.
Childress, J. J. (1975). The respiratory rate of midwater crustaceans as a function of depth of occurrence and relation to the oxygen minimum layer off southern California. *Comp. Biochem. Physiol.* **50,** 787–799.
Childress, J. J., and Nygaard, M. H. (1973). The chemical composition of midwater fishes as a function of depth of occurrence off southern California. *Deep-Sea Res.* **20** (12), 1093–1110.
Childress, J. J., and Nygaard, M. H. (1974). Chemical composition of buoyancy of midwater crustaceans as a function of depth of occurrence off southern California. *Mar. Biol.* **27,** 225–238.
Culkin, F., and Morris, R. J. (1970). The fatty acids of some marine teleosts. *J. Fish Biol.* **2,** 107–112.
Dahl, E. (1979). Deep sea carion feeding amphipods: Evolutionary patterns in Miche adaptation *Oikos* **33,** 167–175.
Dayton, P. K., and Hessler, R. R. (1972). Role of biological disturbance in maintaining diversity in the deep sea. *Deep-Sea Res.* **19,** 199–208.
Degens, E. T., and Ross, D. A. (1970). The red sea hot brines. *Sci. Am.* **222,** 32–42.
Denton, E. J., and Marshall, N. B. (1958). The buoyancy of bathypelagic fishes without a gas-filled swimbladder. *J. Mar. Biol. Assoc. U. K.* **37,** 753–767.
Denton, E. J., and Warren, F. J. (1956). Visual pigments of deep sea fish. *Nature (London)* **178,** 1059.
Denton, E. J., Gilpin-Brown, J. B., and Shaw, T. I. (1969). A buoyancy mechanism found in cranchid squid. *Proc. R. Soc. B London, Ser.* **174,** 271–279.
Dobzhansky, T. (1970). "Genetics of the Evolutionary Process." Columbia Univ. Press, New York.
Ebbecke, U. (1935). Uber die Wirkungen hoher Drucke auf marine Lebewesen. *Pfluegers Arch. Gesamte Physiol. Menschen Tiere* **236,** 648–657.
Ebeling, A. W. (1967). Zoogeography of tropical deep sea animals. *Stud. Trop. Oceanogr.* **5,** 593–613.
Fell, J. W. (1970). *In* "Recent Trends in Yeast Research" (D. G. Ahearn, ed.), Georgia State University, Atlanta.
Filatova, Z. A. (1960). On the quantitative distribution of the bottom fauna in the central Pacific. *Tr. Inst. Okeanol., Akad. Nauk SSSR* **41,** 85–97.
Flugel, H., and Schlieper, C. (1970). The effects of pressure on marine invertebrates and fishes. *In* "High Pressure Effects on Cellular Processes" (A. M. Zimmerman, ed.), pp. 211–234, Academic Press, New York.
Fontaine, M. (1930). Recherches expérimentales sur les réactions des êtres vivants aux fortes pression. *Ann. Inst. Oceanogr., (Monaco)* **8,** 1–99.
Fournier, R. O. (1966). North Atlantic deep sea fertility. *Science* **153,** 1250–1252.
Fuzessery, Z. M., and Childress, J. J. (1976). Comparative chemosensitivity to amino acids and their role in the feeding activity of bathypelagic and littoral crustaceans. *Biol. Bull. (Woods Hole, Mass.)* **150,** 15–37.
George, R. Y. (1979a). What adaptive strategies promote immigration and speciation in deep sea environment. *Sarsia* **64,** 61–65.
George, R. Y. (1979b). Behavioral and metabolic adaptation of polar and deep sea crustaceans: A hypothesis concerning physiological basis for evolution of cold adapted crustaceans. *Bull. Biol. Soc. Wash.* **3,** 283–296.

George, R. Y., and Higgins, R. P. (1979). Eutrophic hadal benthic community in the Puerto Rico Trench. *AMBIO* **6,** 51–58.

George, R. Y., and Marum, J. P. (1974). Behavior and tolerance of euplanktonic organisms to increased hydrostatic pressure. *Int. Rev. Gesamten Hydro.biol.* **59,** (2), 171–186.

George, R. Y., and Menzies, R. J. (1967). Indication of cyclic reproductive activity in abyssal organisms. *Nature (London)* **215,** 878.

George, R. Y., and Menzies, R. J. (1968a). Further evidence for seasonal breeding cycles in deep sea. *Nature (London)* **220,** 80–81.

George, R. Y., and Menzies, R. J. (1968b). Additions to the Mediterranean deep sea fauna (VEMA cruise 14). *Rev. Roum. Biol., Ser. Zool.* **13** (6), 367–384.

George, R. Y., and Nesbitt, D. (1980). Asynchronous breeding and seasonal spawning in deep sea decapod *Parapagurus pilosimanus. Jour. Crust. Biol. (mss. submitted)*

George, R. Y., and Wright, S. L. (1977). Cayman trench hadal community: New information from the Caribbean subtropical trench (Abstract). *ASLO. Deep-Sea Biology Symposium,* Seattle, 1976.

Gillen, R. G. (1971). The effect of pressure on muscle lactate dehydrogenase activity of some deep sea and shallow water fishes. *Mar. Biol.* **8,** 7–11.

Gooch, J. L., and Schopf, T. J. M. (1972). Genetic variability in the deep sea: Relation to environmental variability. *Evolution* **26,** 545–552.

Gordon, D. C. (1970). Some studies on the distribution and composition of particulate organic carbon in the North Atlantic ocean. *Deep-Sea Res.* **17,** 233–243.

Grice, D. G., and Hulseman, K. (1965). Abundance, vertical distribution and taxonomy of calanoid copepods at selected stations in the north-east Atlantic. *J. Zool.* **146,** 213–262.

Hamilton, R. D., Holm-Hansen, D., and Strickland, J. D. H. (1968). Notes on the occurrence of microscopic organisms in deep water. *Deep-Sea Res.* **15,** 651–656.

Hedgpeth, J. W. (1957). Classification of marine environments. Treatise of marine ecology and paleoecology. *Mem., Geol. Soc. Am.* **67** (1), 17–27.

Hemmingsen, E. A. (1975). Clathrate hydrate of oxygen: Does it occur in deep sea fish. *Deep-Sea Res.* **22,** 145–149.

Herring, P. J. (1974). New observations on the bioluminescence of echinoderms. *J. Zool.* **172,** 401–418.

Hochachka, P. W., ed. (1975a). Pressure effects on biochemical systems of abyssal and midwater organisms: The 1973 Kona Expedition of the alpha helix. *Comp. Biochem. Physiol. B* **52,** 1–199.

Hochachka, P. W. (1975b). Why study protein of abyssal organisms. *Comp. Biochem. Physiol. B* **52,** 1–2.

Hochachka, P. W. (1975c). How abyssal organisms maintain enzymes of the right size. *Comp. Biochem. Physiol. B* **52,** 39–42.

Hochachka, P. W., and Somero, G. N. (1973). "Strategies of Biochemical Adaptation." Saunders Philadelphia, Pennsylvania.

Hochachka, P. W., Schneider, D. E., and Kuznetsov, A. (1970). Interacting pressure and temperature effects on enzymes of marine poikilotherms: Catalytic and regulatory properties of FDPase from deep and shallow water fishes. *Mar. Biol.* **7,** 285–293.

Hulbert, W. C., and Moon, T. W. (1975). Tissue ultra structure and alterations as a result of applied hydrostatic pressure in two marine teleosts. *Comp. Biochem. Physiol. B* **52,** 117–126.

Ivanov, A. V. (1963). "Pogonophora." Academic Press, New York.

Jannasch, H. W., and Wirsen, C. O. (1973). Deep sea microorganisms: *In situ* response to nutrient enrichment. *Science* **180,** 641–643.

Jannasch, H. W., Eimhjillen, K., Wirsen, C. O., and Formanfarmaian, A. (1971). Microbial degradation of organic matter in the deep sea. *Science* **171,** 672–675.

Johnson, R. M., Schwent, R. M., and Press, W. (1968). The characteristics and distribution of marine bacteria isolated from the Indian Ocean. *Limnol. Oceanogr.* **13,** 656–664.

Johnston, R. (1972). The theories of August Pütter. *Trans. R. Soc. Edinburgh, Ser. B.* **72,** 401–410.

Karl, D. M., LaRock, P. A., Morse, J. W., and Sturges, W. (1976). Adenosine triphosphate in the North Atlantic Ocean and its relationship to the oxygen minimum. *Deep-Sea Res.* **23,** (1), 81–89.

Karl, D. M., Wirsen, C. O., and Jannasch, H. W. (1980). Deep-sea primary production at the Galapagos hydrothermal vents. *Science,* **207,** 1345–1347.

Kim, J., and ZoBell, C. E. (1972). Agarase, amylase, cellulase and chitinase activity at deep sea pressure. *J. Oceanogr. Soc. Jpn.* **28,** 131–137.

Kohlmeyer, J. (1968). The first ascomycete from the deep sea. *J. Elisha Mitchell Sci. Soc.* **84,** 239–241.

Kriss, A. E. (1960) Microorganisms as indicators of hydrological phenomena in seas and oceans I. *Deep-Sea Res.* **6,** 88–94.

Krogh, A. (1941). "Comparative Physiology of Respiratory Mechanisms." University of Pennsylvania, Philadelphia.

Lewis, R. W. (1962). Temperature and pressure effects on fatty acids of some marine ectotherms. *Comp. Biochem. Physiol.* **6,** 75–89.

Lewis, R. W. (1967). Fatty acid composition of some marine animals from various depths. *J. Fish. Res. Board Can.* **24,** 1101–1115.

Little, C., and Gupta, B. L. (1970). Studies on pogonophora. 3. Uptake of nutrients. *J. Exp. Biol.* **51,** 759–773.

MacDonald, A. G. (1972). The role of high hydrostatic pressure in the physiology of marine animals. *Sym. Soc. Exp. Biol.* **26,** 209–231.

MacDonald, A. G. (1973). Locomotor activity and oxygen consumption in shallow and deep sea invertebrates exposed to high hydrostatic pressures and low temperature. *In* "Fifth Symposium on Underwater Physiology" (C. J. Lambertsen, ed.), pp. 405–419.

MacDonald, A. G. (1975). "Physiological Aspects of Deep Sea Biology." Cambridge Univ. Press, London and New York.

MacDonald, A. G., and Teal, J. M. (1975). Tolerance of oceanic and shallow water crustacea to high hydrostatic pressure *Deep-Sea Res.* **22,** (3), 131–144.

MacDonald, A. G., Gilchrist, I., and Teal, J. M. (1972). Some observations on the tolerance of oceanic plankton to high hydrostatic pressure. *J. Mar. Biol. Assoc. U. K.* **52,** 213–223.

Magnum, C. P. (1972). Temperature sensitivity of metabolism in offshore and intertidal onuphid polychaetes. *Mar. Biol.* **17,** 108–114.

Marshall, N. B. (1960). Swimbladder structure of deep sea fishes in relation to their systematic and biology. *"Discovery" Rep.* **31,** 1–122.

Marshall, N. B. (1971). "Exploration in the Life of Fishes." Harvard Univ. Press, Cambridge, Massachusetts.

Marshland, D. A. (1970). Pressure-temperature studies on the mechanisms of cell division. *In* "High Pressure Effects on Cellular Processes" (A. M. Zimmerman, ed.), pp. 259–312. Academic Press, New York.

Mead, G. W., Bertelsen, E., and Cohen, D. M. (1964). Reproduction among deep sea fishes. *Deep-Sea Res.* **11,** (4), 569–596.

Meek, R. P., and Childress, J. J. (1973). Respiration and the effect of pressure in the mesopelagic fish *Anoplogaster cornuta* (Beryciformes). *Deep-Sea Res.* **20** (12), 1111–1118.

Menzel, D. W., and Ryther, J. H. (1961). Zooplankton in the Sargasso Sea off Bermuda and its relation to organic production. *J. Cons., Cons. Int. Explor. Mer* **26,** 250–258.

Menzies, R. J., and George, R. Y. (1972). Temperature effects on behavior and survival of marine invertebrates exposed to variations in hydrostatic pressure. *Mar. Biol.* **13** (2), 155–159.

Menzies, R. J., George, R. Y., and Rowe, G. T. (1973). "Abyssal Environment and Ecology of the World Oceans." Wiley (Interscience), New York.

Moon, T. W. (1975). Effects of hydrostatic pressure on gill Na-K-ATP in an abyssal and a surface dwelling teleost. *Comp. Biochem. Physiol. B* **52,** 59–66.

Morita, R. Y., and ZoBell, C. E. (1955). Occurrence of bacteria in the pelagic sediments collected during the MidPacific Expedition. *Deep-Sea Res.* **3,** 66–73.

Murphy, L. S., Rowe, G. T., and Haedrick, R. L. (1976). Genetic variability in deep sea echinoderms. *Deep-Sea Res.* **23,** 339–348.

Napora, T. A. (1964). The effect of hydrostatic pressure on the prawn *Systellaspis debilis. Proc. Symp. Exp. Mar. Ecol.* **2,** pp. 92–94.

Naroska, V. V. (1968). Vergeichende Untersuchungen über den Einfluss des hydrostatischen Druckes auf Überlebensfähigkeit and Stoffwechselintensitat Mariner Evertebraten and Teleosteer. *Kiel. Meeresforsch* **24,** 95–123.

Nicol, J. A. C. (1958). Observations on luminescence in pelagic animals. *J. Mar. Biol. Assoc. U. K.* **37,** 25–36.

Nielsen, J. G., and Munk, O. (1964). A Hadal fish (*Bassogigas profundissimus*) with a functional swimbladder. *Nature* (*London*) **204,** 594–595.

Patton, J. S. (1975). The effects of pressure and temperature on phospholipid and triglyceride fatty acids of fish white muscle: A comparison of deep water and surface marine species. *Comp. Biochem. Physiol. B* **52,** 105–110.

Penniston, J. T. (1971). High hydrostatic pressure and enzymatic activity: Inhibition of multimeric enzymes by dissociation. *Arch. Biochem. Biophys.* **142,** 322–332.

Pfeiler, E. (1978). Effects of hydrostatic pressure on ($Na^+ + K^+$) ATPase and Mg^{2+} ATPase in gills of Teleost fish. *J. Exp. Zool.* **205,** 393–402.

Prosser, C. L., Weems, W., and Meiss, R. (1975). Physiological state, contractile properties of heart and lateral muscles of fishes from different depths. *Comp. Biochem. Physiol.* **52,** (1), 127–132.

Regnard, P. (1891). "Recherches expérimentales sur les conditions physiques de la vie dans les eaux." Masson, Paris.

Rokop, F. J. (1974). Reproductive patterns in the deep sea benthos. *Science* **186,** 743–745.

Salmon, E. D. (1975). Pressure induced depolymerization of brain microtubules in vitro. *Science* **189,** 884–886.

Sanders, H. L. (1968). Marine benthic diversity: A comparative study. *Am. Nat.* **102,** 243–282.

Schlieper, C. (1972). Comparative investigations on the pressure tolerance of marine invertebrates and fishes. *Symp. Soc. Exp. Biol.* **26,** 197–207.

Schoener, A. (1968). Evidence for reproductive periodicity in the deep sea. *Ecology* **49,** 81–87.

Schölander, P. F., and Van Dam, L. (1953). Composition of the swimbladder gas in deep sea fishes. *Biol. Bull.* (*Woods Hole, Mass.*) **104,** 75–86.

Schwarz, J. R., Yayanos, A. A., and Colwell, R. R. (1975). Metabolic activity of the intestinal microflora of a deep sea invertebrate. *Appl. Environ. Microbiol.* **31,** 46–48.

Shulenberger, E., and Hessler, R. R. (1974). Scavenging abyssal benthic amphipods trapped under oligotrophic central North Pacific Gyre waters. *Mar. Biol.* **28,** 185–187.

Sieburth, J. M. (1971). Distribution and activity of oceanic bacteria. *Deep-Sea Res.* **18,** 1111–1121.

Smith, K. L. (1978). Benthic community respiration in the N. W. Atlantic Ocean: *In situ* measurements from 40 to 5200 meters. *Mar. Biol.* **47,** 337–347.

Smith, K. L., and Hessler, R. R. (1974). Respiration of Benthopelagic fishes: *In situ* measurements at 1230 m. *Science* **184,** 72–73.

Smith, K. L., and Teal, J. M. (1973). Deep sea benthic community respiration, an *in situ* study at 1850 m. *Science* **170,** 283.

Sokolova, M. N. (1959). On the distribution of deep water bottom animals in relation to their feeding habits and the character of sedimentation. *Deep-Sea Res.* **6,** 1–4.

Sokolova, M. N. (1972). Trophic structure of deep sea macrobenthos. *Mar. Biol.* **16,** 1–12.

Southward, A. J., and Southward, E. C. (1968). Uptake and incorporation of labelled glycerine by pogonophores. *Nature* (*London*) **218,** 875–876.

Southward, A. J., and Southward, E. C. (1970). Observation on the role of dissolved organic compounds in the nutrition of benthic invertebrates. *Sarsia* **45,** 69–96.

Southward, E. C., and Southward, A. J. (1967). The distribution of pogonophora in the Atlantic Ocean. *Symp. Zool. Soc. London* **19,** 145–158.

Sutcliffe, W. H., Baylor, E. R., and Menzel, D. W. (1963). Sea surface chemistry and langmuir circulation. *Deep-Sea Res.* **10,** 233–243.

Teal, J. M. (1971). Pressure effects on the respiration of vertical migrating decapod crustacea. *Am. Zool.* **11,** 571–576.

Teal, J. M., and Carey, F. G. (1967). Effects of pressure and temperature on the respiration of euphausids. *Deep-Sea Res.* **14,** 725–733.

Thiel, H. (1975). The size structure of the deep sea benthos. *Int. Rev. Gesamten Hydrobiol.* **60** (5), 575–606.

Torres, J. J., Gluck, D. L., and Childress, J. J. (1977). Activity and physiological significance of the pleopods in the respiration of *Callianassa californiensis* (Dana), (Crustacea: Thalassinidea). *Biol. Bull.* (*Woods Hole, Mass.*) **152,** 134–146.

Turekian, K. K., Cochran, J. K., Kharkar, D. P., Cerrato, R. M., Vaisnys, J. R., Sanders, H. L., Grassle, J. F., and Allen, J. A. (1975). Slow growth rate of a deep sea clam determined by ^{228}Ra chronology. *Proc. Natl. Acad. Sci. U.S.A.* **72,** 2829–2832.

Valentine, J. W., and Ayala, F. J. (1975). Genetic variation in *Frieleia halli,* a deep sea brachiopod. *Deep-Sea Res.* **22,** 37–44.

Vernberg, W. B., and Vernberg, F. J. (1972). "Environmental Physiology of Marine Animals." Springer-Verlag, Berlin and New York.

Vinogradov, M. E. (1962). Feeding of deep sea zooplankton. *Rapp. P.-V. Reun. Cons. Int. Explor. Mer.* **153,** (18), 114–120.

Wieser, W., ed. (1973). "Effects of Temperature on Ectothermic Organisms." Springer-Verlag, Berlin and New York.

Williams, P. M. (1971). *In* "Organic Compounds in Aquatic Environment" (eds.), (S. D. Faust and J. V. Hunter, eds.), pp. 200–220, Dekker, New York.

Yayanos, A. A. (1978). Recovery and maintenance of live amphipods at a pressure of 580 bars from an ocean depth of 5700 meters. *Science* **200,** 1056–1059.

Yayanos, A. A., Dietz, A. S., and Boxtel, R. V. (1979). Isolation of a deep sea barophilic bacterium and some of its growth characteristics. *Science* **205,** 808–810.

Zimmerman, A. M. ed. (1970). "High Pressure Effects on Cellular Processes." Academic Press, New York.

Zimmerman, A. M., and Marsland, D. (1964). Cell division: Effects of pressure on the mitotic mechanisms of marine eggs (*Arbacia punctulata*). *Exp. Cell Res.* **35,** 293–302.

ZoBell, C. E., and Morita, R. Y. (1957). Barophilic bacteria in some deep sea sediments. *J. Bacteriol.* **73,** 563–568.

Index

N

T